Lecture Notes in Physics

Lecture Notes in Physics

Edited by H. Araki, Kyoto, J. Ehlers, München, K. Hepp, Zürich
R. Kippenhahn, München, H. A. Weidenmüller, Heidelberg
and J. Zittartz, Köln

166

Computer Simulation of Solids

Edited by C.R.A. Catlow and W.C. Mackrodt

Springer-Verlag
Berlin Heidelberg GmbH 1982

Editors

C.R.A. Catlow
Department of Chemistry, University College London
20 Gordon Street, London WC1H 0AJ, England

W.C. Mackrodt
ICI New Science Group, The Heath
Runcorn, Cheshire WA7 4QE, England

ISBN 978-3-540-11588-5   ISBN 978-3-540-39347-4 (eBook)
DOI 10.1007/978-3-540-39347-4

Originally published by Springer-Verlag Berlin Heidelberg New York in 1982

2153/3140-543210

PREFACE

Over the last few years simulation methods have become an established part of computational solid state physics and chemistry.   In May 1980 a meeting was held at the Daresbury Laboratory of the Science and Engineering Research Council to discuss recent progress in certain areas of this expanding subject.   The present volume is derived from the presentations and discussions at that meeting. The intention is to give research workers in solid state chemistry and physics and in materials science generally an informed view of the current state of the chosen topics, and perhaps more important, a glimpse of future developments.

Acknowledgements

We would like to thank Dr. V.R. Saunders for his collaboration in the
organisation of the original meeting at Daresbury, on which the present volume
is based, and Dr. B.E.F. Fender for his advice on the preparation and presentation
of this book. We would also like to thank Ms. R.A. Rosier, for her efficiency in
preparing the manuscript.

* * * * * * * *

INTRODUCTION

The techniques of computer simulation of solids have, in recent years, been developed to the level that they now provide a good predictive tool - one which may be used in the study of a diverse range of solid state properties, including thermodynamic, structural and transport properties. This has been due both to developments in numerical and computational methods and to improved understanding of interatomic forces in solids, which are the basic physical input of any simulation. This book has therefore three main aims: first to review the present state of methodology in this field; second, to describe the current state of our understanding of interatomic potentials; and third to survey the application of the techniques. Throughout the book, the limitations of the present methods will be stressed, with a view to identifying those areas where an improved understanding at a <u>fundamental</u> level is required.

Much of the earlier work in the simulation field was concerned with the defective solid. The contributions here of the Theoretical Physics group at Harwell should be stressed. In particular the development by Norgett[1] of a generalised Mott-Littleton program (HADES) for the calculation of defect energies in cubic ionic solids led to a number of notable successes in calculating accurate values of the energies governing the concentration and mobilities of defects in strongly ionic materials[2,3,4,5]. Simulations using this program have also made a useful contribution to our understanding of heavily defective materials - in particular the complex types of defect clustering that occur in non-stoichiometic and heavily doped solids[6,7]. The scope of the method has been extended with the development by the present authors and their collaborators of a completely generalised Mott-Littleton defect simulation package (HADES III)[8] which may be applied to crystals of any symmetry class. Defect simulation thus remains an active area for the future, and the initial chapter of section A (Techniques) is devoted to the methodology of these calculations.

Defect calculations have, of course, been applied to other classes of materials, including metals and semi-conductors. These applications are discussed in Chapters (13) and (14); less success has, in general, been enjoyed than in the study of ionic materials, owing to the greater difficulties in obtaining adequate potentials. Extensions of the simulation methods to treat two dimensional defects (including surfaces) is an active field of research. Results are described in Chapter (19).

The defect codes essentially minimise the energy of a static lattice (i.e. one in which no thermal vibrations are included) with respect to the coordinates of the ions surrounding the defect species. Energy minimisation may also be used in a second group of programs (PLUTO) developed by Catlow and Norgett[9] which perform calculations on perfect lattices and which are applied to the study of structural and other properties of complex crystals. By performing minimisations of energy with respect to the coordinates they may be used to predict crystal structures - an application discussed in Chapter (15). Further applications to long range order in defect super-cells are discussed by Cormack in Chapter (20).

A complete understanding of the thermodynamic properties of solids requires, of course, the calculation of entropies as well as energies. We include therefore in Chapter (2) an account of a recent successful work on the formation entropies of defects. Advances have also been made recently in the theory of defect volumes which are discussed by Gillan and Lidiard in Chapter (3).

Static simulation methods of the type referred to above are limited to the modelling of processes with energy parameters >> kT. For many areas, including most classical defect studies, this limitation is unimportant. However, interest in the last few years in 'superionic' solids - that is materials which show transport coefficients of a liquid-like magnitude for at least one sub-lattice - has led to the application to these materials of simulation methods whose use had previously been largely confined to liquids. We refer in particular to the molecular dynamics techniques discussed also by Walker in Chapter (5) and by Dixon and Gillan in Chapter (18). The Monte-Carlo method is also discussed in Chapter (4), for although it cannot yield dynamical information, it supplements molecular dynamics by providing calculated values of thermodynamic quantities.

The reliability of all these calculations, both static and dynamic is essentially limited by the quality of the interatomic potentials that are used in the simulation. For this reason a whole section of the book is devoted to a discussion of the derivation of interatomic potential parameters and of the models that have been implemented in computer codes. The fundamental aspects of potential development from electron densities is discussed by March, while Inglesfield considers the special problems presented by interatomic potentials in metals. Detailed consideration is given to modelling of ionic solids in the chapter by Catlow, Dixon and Mackrodt, which stresses the limitations of potential models derived purely from empirical data, and of an accurate treatment of polarisability in modelling polar solids. The article by Stoneham and Harding, discussing the problem of the effects of covalence in solids is also highly relevant here. Indeed, we believe that future progress in this field is very largely dependent upon an improved understanding of the contribution of covalence to the cohesion in heteropolar solids and on the development of models which allow covalence effects to be included in a viable manner into computer codes.

The diversity of applications of the techniques is illustrated by the work reported in Chapters (12)-(20). The materials studied range from simple oxides and halides to highly complex solids including mineral systems. The problems discussed include complex structural questions (see Chapters (15),(16) and (17)), as well as transport and thermodynamic properties. An important feature of all this work is, we believe, the way in which it interacts with and complements experiment. This feature is stressed in all the chapters discussing applications.

Future developments in the field of interatomic potential studies will, we believe, rely increasingly on the use of quantum mechanical methods. Indeed, quantum mechanical calculations are playing an increasingly important role in solid state studies. Application of quantum mechanical cluster methods are described in Chapter (6), while Harker in Chapter (7) discusses calculation of the electronic energy levels of defects. Good simulation studies are indeed a bridge between fundamental physics and experimental work. We hope that the book introduces the reader to the wide range of activities and techniques involved, and to the role which simulation studies can play in elucidating structural, transport and thermodynamic properties of solids.

References

1. Norgett, M.J. UKAEA Report, AERE-R.7650 (1974).

2. Catlow, C.R.A., Corish, J., Diller, K.M., Jacobs, P.W.M. and Norgett, M.J., J. Phys. C. 12, 451 (1979).

3. Mackrodt, W.C. and Stewart, R.F., J. Phys. C. 10, 1431 (1977).

4. Mackrodt, W.C. and Stewart, R.F., J. Phys. C. 12, 431 (1979).

5. Catlow, C.R.A., Norgett, M.J., Mackrodt, W.C. and Stoneham, A.M., Phil. Mag. 40, 161 (1979).

6. Catlow, C.R.A. and Fender, B.E.F., J. Phys. C. 8, 3267 (1975).

7. Catlow, C.R.A. in 'Non-stoichiometric Oxides' (Ed. O.T. Sorensen), Academic Press, 1981.

8. Catlow, C.R.A., James, R., Mackrodt, W.C. and Stewart, R.F., Phys. Rev. B25, 1006, (1982).

9. Catlow, C.R.A. and Norgett, M.J., UKAEA Report AERE-M.2763 (1978).

CONTENTS

Section A: Techniques

Section B: Potentials

Section C: Applications

<u>SECTION A</u>

TECHNIQUES

CHAPTER 1

THEORY OF SIMULATION METHODS FOR LATTICE
AND DEFECT ENERGY CALCULATIONS IN CRYSTALS

by

C.R.A. Catlow
Department of Chemistry, University College London,
20 Gordon Street, London WC1

and

W.C. Mackrodt
I.C.I., PLC, The Heath, Runcorn, Cheshire WA7 4QE

1. Introduction

We are concerned in this chapter, first with the calculation of two distinct types of energy, that is _lattice_ and _defect_ energies. The former is the binding or cohesive energy of the perfect crystal per unit cell; the latter is the energy to create a defect (defined in the most general sense, to include point defects, dislocations, planar defects and surfaces) within the perfect lattice. The former quantity is evidently of central importance in treating thermochemical properties of solids and in assessing relative stabilities of different structures. Moreover its derivatives with respect to elastic strain and displacement are related to dielectric, piezo-electric and elastic constants and phonon dispersion curves. The defect energy, as will be clear from later chapters in the book, controls many of the thermodynamic and transport properties of the crystal. Related to the energetic calculations are corresponding structural properties: in the first case the minimum energy of the perfect lattice and in the latter the structure of the relaxed lattice about the defect.

In many instances notably for ionic and semi-ionic materials the lattice and defect energies and the corresponding structural properties can be calculated directly from suitable interatomic potentials such as those discussed later in this volume, and it is with the basis of these calculations that the present chapter is concerned. Our discussion relates to three main topics: the stability condition for non-defective lattices and the calculation of dielectric, piezo-electric and elastic constants, the treatment of defective lattices and minimization methods in defect calculations and crystal structure prediction. Our discussion of defect systems is confined to the treatment of point defects; extension to two dimensional defects (e.g. surfaces) is considered in Chapter (21). In addition, our treatment is, as noted, particularly suitable for ionic and semi-ionic systems. The methods appropriate to metals and covalent solids are discussed in Chapters (13) and (14).

2. The stability conditions for non-defective lattices

We begin our discussion by considering the stability conditions that determine the equilibrium configuration of perfect or non-defective lattices. The treatment given here follows that outlined previously by Catlow and Norgett[1]. For a lattice containing s atoms per unit cell near its equilibrium configuration the lattice energy or energy density, to use the terminology of Born and Huang[2], u_L, can be written to second order in the total strain as

$$u_L(\underline{R}') = u_L(\underline{R}) + \underline{g}^T.\underline{\delta} + \underline{\delta}^T.\underline{\underline{W}}.\underline{\delta} \tag{1}$$

in which $\underline{\delta}$ is a generalised $(3s + 6)$-dimensional strain vector consisting of $3s$ internal components, $\underline{\delta r}$, and 6 bulk components, $\underline{\delta \epsilon}$, thus

$$\underline{\delta} = [\underline{\delta r}, \underline{\delta \epsilon}] \tag{2}$$

$\underline{g}$ is a vector of first-derivatives of the energy

$$\underline{g} = [\partial u_L/\partial \underline{r}, \; \partial u_L/\partial \underline{\epsilon}] \tag{3}$$

and $\underline{\underline{W}}$ the corresponding matrix of second derivatives

$$\underline{\underline{W}} = \begin{bmatrix} \partial^2 u_L/\partial \underline{r}\partial \underline{r} & \partial^2 u_L/\partial \underline{r}\partial \underline{\epsilon} \\ \partial^2 u/\partial \underline{\epsilon}\partial \underline{r} & \partial^2 u_L/\partial \underline{\epsilon}\partial \underline{\epsilon} \end{bmatrix}. \tag{4}$$

The configuration, $\underline{R}' \equiv \{\underline{r}'_s\}$, is related to $\underline{R}$ by the transformation

$$\underline{r}'_s = \Delta\epsilon.(\underline{r}_s + \underline{\delta r}_s) \tag{5}$$

in which $\Delta\epsilon$ is the symmetric strain matrix formed from the components of $\underline{\delta \epsilon}$

$$\Delta\epsilon = \begin{bmatrix} \delta\epsilon_1 & \tfrac{1}{2}\delta\epsilon_6 & \tfrac{1}{2}\delta\epsilon_5 \\ \tfrac{1}{2}\delta\epsilon_6 & \delta\epsilon_2 & \tfrac{1}{2}\delta\epsilon_4 \\ \tfrac{1}{2}\delta\epsilon_5 & \tfrac{1}{2}\delta\epsilon_4 & \delta\epsilon_3 \end{bmatrix}. \tag{6}$$

The numbering of the components here, as elsewhere, follows the normal Voight convention. $\underline{r}'_s$, $\underline{r}_s$ and $\underline{\delta r}_s$ are 3-vectors with Cartesian components for the s ions per unit cell.

In the absence of external fields, u_L is an harmonic function of $\underline{\delta}$; thus from equation (1), the condition

$$\underline{\delta} = -\underline{\underline{W}}^{-1}.\underline{g} \tag{7}$$

defines a stationary point in the energy hypersurface which is a minimum provided W is positive definite. Even when u_L is not an harmonic function of the strain, the use of equation (7) gives a new configuration that is closer to the minimum provided that the initial W matrix is positive definite. There is, however, a basic ambiguity in the calculation of the optimum lattice configuration because the position of the lattice is undefined up to a constant translation of the constituent atoms. This does not affect possible rotations of the lattice since the strain matrix, $\Delta\varepsilon$, is symmetric; but the inherent ambiguity is such that the rank of W is three less than the dimension and $\underline{\underline{W}}^{-1}$ is not defined. However, the singularity in W can be resolved by fixing the position of a single atom, either by deleting the appropriate rows and columns of $\underline{\underline{W}}$ or, equivalently, by adding large diagonal terms to the matrix at the location of the fixed atom. It is in this sense that $\underline{\underline{W}}$ and $\underline{\underline{W}}^{-1}$ in equations (1) and (7) are interpreted.

Turning now to an explicit representation of $\underline{g}$ and $\underline{\underline{W}}$, in general the coordinate derivatives of the energy, $\partial u_L/\partial \underline{r}$ and $\partial^2 u_L/\partial \underline{r}\partial \underline{r}$ are quite straightforward to evaluate. Thus for lattice models based on pair potentials

$$u_L = \sum_{i>j} \Phi_{ij}(|\underline{r}_i-\underline{r}_j|) \tag{8}$$

$$\partial u_L/\partial \underline{r}_i^\alpha = \sum_j [(\underline{r}_i-\underline{r}_j)^\alpha/|\underline{r}_i-\underline{r}_j|] \; \Phi'_{ij}(|\underline{r}_i-\underline{r}_j|) \tag{9}$$

$$\partial^2 u_L/\partial \underline{r}_i^\alpha \partial \underline{r}_j^\beta = -\delta_{\alpha\beta} \; \Phi'_{ij}(|\underline{r}_i-\underline{r}_j|)/|\underline{r}_i-\underline{r}_j|$$

$$- [(\underline{r}_i-\underline{r}_j)^\alpha(\underline{r}_i-\underline{r}_j)^\beta/|\underline{r}_i-\underline{r}_j|^2][\Phi''_{ij}(|\underline{r}_i-\underline{r}_j|) - \Phi'_{ij}(|\underline{r}_i-\underline{r}_j|)$$

$$/|\underline{r}_i-\underline{r}_j|] \qquad \text{for } i\neq j \tag{10}$$

and

$$\partial^2 u_L/\partial \underline{r}_i^\alpha \; \partial \underline{r}_i^\beta = \sum_{j\neq i} \{\delta_{\alpha\beta} \; \Phi'_{ij}(|\underline{r}_i-\underline{r}_j|)/|\underline{r}_i-\underline{r}_j|$$

$$+ [(\underline{r}_i-\underline{r}_j)^\alpha(\underline{r}_i-\underline{r}_j)^\beta/|\underline{r}_i-\underline{r}_j|^2][\Phi''_{ij}(|\underline{r}_i-\underline{r}_j|) - \Phi'_{ij}(|\underline{r}_i-\underline{r}_j|)/|\underline{r}_i-\underline{r}_j|]$$

$$\tag{11}$$

in which $\delta_{\alpha\beta}$ is the Kronker delta, α is the coordinate index and Φ'_{ij} and Φ''_{ij} the first and second derivatives of Φ_{ij}. Hereafter, we use the following abbreviations

$$r_{ij} = |\underline{r}_i-\underline{r}_j| , \tag{12a}$$

$$r_{ij}^\alpha = (\underline{r}_i-\underline{r}_j)^\alpha, \tag{12b}$$

$$v_{ij}^{(1)} = \Phi'_{ij}(|\underline{r}_i-\underline{r}_j|)/|\underline{r}_i-\underline{r}_j| \tag{12c}$$

$$v_{ij}^{(2)} = [\Phi''_{ij}(|\underline{r}_i-\underline{r}_j|) - v_{ij}^{(1)}]/|\underline{r}_i-\underline{r}_j|^2. \tag{12d}$$

The strain derivatives of the energy, on the other hand, are somewhat more complicated to derive. Following Born and Huang[2], it is simplest to regard the potential function as dependent on the square of the separation. From the definition of the bulk strains, $\{\varepsilon_\kappa\}$, $\kappa=1,\ldots,6$, then

$$\underline{r}' = (\underline{\underline{I}}+\underline{\underline{\varepsilon}})\cdot\underline{r} \tag{13}$$

in which I is a unit matrix and $\underline{\underline{\varepsilon}}$ the symmetric strain tensor formed from $\{\varepsilon_\kappa\}$ (see equation (6)),

$$r'^2 = r^2 + 2\underline{r}^T \cdot \underline{\underline{\epsilon}} \cdot \underline{r} + \underline{r}^T \cdot \underline{\underline{\epsilon}}^2 \cdot \underline{r} \tag{14}$$

in which $r = |\underline{r}|$. At zero strain we obtain the following identities

$$\tfrac{1}{2}\partial r^2/\partial \epsilon_\kappa = r^\alpha r^\beta \qquad \text{for} \quad \left\{ \begin{array}{l} \kappa=1,2,3,4,5,6 \\ \alpha=x,y,z,y,x,x \\ \beta=x,y,z,z,z,y; \end{array} \right. \tag{15}$$

$$\tfrac{1}{2}\partial^2 r^2/\partial r^\alpha \partial \epsilon_\kappa = 2r^\alpha \quad \text{for} \left\{ \begin{array}{l} \kappa=1,2,3 \\ \alpha=x,y,z; \end{array} \right. \tag{16}$$

$$= r^\beta \quad \text{for} \left\{ \begin{array}{l} \kappa=4,4,5,5,6,6 \\ \alpha=y,z,x,z,x,y \\ \beta=z,y,z,x,y,x; \end{array} \right. \tag{17}$$

$$\tfrac{1}{2}\partial^2 r^2/\partial \epsilon_\kappa \partial \epsilon_\kappa = r^\alpha r^\alpha \quad \text{for} \left\{ \begin{array}{l} \kappa=1,2,3 \\ \alpha=x,y,z; \end{array} \right. \tag{18}$$

$$= \tfrac{1}{4}(r^\alpha r^\alpha + r^\beta r^\beta) \quad \text{for} \left\{ \begin{array}{l} \kappa=4,5,6 \\ \alpha=y,x,x \\ \beta=z,z,y; \end{array} \right. \tag{19}$$

$$\tfrac{1}{2}\partial^2 r^2/\partial \epsilon_\kappa \partial \epsilon_\ell = \tfrac{1}{2}r^\alpha r^\beta \quad \text{for} \left\{ \begin{array}{l} \kappa=1,1,2,2,3,3 \\ \ell=5,6,4,6,4,5 \\ \alpha=x,x,y,x,y,x \\ \beta=z,y,z,y,z,z; \end{array} \right. \tag{20}$$

$$= \tfrac{1}{4}r^\alpha r^\beta \quad \text{for} \left\{ \begin{array}{l} \kappa=4,4,5 \\ \ell=5,6,6 \\ \alpha=x,x,y \\ \beta=y,z,z; \end{array} \right. \tag{21}$$

$$= 0 \quad \text{for} \left\{ \begin{array}{l} \kappa=1,1,2,1,2,3 \\ \ell=2,3,3,4,5,6 \end{array} \right. . \tag{22}$$

Thus,

$$\partial u_L/\partial \epsilon_\kappa = \sum_{i>j} (\tfrac{1}{2}\partial r_{ij}^2/\partial \epsilon_k) v_{ij}^{(1)} \tag{23}$$

$$= \sum_{i>j} r_{ij}^\alpha r_{ij}^\beta v_{ij}^{(1)} \tag{24}$$

with κ, α and β corresponding to the assignment given in equation (15). Similarly

$$\partial^2 u_L/\partial\epsilon_\kappa\partial\epsilon_\ell = \sum_{i>j} (\tfrac{1}{2}\partial^2 r_{ij}^2/\partial\epsilon_\kappa\partial\epsilon_\ell) \ V_{ij}^{(1)} + (\tfrac{1}{2}\partial r_{ij}^2/\partial\epsilon_\kappa)(\tfrac{1}{2}\partial r_{ij}^2/\partial\epsilon_\ell) \ V_{ij}^{(2)} \tag{25}$$

$$= \sum_{i>j} \{C_{\kappa\ell}^{\alpha\beta} \ V_{ij}^{(1)} + D_{\kappa\ell}^{\alpha\beta\gamma\lambda} \ V_{ij}^{(2)}\} . \tag{26}$$

For $\kappa=\ell$,

$$\left.\begin{matrix} C_{\kappa\ell}^{\alpha\beta} \\ D_{\kappa\ell}^{\alpha\beta\gamma\lambda} \end{matrix}\right\} = \begin{cases} r_{ij}^\alpha \ r_{ij}^\beta \\ r_{ij}^\alpha r_{ij}^\beta r_{ij}^\gamma r_{ij}^\lambda \end{cases} \quad \text{for} \begin{cases} \kappa=1,2,3 \\ \alpha \text{ or } \gamma=x,y,z \\ \beta \text{ or } \lambda=x,y,z; \end{cases} \tag{27a}$$

$$= \begin{cases} \tfrac{1}{4}(r_{ij}^\alpha r_{ij}^\alpha + r_{ij}^\beta r_{ij}^\beta) \\ r_{ij}^\alpha r_{ij}^\beta r_{ij}^\gamma r_{ij}^\lambda \end{cases} \quad \text{for} \begin{cases} \kappa=4,5,6 \\ \alpha \text{ or } \gamma=y,x,x \\ \beta \text{ or } \lambda=z,z,y; \end{cases} \tag{27b}$$

For $\kappa\neq\ell$

$$= \begin{cases} 0 \\ r_{ij}^\alpha r_{ij}^\beta r_{ij}^\gamma r_{ij}^\lambda \end{cases} \quad \text{for} \begin{cases} \kappa \text{ or } \ell=1,2,3,4,5,6 \\ \alpha \text{ or } \gamma=x,y,z,y,x,x \\ \beta \text{ or } \lambda=x,y,z,z,z,y; \end{cases} \tag{27c}$$

$$\text{except } \kappa=1,1 \\ \ell=5,6$$

$$= \begin{cases} \tfrac{1}{2}r_{ij}^\alpha r_{ij}^\beta \\ 0 \end{cases} \quad \text{for} \begin{cases} \kappa=1,1,2,2,3,3 \\ \ell=5,6,4,6,4,5 \\ \alpha=x,x,y,x,y,x \\ \beta=z,y,z,y,z,z; \end{cases} \tag{27d}$$

$$= \begin{cases} \tfrac{1}{4}r_{ij}^\alpha r_{ij}^\beta \\ 0 \end{cases} \quad \text{for} \begin{cases} \kappa=4,4,5 \\ \ell=5,6,6 \\ \alpha=x,x,y \\ \beta=y,z,z . \end{cases} \tag{27e}$$

The mixed derivatives, $\partial^2 u_L/\partial\underline{r}\partial\underline{\epsilon}$ are obtained from equations (24) and (15) to (17). Thus

$$\partial^2 u_L/\partial\underline{r}_i^\alpha\partial\epsilon_\kappa = \sum_j (\tfrac{1}{2}\partial^2 r_{ij}^2/\partial\underline{r}_i^\alpha\partial\epsilon_\kappa) \ V_{ij}^{(1)} + (\tfrac{1}{2}\partial r_{ij}^2/\partial\underline{r}_i^\alpha)(\tfrac{1}{2}\partial r_{ij}^2/\partial\epsilon_\kappa) \ V_{ij}^{(2)} \tag{28}$$

$$= \sum_j \{C_\kappa^\alpha \ V_{ij}^{(1)} + D_\kappa^{\alpha\beta\gamma} \ V_{ij}^{(2)}\} \tag{29}$$

with

$$\left.\begin{matrix} C_\kappa^\alpha \\ D_\kappa^{\alpha\beta\gamma} \end{matrix}\right\} = \begin{cases} 2r_{ij}^\alpha \\ r_{ij}^\alpha r_{ij}^\beta r_{ij}^\gamma \end{cases} \quad \text{for} \quad \begin{cases} & \kappa=1,2,3 \\ \alpha,\beta \text{ or } & \gamma=x,y,z; \end{cases} \tag{30a}$$

$$= \begin{cases} 0 \\ r_{ij}^\alpha r_{ij}^\beta r_{ij}^\gamma \end{cases} \quad \text{for} \quad \begin{cases} & \kappa=1,1,2,2,3,3 \\ & \alpha=y,z,x,z,x,y \\ \beta \text{ or } & \gamma=x,x,y,y,z,z; \end{cases} \tag{30b}$$

$$= \begin{cases} r_{ij}^\alpha \\ r_{ij}^\alpha r_{ij}^\beta r_{ij}^\gamma \end{cases} \quad \text{for} \quad \begin{cases} \kappa=4,4,5,5,6,6 \\ \alpha=y,z,x,z,x,y \\ \beta=y,y,x,x,x,x \\ \gamma=z,z,z,z,y,y \end{cases} . \tag{30c}$$

We have, therefore, a complete set of expressions for all the terms occurring in equations (1) and (7) from which the zero strain lattice energy and equilibrium configuration can be determined.

To complete the treatment of the non-defective or perfect lattice we consider the way in which the strain derivatives of the lattice energy relate to the dielectric, piezoelectric and elastic constants. In the absence of external fields the equilibrium condition for lattices without a permanent moment is $\underline{g}=0$. In the presence of an external field, $\underline{E}_{ext}$, therefore, we can write

$$u_L(\underline{R}) = u_L(\underline{R}^e) + \tfrac{1}{2}\underline{\delta}^T.\underline{\underline{W}}.\underline{\delta} - \underline{q}^T.\delta\underline{r}^\alpha \ E_{ext}^\alpha \tag{31}$$

in which $\underline{R}^e$ denotes the field free equilibrium configuration and q an s-dimensional vector of charges that couple to the external field. The repeated coordinate index, α, implies a summation over all components. Expanding equation (31) we have

$$u_L(\underline{R}) = u_L(\underline{R}^e) + \tfrac{1}{2}\delta\underline{r}^T.\underline{\underline{W}}_{rr}.\delta\underline{r} + \delta\varepsilon.\underline{\underline{W}}_{\varepsilon r}.\delta\underline{r} + \tfrac{1}{2}\delta\varepsilon.\underline{\underline{W}}_{\varepsilon\varepsilon}.\delta\varepsilon$$
$$- \underline{q}^T.\delta\underline{r}^\alpha \ E_{ext}^\alpha \tag{32}$$

in which $\underline{\underline{W}}_{rr}=\partial^2 u_L/\partial\underline{r}\partial\underline{r}$, etc. From the equilibrium condition

$$\partial u_L/\partial\delta\underline{r} = 0 \tag{33}$$

we get

$$\delta\underline{r}^\alpha = - [\underline{\underline{W}}_{rr}^{-1}.\underline{\underline{W}}_{r\varepsilon}.\delta\underline{\varepsilon}]^\alpha + [\underline{\underline{W}}_{rr}^{-1}]^{\alpha\beta}.\underline{q} \ E_{ext}^\beta \tag{34}$$

corresponding to the equilibrium displacement of the atoms in the presence of an external field $\underline{E}_{ext}$. From the definition of the electric displacement field $\underline{D}$, we have

$$\underline{D}^\alpha = \underline{E}_{ext}^\alpha + (4\pi/v_c) \ \underline{q}^T.\delta\underline{r}^\alpha \tag{35}$$

which on substituting for $\delta \underline{r}^\alpha$ from equation (34) gives

$$\underline{D}^\alpha = -(4\pi/v_c) \; \underline{q}^T \cdot [\underline{\underline{W}}_{rr}^{-1} \cdot \underline{\underline{W}}_{r\epsilon} \cdot \delta \underline{\epsilon}]^\alpha + (\delta_{\alpha\beta} + (4\pi/v_c) \; \underline{q}^T \cdot [\underline{\underline{W}}_{rr}^{-1}]^{\alpha\beta} \cdot \underline{q}) \; \underline{E}_{ext}^\beta \quad . \tag{36}$$

The piezoelectric and dielectric constant can be identified immediately by comparison with the definition

$$\underline{D}^\alpha = \sum_i \; \underline{\lambda}_i^\alpha \; \epsilon_i + \sum_\beta \; \underline{\underline{\kappa}}^{\alpha\beta} \; \underline{E}_{ext}^\beta \quad . \tag{37}$$

where $\underline{\lambda}$ and $\underline{\kappa}$ are the piezoelectric and dielectric tensors respectively. Thus, the piezoelectric tensor is given by

$$\underline{\lambda}_i^\alpha = -(4\pi/v_c) \; \underline{q}^T \cdot [\underline{\underline{W}}_{rr}^{-1} \cdot \underline{\underline{W}}_{r\epsilon}]_i^\alpha \tag{38}$$

and the dielectric tensor by

$$\underline{\underline{\kappa}}^{\alpha\beta} = \delta^{\alpha\beta} + (4\pi/v_c) \; \underline{q}^T \cdot [\underline{\underline{W}}_{rr}^{-1}]^{\alpha\beta} \cdot \underline{q} \tag{39}$$

Elastic constants are defined as the second-derivatives of the lattice energy with respect to strain, with the lattice energy normalised to unit volume. At zero electric field, substitution of equation (34) in (32) gives

$$u_L(\underline{R}) = u_L(\underline{R}^e) + \tfrac{1}{2}\delta\epsilon \cdot [\underline{\underline{W}}_{\epsilon\epsilon} - \underline{\underline{W}}_{\epsilon r} \cdot \underline{\underline{W}}_{rr}^{-1} \cdot \underline{\underline{W}}_{r\epsilon}] \cdot \delta\underline{\epsilon} \tag{40}$$

from which the elastic constraint tensor $\underline{\underline{C}}$ (with Voight components) can immediately be identified as

$$\underline{\underline{C}} = (1/v_c) \; [\underline{\underline{W}}_{\epsilon\epsilon} - \underline{\underline{W}}_{\epsilon r} \cdot \underline{\underline{W}}_{rr}^{-1} \cdot \underline{\underline{W}}_{r\epsilon}] \quad . \tag{41}$$

3. The equations for the defective lattice

The most convenient general formulation for treating the defective lattice is that developed by Lidiard and Norgett[3] and Norgett[4]. It is based on the notion that the total energy of the system is minimised by a relaxation of the ions surrounding a defect, and that this relaxation decreases fairly rapidly for distances away from the defect. As a result, the crystal can be formally partitioned into an inner region I in which the lattice configuration is evaluated explicitly, and an outer region II, which can be viewed from the defect as a continuum, within which the displacements can be calculated on the basis of some suitable approximation. Quite formally, then, the total energy of the system is written as

$$E = E_1(\underline{x}) + E_2(\underline{x},\underline{\zeta}) + E_3(\underline{\zeta}) \tag{42}$$

in which $E_1(\underline{x})$ is the energy of the inner region, $E_3(\underline{\zeta})$, the energy of the outer region and $E_2(\underline{x},\underline{\zeta})$, the interaction energy between regions I and II. $\underline{x}$ is a vector of the independent coordinates describing the inner region, while $\underline{\zeta}$ is a corresponding vector of the displacements in the outer region and is both formally distinguished from $\underline{x}$ and assumed to be an implicit function of it. Thus

$$\underline{\zeta} \equiv \underline{\zeta}(\underline{x}) \tag{43}$$

though, in practice, $\underline{\zeta}$ is more often than not determined solely by the position of the defects and not by the surrounding ions in region I. Now $E_3(\underline{\zeta})$ is _defined_ to be a quadratic function of $\underline{\zeta}$

$$E_3(\underline{\zeta}) = \tfrac{1}{2}\underline{\zeta}^T \cdot \underline{\underline{A}} \cdot \underline{\zeta} \tag{44}$$

which together with the equilibrium condition

$$\partial E/\partial\underline{\zeta} = \partial E_2(\underline{x},\underline{\zeta})/\partial\underline{\zeta} \bigg|_{\underline{\zeta}=\overline{\underline{\zeta}}} + \underline{\underline{A}} \cdot \overline{\underline{\zeta}} \tag{45}$$

$$= 0 \tag{46}$$

lead to an alternative expression for $E_3(\underline{\zeta})$, and hence for the total energy E. Thus

$$E_3(\underline{\zeta}) = - \tfrac{1}{2}\partial E_2(\underline{x},\underline{\zeta})/\partial\underline{\zeta} \bigg|_{\underline{\zeta}=\overline{\underline{\zeta}}} \cdot \overline{\underline{\zeta}} \tag{47}$$

and

$$E = E_1(\underline{x}) + E_2(\underline{x},\underline{\zeta}) - \tfrac{1}{2} E_2(\underline{x},\underline{\zeta})/\partial\underline{\zeta} \bigg|_{\underline{\zeta}=\overline{\underline{\zeta}}} \cdot \overline{\underline{\zeta}} \tag{48}$$

$\overline{\underline{\zeta}}$ are the equilibrium values $\underline{\zeta}$ corresponding to arbitrary values of $\underline{x}$ determined from equation (45), and hence are also implicit functions of the inner region coordinates.

Still proceeding formally, the defect energy, E, can in principle be found by minimising the energy directly, i.e. by solving the equations

$$dE/d\underline{x} = 0 \tag{49}$$

in which the total derivative implies an _explicit_ differentiation of $\overline{\underline{\zeta}}$, the outer region displacements, with respect to $\underline{x}$. Now as Norgett[4] has pointed out, while this is consistent, it is difficult to apply in practice since the complicated nature of E precludes an analytic evaluation of $dE/\partial\underline{x}$. Alternatively, E can be found by requiring that the force on each ion in the inner region is zero, i.e. by solving the partial differential equations

$$\partial E/\partial\underline{x} \bigg|_{\underline{\zeta} = \text{constant}} = 0 \tag{50}$$

Writing

$$dE/d\underline{x} = \partial E/\partial\underline{x} \bigg|_{\underline{\zeta}=\overline{\underline{\zeta}}} + \partial E/\partial\zeta \bigg|_{\underline{x}} \cdot \partial\underline{\zeta}/\partial\underline{x} \tag{51}$$

it can be seen that the two approaches are equivalent if

$$\partial E/\partial\underline{\zeta} \bigg|_{\underline{x}} = 0 \tag{52}$$

i.e. if region II is in equilibrium. The complete force-balance equations, then, are

$$\partial E_1/\partial \underline{x} + \partial E_2(\underline{x},\underline{\varsigma})/\partial \underline{x}\Big|_{\underline{\varsigma}=\underline{\bar{\varsigma}}} - \tfrac{1}{2}\partial^2 E_2(\underline{x},\underline{\varsigma})/\partial \underline{x}\partial \underline{\varsigma}\Big|_{\underline{\varsigma}=\underline{\bar{\varsigma}}} \cdot \underline{\bar{\varsigma}} = 0. \qquad (53)$$

As they stand equations (42) to (53) represent a purely formal approach to the problem. To develop them further we need to consider an explicit representation for the energy, which as in section 1 we take to be the sum of two-body interactions. Thus the energy of the perfect or non-defective lattice, E_L we write as

$$E_L = \sum_{i>j} \Phi_{ij}(|\underline{R}_i - \underline{R}_j|) . \qquad (54)$$

in which $\{\underline{R}_i\}$ represents the equilibrium configuration discussed in the previous section. Similarly, the energy of the defective lattice, E_D, is written as

$$E_D = \sum_{i>j}{}' \Phi_{ij}(|\underline{r}_i - \underline{r}_j|) \qquad (55)$$

in which $\{\underline{r}_i\}$ are the displaced lattice coordinates. The defect energy, E, therefore, is given by

$$E = \sum_{i>j}{}' \Phi_{ij}(|\underline{r}_i - \underline{r}_j|) - \sum_{i>j} \Phi_{ij}(|\underline{R}_i - \underline{R}_j|) \qquad (56)$$

in which it is important at this stage to distinguish between the summation over the perfect and defective lattices, since defects involving vacancies and interstitials have no immediate correspondence from one summation to the other in the usual sense. However, by defining a summation convention in which,

    i) the initial position of an interstitial ion is taken to be at infinity,

    ii) the final position of an ion that is removed from the lattice to form a vacancy is also taken to be at infinity,

and

    iii) substitutions are treated as a combination of a vacancy plus an interstitial

we can use a single summation, $\sum''_{i>j}$, for both the perfect and defective lattice. Thus, E, the defect energy can be written as a single sum

$$E = \sum_{i>j}{}'' \{\Phi_{ij}(|\underline{r}_i - \underline{r}_j|) - \Phi_{ij}(|\underline{R}_i - \underline{R}_j|)\} \qquad (57)$$

in which the summation extends over all ions and over all space. For an interstitial ion, then, $\underline{R}_{int} = \infty$ in equation (57), whereas for a vacancy (ion), $\underline{r}_{vac} = \infty$.

Now the identification of equation (57) with equation (42) or its alternative form, equation (48) is not quite as straightforward as it might seem. For while, E_1 is evidently given by

$$E_1 = \sum_{\substack{i,j\in I\\ i>j}}{}'' \{\Phi_{ij}(|\underline{r}_i - \underline{r}_j|) - \Phi_{ij}(|\underline{R}_i - \underline{R}_j|)\} \qquad (58)$$

and E_2 can be written as

$$E_2 = \sum_{\substack{i \in I \\ i \in II}}^{"} \{\Phi_{ij}(|\underline{r}_i - \underline{r}_j|) - \Phi_{ij}(|\underline{R}_i - \underline{R}_j|)\} \tag{59}$$

a simple summation of the type $\sum_{\substack{i,j \in II \\ i>j}}^{"}$, for the outer region is not an adequate

description for E_3, since it is not a quadratic function of the displacements, $\{(\underline{r}_i - \underline{R}_i)\}$, as required by equation (44). To see this we write $\underline{r}_i$ and $\underline{r}_j$ in the form

$$\underline{r}_i = \underline{R}_i + \underline{\zeta}_i \tag{60a}$$

and

$$\underline{r}_j = \underline{R}_j + \underline{\zeta}_j \tag{60b}$$

and expand the function, E_3', to second order in $\underline{\zeta}_i$ and $\underline{\zeta}_j$, the outer region displacements. Thus

$$E_3' = \sum_{\substack{i,j \in II \\ i>j}}^{"} \{\Phi_{ij}(|\underline{r}_i - \underline{r}_j|) - \Phi_{ij}(|\underline{R}_i - \underline{R}_j|) \tag{61}$$

$$= \sum_{\substack{i,j \in II \\ i>j}}^{"} \{(\partial \Phi_{ij}(|\underline{R}_i - \underline{r}_j|)/\partial \underline{r}_j) \cdot \underline{\zeta}_j + (\partial \Phi_{ij}(|\underline{r}_i - \underline{R}_j|)/\partial \underline{r}_i) \cdot \underline{\zeta}_i$$

$$+ \tfrac{1}{2}[\underline{\zeta}_j \cdot (\partial^2 \Phi_{ij}(|\underline{R}_i - \underline{r}_j|)/\partial \underline{r}_j \partial \underline{r}_j) \cdot \underline{\zeta}_j]$$

$$+ \tfrac{1}{2}[\underline{\zeta}_i \cdot (\partial^2 \Phi_{ij}(|\underline{r}_i - \underline{R}_j|)/\partial \underline{r}_i \partial \underline{r}_i) \cdot \underline{\zeta}_i] + \text{cross terms} \tag{62}$$

which is evidently linear in the displacements. To eliminate this linear dependence we consider the auxiliary function F, which we expand, as before, as a power series in $\underline{\zeta}_j$

$$F = \sum_{\substack{i \in I \\ j \in II}}^{"} \{\Phi_{ij}(|\underline{R}_i - \underline{r}_j|) - \Phi_{ij}(|\underline{R}_i - \underline{R}_j|)\} \tag{63}$$

$$= \sum_{\substack{i \in I \\ j \in II}}^{"} \{(\partial \Phi_{ij}(|\underline{R}_i - \underline{r}_j|)/\partial \underline{r}_j) \cdot \underline{\zeta}_j$$

$$+ \tfrac{1}{2} [\underline{\zeta}_j \cdot (\partial^2 \Phi_{ij}(|\underline{R}_i - \underline{r}_i|)/\partial \underline{r}_j \partial \underline{r}_j) \cdot \underline{\zeta}_j] \quad . \tag{64}$$

Adding F+E$_3'$ and collecting the <u>linear terms only</u>, we get

$$F+E_3^1 \text{ (linear terms)} = \sum_{\substack{i\in I\& II \\ j\in II \\ i\neq j}}'' (\partial\Phi_{ij}(|\underline{R}_i-\underline{r}_j|)/\partial\underline{r}_j)\cdot \underline{\varsigma}_j \qquad (65)$$

in which we have made use of the identity

$$\sum_{\substack{i,j\in II \\ i>j}} \{(\partial\Phi_{ij}(|\underline{R}_i-\underline{r}_i|)/\partial\underline{r}_j)\cdot \underline{\varsigma}_j + (\partial\Phi_{ij}(|\underline{r}_i-\underline{R}_j|)/\partial\underline{r}_i)\cdot \underline{\varsigma}_i\}$$

$$= \sum_{\substack{i,j\in II \\ i\neq j}} (\partial\Phi_{ij}(|\underline{R}_i-\underline{r}_j|)/\partial\underline{r}_j)\cdot \underline{\varsigma}_j \qquad (66)$$

since the summations are symmetric in i and j. The linear terms of $F+E_3^1$ can be re-written as

$$F+E_3^1 \text{ (linear terms)} = \sum_{j\in II}'' \underline{\varsigma}_j \cdot \sum_{\substack{i\in \text{ all space} \\ i\neq j}}'' (\partial\Phi(|\underline{R}_i-\underline{r}_j|)/\partial\underline{r}_j). \qquad (67)$$

Now for a lattice at equilibrium, the interaction energy, e_j of a single atom at $\underline{r}_j$ with the remaining lattice, is

$$e_j = \sum_{\substack{i\in \text{ all space} \\ i\neq j}}'' \Phi_{ij}(|\underline{R}_i-\underline{r}_j|). \qquad (68)$$

and the force on j

$$\partial e_j/\partial\underline{r}_j = \sum_{\substack{i\in \text{ all space} \\ i\neq j}}'' \partial\Phi_{ij}(|\underline{R}_i-\underline{r}_j|)/\partial\underline{r}_j \qquad (69)$$

which at equilibrium is zero. The summation over i, therefore, in (69) is zero, so that the function $F+E_3^1$ has no linear dependence, the lowest order terms being those quadratic in $\underline{\varsigma}_i$ and $\underline{\varsigma}_j$ the displacements of the outer region. Thus we write

$$E_3 = F+E_3^1 \qquad (70)$$

so that E_3 as defined by equation (70) is given by

$$E_3 = \sum_{\substack{i,j\in II \\ i>j}}'' \{\Phi_{ij}(|\underline{r}_i-\underline{r}_j|) - \Phi_{ij}(|\underline{R}_i-\underline{R}_j|)\}$$

$$+ \sum_{\substack{i\in I \\ i\in II}} \{\Phi_{ij}(|\underline{R}_i-\underline{r}_j|) - \Phi_{ij}(|\underline{R}_i-\underline{R}_j|)\}. \qquad (71)$$

To compensate for the additional terms, via. $\sum_{\substack{i\in I \\ j\in II}}'' \{\Phi_{ij}(|\underline{R}_i-\underline{r}_j|)-\Phi_{ij}(|\underline{R}_i-\underline{R}_j|)\}$

we subtract them from E_2, which now becomes

$$E_2 = \sum_{\substack{i\in I \\ j\in II}}^{\prime\prime} \{\Phi_{ij}(|\underline{r}_i-\underline{r}_j|) - \Phi_{ij}(|\underline{R}_i-\underline{r}_j|)\} \tag{72}$$

from which E_3 can be re-written as

$$E_3 = -\tfrac{1}{2} \sum_{\substack{i\in I \\ j\in II}}^{\prime\prime} \{\partial\Phi_{ij}(|\underline{r}_i-\underline{r}_j|)/\partial\underline{r}_j -\partial\Phi_{ij}(|\underline{R}_i-\underline{r}_j|)/\partial\underline{r}_j\}.(\underline{r}_j-\underline{R}_j) . \tag{73}$$

The complete expression for E, the energy of lattice defect, then, is

$$E = \sum_{\substack{i,j\in I \\ i>j}}^{\prime\prime} \{\Phi_{ij}(|\underline{r}_i-\underline{r}_j|) - \Phi_{ij}(|\underline{R}_i-\underline{R}_j|)\}$$

$$+ \sum_{\substack{i\in I \\ j\in II}}^{\prime\prime} \{\Phi_{ij}(|\underline{r}_i-\underline{r}_j|) - \Phi_{ij}(|\underline{R}_i-\underline{r}_j|)\}$$

$$-\tfrac{1}{2} \sum_{\substack{i\in I \\ j\in II}}^{\prime\prime} \{\partial\Phi_{ij}(|\underline{r}_i-\underline{r}_j|)/\partial\underline{r}_j - \partial\Phi_{ij}(|\underline{R}_i-\underline{r}_j|)/\partial\underline{r}_j\}. (\underline{r}_j-\underline{R}_j) . \tag{74}$$

It involves interactions only between ions in region I and between those in region I and region II: interactions between ions in region II have been eliminated. The force balance equations corresponding to equation (74) follow from equation (53). Thus for the α-component of the displacement of the ith atom in region I we have

$$\sum_{j\in I\&II}^{\prime\prime} [(\underline{r}_i-\underline{r}_j)^\alpha/|\underline{r}_i-\underline{r}_j|] \Phi_{ij}^{\prime}(|\underline{r}_i-\underline{r}_j|) + \tfrac{1}{2} \sum_{j\in II}^{\prime\prime} \sum_{\beta} \{\delta_{\alpha\beta} \Phi_{ij}^{\prime}(|\underline{r}_i-\underline{r}_j|)/|\underline{r}_i-\underline{r}_j|$$

$$+ [(\underline{r}_i-\underline{r}_j)^\alpha(\underline{r}_i-\underline{r}_j)^\beta/|\underline{r}_i-\underline{r}_j|^2][\Phi_{ij}^{\prime\prime}(|\underline{r}_i-\underline{r}_j|) - \Phi_{ij}^{\prime}(|\underline{r}_i-\underline{r}_j|)/|\underline{r}_i-\underline{r}_j|]\}$$

$$\times (\underline{r}_j-\underline{R}_j)^\beta = 0. \tag{75}$$

In equations (74) and (75), therefore, we have complete expressions for the total defect energy, E, and the force on each atom in the inner region. It is important to emphasise that they are perfectly general and apply to all types. They derive solely from

(i) the <u>partition</u> of the total defect energy, E, in the form given in equation (42)

(ii) the assumption that the energy of region II is a quadratic function of the displacements and

(iii) the pair-wise additivity of the total energy.

An extension to include higher-order interactions is quite straightforward in principle,though the resulting equations are somewhat unwieldly.

Now the precise way in which equations (74) and (75) are implemented depends on the nature of the potential, Φ_{ij}. For potentials of all types the interaction between atoms within region I is purely local and is best evaluated by explicit summation. The interaction between regions I and II, however, require special attention. In the case of short-range potentials of the type involved in exchange interactions for example, it is only the inner part of region II that makes a significant contribution to E and the forces on region I. In which case we define a truncated region II, viz. II', and carry out the various summations explicitly as before. In the case of long-range interactions which predominate in ionic materials, on the other hand, alternative methods need to be devised. In Norgett's HADES* formulation for ionic materials, for example[4-6], the Coulombic contribution to E_2 (and E_3) corresponding to the outer part of region II which formally extends to infinity, reduces to the form

$$E_2^{outer}(\text{Coulombic}) = -Q \sum_{j \in II(outer)} q_j (\underline{\zeta}_j \cdot \underline{R}_j)/|\underline{R}_j|^3 \tag{76}$$

in which Q is the total **effective** charge of the defect which is taken at the origin, and q_j, $\underline{\zeta}_j$ and $\underline{R}_j$ the charge, displacement and equilibrium position of the jth ion in region II. In a Mott-Littleton approximation, $\underline{\zeta}_j$ can be obtained from equation (35) in which $\underline{E}_{ext}$ is taken to be the electric field due to the charge Q at the origin. For cubic materials, $E_2^{outer}(\text{Coulombic})$ reduces to

$$E_2^{outer}(\text{Coulombic}) = -\tfrac{1}{2} Q^2 \sum_{j \in II \; outer} q_j \, K_j/|\underline{R}_j|^4 \tag{77}$$

whereas for non-cubic materials it takes the more general form

$$E_2^{outer}(\text{Coulombic}) = -\tfrac{1}{2} Q^2 \sum_{j \in II \; outer} \sum_{\alpha,\beta} M_{ij}^{\alpha\beta} R_j^\alpha R_j^\beta/|\underline{R}_j|^6 \tag{78}$$

in which the constants K_j and $M_j^{\alpha\beta}$ are derived from the $\underset{=}{W}$ matrix. The most straightforward method of evaluating the partial sums occurring in equations (77) and (78) is to evaluate the complete sums analytically and then subtract the inner part which can be obtained by explicit summation. Analytic sums for the general expression have recently been derived[7] and implemented in the most general version of HADES and other computational procedures currently in use[8,9].

---

*Harwell Automatic Defect Evaluation System

4. Minimisation methods

As mentioned in section 1 and 2 lattice energy calculations for both the non-defective and defective lattice require function minimisation. In this final section, therefore, we briefly review the methods that have been found to be best suited to lattice calculations, and to ionic materials in particular. Quite generally, the function $F(\underline{x})$, which can be either the lattice energy, equation (1) or the force on a particular atom in the inner region, equation (50) is expanded to second order thus

$$F(\underline{x}) = F(\underline{x}') + \underline{g}^T.\underline{\delta} + \tfrac{1}{2}\,\underline{\delta}^T.\underline{\underline{H}}.\underline{\delta} \tag{79}$$

in which

$$\underline{\delta} = \underline{x} - \underline{x}' \tag{80a}$$

$$\underline{g} = \partial F(\underline{x})/\partial \underline{x} \tag{80b}$$

and

$$\underline{\underline{H}} = \partial^2 F(\underline{x})/\partial \underline{x}\partial \underline{x} \tag{80c}$$

$F(\underline{x})$ is clearly a minimum for

$$\underline{\delta} = -\underline{\underline{H}}^{-1}.\underline{g} \tag{81}$$

but since $F(\underline{x})$ in most cases is not really a quadratic function in $\underline{x}$, the repeated use of equation (81) is necessary for convergence. The difficulty with such an approach, however, is that for problems involving a large number of variables, which in lattice calculations can quite easily be of the order of 200-300, the evaluation of $\underline{\underline{H}}^{-1}$ is extremely time-consuming. The process of matrix inversion requires $O(n^3)$ operations, where n is the number of variables, whereas the evaluation of F for pair potentials, for example, requires $O(n^2)$ operations. Large lattice calculations if they are to be efficient, therefore, require alternative methods. The most useful to emerge so far is that based on the so-called variable-metric method introduced by Davidon[10] and developed by Fletcher and Powell[11], and applied to lattice calculations by Fletcher and Norgett[12]. The essence of this method is to replace equation (81) by a similar equation of the form

$$\underline{\delta} = -\lambda\,\underline{\underline{G}}.\underline{g} \tag{82}$$

in which $\underline{\underline{G}}$ is an approximation to $\underline{\underline{H}}^{-1}$ and λ is a linear parameter. Initially $\underline{\underline{G}}$ is set equal to $\underline{\underline{H}}^{-1}$ (with $\lambda=1$): thereafter it is updated using one of a number of formulae that have been suggested (11,13,14). Within an iterative procedure, then, we have for the kth iteration

$$\underline{\delta}_k = -\lambda_k\,\underline{\underline{G}}_k.\underline{g}_k \tag{83}$$

in which

$$\underline{\delta}_k = \underline{x}_{k+1} - \underline{x}_k \tag{84}$$

g_k is evaluated at each iteration. Writing

$$Y_k = g_{k+1} - g_k \tag{85}$$

the most useful updating formulae for $\underline{\underline{G}}$ have been found to be:

Davidon-Fletcher-Powell (11)

$$\underline{\underline{G}}_{k+1} = \underline{\underline{G}}_k - \frac{(\underline{\underline{G}}_k \cdot Y_k) \cdot (Y_k^T \cdot \underline{\underline{G}}_k)}{(Y^T \cdot \underline{\underline{G}}_k \cdot Y)} - \frac{(\delta_k \cdot \delta_k^T)}{(\delta_k^T \cdot Y_k)} \tag{86}$$

Davidon (13)

$$\underline{\underline{G}}_{k+1} = \frac{\underline{\underline{G}}_k + (\delta_k - \underline{\underline{G}}_k \cdot Y_k)(\delta_k - \underline{\underline{G}}_k \cdot Y_k)^T}{(\delta^T \cdot Y_k) - (Y_k^T \cdot \underline{\underline{G}} \cdot Y_k)} \tag{87}$$

Broyden, Fletcher and Shanno (14,15)

$$\underline{\underline{G}}_{k+1} = \left(1 - \frac{\delta_k \cdot Y_k^T}{\delta_k^T \cdot Y_k}\right) \cdot \underline{\underline{G}}_k \cdot \left(1 - \frac{Y_k \cdot \delta_k^T}{\delta_k^T \cdot Y_k}\right) + \frac{(\delta_k \cdot \delta_k^T)}{(\delta_k^T \cdot Y_k)} \tag{88}$$

To ensure quadratic termination, the Davidon-Fletcher-Powell and Broyden-Fletcher-Shanno formulae require a linear search to determine λ_k. However, Fletcher[16] has suggested an algorithm which requires only occasional linear search and this is implemented in recent defect lattice procedures[16]. In the majority of defect calculations an initial matrix inversion ($\underline{\underline{G}}_o = \underline{\underline{H}}^{-1}$) followed by updating seems to give rapid convergence; but there are occasions when a second inversion after four or five iterations is preferable.

Despite the effectiveness of Newton and related methods, a basic computational problem arises in the necessity of storing the Hessian matrix. Thus for large low symmetry crystal structures calculations may not prove possible even with the largest memories available on modern computers. For this reason we have implemented for such calculations, the somewhat simpler conjugate gradient methods, which require only the storage of the first derivative. While these methods are undoubtedly more slowly convergent than those discussed above, they do enable exceptionally complex crystal structure to be examined; for example, the Na zeolite A structure (containing 700 atoms) has been investigated in recent work of Catlow, Parker and Sanders (to be published).

5. Computer codes

For crystals in which the interatomic potentials may be adequately represented by 'pair potential' models (the definition of which is given in Chapter (8)) the methods described above may be used to construct general purpose computer programs. Since pair potential models are most applicable to ionic materials, the presently available programs are adapted specifically for these systems. Long range Coulomb summations are generally handled by the Ewald procedure[17], the adaptation of which for defect systems is discussed by Norgett[5].

The principle codes available at present are as follows:

1. HADES (Harwell Automatic Defect Evaluation System). This is a general purpose program for the calculation of defect energies, developed originally for cubic crystals by Norgett[5], but generalised for crystals of any crystal symmetry by Catlow, James, Mackrodt and Stewart[8].

2. PLUTO (Perfect Lattice Unrestricted Testing Operation) which performs calculations of the lattice energy and elastic and dielectric properties of a specified perfect lattice structure. The program (which was written by Catlow and Norgett[18]) may be extended to calculate phonon dispersion curves, and also may be used in a 'reverse mode' to fit potential parameters as discussed in Chapter (10).

3. Cascade

This program (written by Leslie and Smith[20]) is essentially a fusion of the defect and perfect lattice simulation techniques, an important feature of which is the adaptation of the coding to exploit the parallel processor freely available on, for example, the CRAY series of computers.

4. METAPOCS[19] (Minimisation of Energy Techniques Applied to the Prediction of Crystal Structures). This program is an adaptation of the techniques used in the given PLUTO code, specified interaction potentials minimises the energy of a trial structure with respect to structural parameters. Two versions are available. METAPOCS(1), which adjusts atomic coordinates within a unit cell of given fixed dimensions; and METAPOCS(2) which varies unit-cell, in addition to atomic coordinates.

Applications of the program will be discussed in the book, especially in Chapters (15) and (20).

6. Summary and Conclusions

The methods described in this section enable us to perform accurate and reliable calculations of the properties of perfect and defective lattices where good potential models of the pair-wise type may be derived. The limitation of the type of potential model which may be effectively used in the calculations has tended to confine their application to the study of ionic and semi-ionic materials; the types of model appropriate for covalent and metallic systems are discussed in Chapters (11) and (19). Additional restrictions follow from the omission of any direct representation of quantum mechanical effects which excludes the applications of the methods to metals and semi conductors. (Treatments of defects in these systems are discussed in Chapters (13) and (14). However, in spite of these limitations, the methods we have described in this chapter have had possibly the greatest degree of quantitative success of any similar procedure in solid state studies - a feature which should become apparent in later chapters.

References

1. Catlow, C.R.A. and Norgett, M.J. AERE Harwell Report M.2763 (1978).

2. Born, M. and Huang, K., Dynamical Theory of Crystal Lattices (Oxford University Press), 1954.

3. Lidiard, A.B. and Norgett, M.J. In 'Computational Solid State Physics' 1972. Ed. F. Herman, N.W. Dalton and T.R. Koehler, p.385 (Plenum Press, New York).

4. Norgett, M.J., AERE Harwell Report R.7015 (1972).

5. Norgett, M.J., AERE Harwell Report R.7650 (1974a).

6. Norgett, M.J., AERE Harwell Report R.7780 (1974b).

7. Mackrodt, W.C. J. Comp. Phys. - in press (1981).

8. Catlow, C.R.A., James, R., Mackrodt, W.C. and Stewart, R.F., Phys. Rev. B25, 1006, (1982).

9. Mackrodt, W.C., Stewart, R.F., Campbell, J.C. and Hillier, I.H., J. de Physique C7, 64 (1980).

10. Davidon, W.C., AEC Report ANL-5990 (Rev)(1959).

11. Fletcher, R. and Powell, M.J.D., Computer J. 6, 16 (1963).

12. Norgett, M.J. and Fletcher, R., J. Phys. C3, L190 (1970).

13. Powell, M.J.D., AERE Harwell Report TP372, 1969.

14. Broyden, C.G., J. Inst. Maths. Appl. 6, 222 (1970).

15. Shanno, D.F., Math. Computing 24, 647 (1970).

16. Kendrick, J., I.C.I., Corporate Laboratory Report, CL-R/81/1640 (1981).

17. Ewald, R.P., Ann. Physik [4], <u>64</u>, 253 (1921); Gottinger Nechr., Math. - Phys. Kl. II3, 55 (1937).

18. Catlow, C.R.A. and Norgett, M.J., AERE Harwell Report M

19. Catlow, C.R.A., Cormack, A.N. and Parker, S.C.   To be published.

20. Lesley, M. and Smith, W., Daresbury Laboratory Report - To be published.

21. Catlow, C.R.A., Parker, S.C. and Saunders, M.J.   Nature - in press.

CHAPTER 2   THEORY AND CALCULATION OF DEFECT ENTROPIES

by

P.W.M. Jacobs[*], M.A. Nerenberg[†]
and J. Govindarajan[**]
Centre for Interdisciplinary Studies in Chemical Physics,
University of Western Ontario, London, Ontario, Canada N6A 5B7

1.   Introduction

The majority of numerical studies of defect parameters have concentrated on energy terms.   Thus the methods discussed in the previous chapter are concerned with calculations of formation, migration and activation energies of defects. However, detailed predictions, of the concentration of defects and the rates of defect migration require knowledge of the appropriate entropy term.   The present chapter therefore describes an approach to the calculation of the vibrational defect entropies, based on a Green function formalism, that has been successfully applied in a number of cases.

The vibrational contribution to the Helmholtz energy of a crystal containing N unit cells with ν atoms per cell is given by

$$F = \sum_{i=1}^{3N\nu} \left[\tfrac{1}{2} \hbar\omega_i + kT \ln\{1 - \exp(-\hbar\omega_i/kT)\} \right] \qquad (1)$$

where the ω_i are the phonon frequencies of the crystal at temperature T.   This gives rise to a vibrational contribution to the entropy of the crystal

$$S = k \sum_{i=1}^{3N\nu} \left[\frac{\hbar\omega_i}{kT} \{\exp(\hbar\omega_i/kT) - 1\}^{-1} - \ln\{1 - \exp(-\hbar\omega_i/kT)\} \right] . \qquad (2)$$

A useful approximation, that is valid when $kT \gg \hbar\omega_i$, $i=1,\ldots,3N\nu$, is

$$S = k \sum_{i=1}^{3N\nu} \{1 - \ln(\hbar\omega_i/kT)\} . \qquad (3)$$

We shall refer to (3) as the high-temperature approximation;   one anticipates that it will be strictly valid only for $T > \sim \Theta_D$, where Θ_D is the Debye temperature.   The practical merit of the high-temperature approximation is that it simplifies enormously the problem of calculating the entropy changes between two different thermodynamic states of the crystal.

---

[*]Department of Chemistry
[†]Departments of Applied Mathematics and Physics
[**]Computing Centre

When a defect forms in a crystal, the crystal entropy increases. This increase consists of two terms, the configurational contribution and the non-configurational contribution, the latter comprising the entropy change caused by any changes which might occur in the vibrational frequencies of the crystal. For example, the site fraction c of Schottky or Frenkel defects in a crystal held at constant T and p is

$$c = \eta \exp(-g/kT) \qquad (4)$$

where η is a numerical factor of order unity and g is the change in the non-configurational Gibbs energy on forming the defect. g may be decomposed into a component enthalpy and entropy

$$g = h - Ts \qquad (5)$$

and it is the calculation of s in equation (5) that is the concern of this article. The entropy difference, per defect, between the imperfect crystal and the corresponding perfect crystal at the same V and T, in the high-temperature approximation (3), is

$$s = S' - S = -k \ln \left[\prod_{i=1}^{3\overline{N}'} \omega_i' \Big/ \prod_{i=1}^{3\overline{N}} \omega_i \right] + 3k(\overline{N}' - \overline{N}) \{1 - \ln(\hbar/kT)\} \qquad (6)$$

where a prime denotes the imperfect crystal and $\overline{N}$ is the number of atoms in a crystal which therefore has $3\overline{N}$ normal modes ($\overline{N} \ggg 1$). When $\overline{N}' = \overline{N}$, the ω_i' in (6) may be expressed in terms of the perfect crystal frequencies by

$$\omega_i' = \omega_i + \Delta\omega_i \quad . \qquad (7)$$

Theimer[1] assumed no change in vibrational degrees of freedom accompanying defect formation ($\overline{N}' = \overline{N}$) and that the frequency changes $\Delta\omega_i$ would be small compared to ω_i. Two approximations in (6) then follow

$$s = -k \sum_{i=1}^{3\overline{N}} \ln(\omega_i'/\omega_i)$$

$$= -k \sum_{i=1}^{3\overline{N}} \Delta\omega_i/\omega_i - \tfrac{1}{2} (\Delta\omega_i/\omega_i)^2$$

$$\approx -\tfrac{1}{2} k \sum_{i=1}^{3\overline{N}} \Delta(\omega_i)^2/\omega_i^2 - \tfrac{1}{2}\{\Delta(\omega_i)^2/\omega_i^2\}^2 \quad . \qquad (8)$$

The expression (8) has been evaluated[2,3,4] by replacing $\Delta(\omega_i)^2/\omega_i^2$ by $\Delta f_i/f$, where Δf_i is the force-constant change estimated from the short-range part of a two body interionic potential and the displaced ionic positions due to the presence of the defect. These preliminary calculations, though useful, could not achieve high accuracy because the fractional changes in force constant, particularly for the vibrations of ions that are nearest neighbours to a vacancy, are not small[5].

A way to avoid Theimer's assumption was pointed out by Mahanty and Sachdev[6,7] and indeed is implicit in the discussion of Maradudin et al[8]. The most detailed formulation and application of this kind of approach is given in a series of papers by the authors[5,9-11]; however these writings parallel a gradual evolution of technique which still continues. A restatement and description of this procedure is therefore useful. This is given below, followed by a more detailed description for Frenkel defects in the fluorite structure. A review of the results is then presented.

2.   The Green function method: theory

The first term on the RS of (6) may be written in the equivalent form

$$-\tfrac{1}{2} k \ell n \left[\prod_{i=1}^{3\overline{N}'} (\omega_i')^2 \Big/ \prod_{i=1}^{3\overline{N}} \omega_i^2 \right] \tag{9}$$

which emphasizes that the calculation of s is essentially the calculation of the ratio of the product of the squares of the vibrational frequencies of the imperfect and perfect crystals. Our theoretical approach involves three steps. First, $\prod_i \omega_i^2$ is expressed in terms of the determinant of a Green matrix; second, the Green matrix for the imperfect crystal is expressed in terms of the corresponding matrix for the perfect crystal and a matrix of force constant changes; third a partitioning technique[6-8] is employed to reduce the size of these matrices to dimensions suitable for practical calculation. (This implies that the presence of the defect is felt mainly by a relatively small number of near neighbours). The computational technique involves (i) the evaluation of perfect lattice Green functions; (ii) the calculation of the relaxations experienced by the ions as a result of the presence of the defects; (iii) the calculation of the force constant changes from the relaxed ion positions and the crystal potential. The whole procedure will be illustrated by treating as an example a Frenkel defect on the anion sub-lattice in the fluorite structure.

The equations of motion for the ions in a crystal in the harmonic approximation are

$$(\underline{\underline{\Phi}} - M\omega^2) \ \underline{u} = \underline{\underline{L}} \underline{u} = 0 \tag{10}$$

where $\underline{u}$ is the column matrix of displacements. $\underline{\underline{\Phi}}$ is the force constant matrix and M the (diagonal) mass matrix. The set of equations (10) may be re-written in the form

$$\underline{\underline{L}} \underline{u} = \underline{\underline{M}}^{\frac{1}{2}} (\underline{\underline{D}} - \omega^2 \underline{\underline{I}}) \ \underline{\underline{M}}^{\frac{1}{2}} \ \underline{u} = 0 \tag{11}$$

where $\underline{\underline{I}}$ denotes the unit matrix of appropriate dimensions, whence it follows that the square of the frequencies are the eigenvalues of the (mass-reduced) dynamical matrix $\underline{\underline{D}}$, so that

$$\prod_i \omega_i^2 = |\underline{\underline{D}}| \tag{12}$$

the determinant of $\underline{\underline{D}}$.   Therefore

$$\prod_i \omega_i^2 = \lim_{\omega \to 0} \prod_i (\omega_i^2 - \omega^2)$$

$$= \frac{1}{|\underline{\underline{M}}|} \lim_{\omega \to 0} |\underline{\underline{M}}^{\frac{1}{2}}(\underline{\underline{D}} - \omega^2 \underline{\underline{I}}) \underline{\underline{M}}^{\frac{1}{2}}|$$

$$= \frac{1}{|\underline{\underline{M}}|} \lim_{\omega \to 0} |\underline{\underline{L}}| \quad . \tag{13}$$

Introducing the Green matrix $\underline{\underline{G}}$ defined by

$$\underline{\underline{G}}\underline{\underline{L}} = \underline{\underline{I}} \tag{14}$$

we have

$$\prod_i \omega_i^2 = \frac{1}{|\underline{\underline{M}}|} \lim_{\omega \to 0} |\underline{\underline{G}}^{-1}| \quad . \tag{15}$$

It follows from (14) and the orthonormal and completeness properties of the eigenvectors of $\underline{\underline{D}}$ that the elements of G are given by

$$G_{\alpha\beta}(\ell\kappa; \ell'\kappa'; \omega^2) = \frac{1}{N(m_\kappa m_{\kappa'})^{\frac{1}{2}}} \times$$

$$\sum_{\underline{q},j} \frac{e_\alpha(\kappa|\underline{q}j) \, e_\beta^*(\kappa'|\underline{q}j)}{\omega^2 - \omega_j^2(\underline{q})} \exp[i\underline{q}.\{\underline{r}(\ell\kappa) - \underline{r}(\ell'\kappa')\}] \tag{16}$$

where $\underline{r}(\ell\kappa)$ is the displacement vector from the origin to the κth atom in the ℓth cell, $\underline{q}$ the wave vector, $\omega_j(\underline{q})$ the vibrational frequency at $\underline{q}$ in the jth branch, $e_\alpha(\kappa|\underline{q}j)$ the αth component of the eigenvector $\underline{e}(\kappa|\underline{q}j)$.   [Notice that the eigenvector components in (16) differ from the $e_\alpha(\kappa|\underline{q}j)$ in the book of Maradudin et al[8] by a phase factor $\exp\{-i\underline{q}.\underline{r}(\kappa)\}$ where $\underline{r}(\kappa)$ is the vector from the origin of any unit cell to the κth atom in that cell].

2.1 An interstitial anion in the fluorite structure

As shown by (6),(9) and (15), the problem of calculating the entropy change associated with a defect process has been reduced to that of calculating the ratio of the determinants of the reciprocal of the Green matrices for the imperfect and perfect crystals.   The dynamical matrix of a crystal with an interstitial is

$$\underline{\underline{L}}_I = \begin{bmatrix} \underline{\underline{L}} + \delta\underline{\underline{L}} & \underline{\underline{A}} \\ \underline{\underline{A}}^T & \underline{\underline{\Lambda}} \end{bmatrix} \tag{17}$$

where Λ is the 3×3 part of $\underline{\underline{L}}_I$ that refers exclusively to the interstitial, I; $\underline{A}$ is the $3\bar{N}\times3$ matrix that describes the coupling of the interstitial to the rest of the atoms in the crystal, $\underline{A}^T$ is its transpose and $\delta\underline{\underline{L}}$ is the matrix of force constant changes due to the relaxations of the ions that are induced by the presence of the interstitial. It is reasonable to suppose, and this is confirmed by detailed calculation, that only those ions in the immediate surroundings of the interstitial will be perturbed significantly by its presence; use will be made of this later in the partitioning of $\underline{\underline{L}}_I$. $\underline{\Lambda}$ is the 3×3 diagonal matrix whose diagonal elements are each equal to $m_I(\omega^2-\omega_I^2)$ where m_I is the mass of the interstitial and ω_I the (hypothetical) frequency of vibration of the interstitial I when all the lattice ions are frozen at their equilibrium positions. Partitioning $\underline{\underline{G}}_I$, the Green matrix for the crystal with an interstitial, (defined by $\underline{\underline{G}}_I = \underline{\underline{L}}_I^{-1}$) in the same way as $\underline{\underline{L}}_I$ in equation (17), we have (after some algebra)

$$\underline{\underline{G}}_I = \begin{bmatrix} \underline{\underline{G}}+\underline{\underline{G}}\underline{\underline{T}}_1\underline{\underline{G}} & -(\underline{\underline{G}}+\underline{\underline{G}}\underline{\underline{T}}_1\underline{\underline{G}})\underline{\underline{A}}\underline{\underline{\gamma}} \\ -(\underline{\underline{\gamma}}+\underline{\underline{\gamma}}\underline{\underline{T}}_2\underline{\underline{\gamma}})\delta\underline{\underline{\lambda}}' & \underline{\underline{\gamma}}+\underline{\underline{\gamma}}\underline{\underline{T}}_2\underline{\underline{\gamma}} \end{bmatrix} \qquad (18)$$

and

$$|\underline{\underline{G}}_I| = |\underline{\underline{I}}-\underline{\underline{G}}\delta\underline{\underline{L}}'|^{-1} |\underline{\underline{G}}||\underline{\underline{\gamma}}| \qquad (19)$$

where

$$\underline{\underline{\gamma}} = \underline{\underline{\Lambda}}^{-1} \qquad (20)$$

$$\underline{\underline{T}}_1 = \delta\underline{\underline{L}}'(\underline{\underline{I}}-\underline{\underline{G}}\delta\underline{\underline{L}}')^{-1} \qquad (21)$$

$$\underline{\underline{T}}_2 = \delta\underline{\underline{\lambda}}(\underline{\underline{I}}-\underline{\underline{\gamma}}\delta\underline{\underline{\lambda}})^{-1} \qquad (22)$$

$$\delta\underline{\underline{\lambda}} = \underline{\underline{A}}^T(\underline{\underline{I}}-\underline{\underline{G}}\delta\underline{\underline{L}})^{-1} \underline{\underline{G}}\underline{\underline{A}} = \delta\underline{\underline{\lambda}}'\underline{\underline{A}} . \qquad (23)$$

Equation (19) expresses the determinant of the Green matrix for the imperfect crystal $\underline{\underline{G}}_I$ in terms of $\underline{\underline{G}}$ which is that for the corresponding perfect crystal (without the interstitial), $\underline{\underline{\gamma}}$ is the Green matrix for the interstitial with the host-lattice ions frozen at their equilibrium positions; and $\delta\underline{\underline{L}}'=\delta\underline{\underline{L}}+\underline{\underline{A}}\underline{\underline{\gamma}}\underline{\underline{A}}^T$, where $\delta\underline{\underline{L}}$ is the matrix of force constant changes. However equation (19) may not be evaluated unless the size of the matrices are reduced substantially; for the interstitial in CaF_2, the eight first neighbour (nn) fluoride ions and the six second neighbour (2n) calcium ions alone need be considered, so that for such systems, in which the interstitial interacts with 14 ions, we have

$$\delta L = \begin{bmatrix} 0 & 0 \\ 0 & \delta \ell_I \end{bmatrix} \quad \text{and} \quad A = \begin{bmatrix} 0 \\ a \end{bmatrix} \tag{24}$$

where the dimensions of $\delta\ell_I$ are 42×42 and those of a are 42×3. It then follows that

$$\delta L' = \begin{bmatrix} 0 & 0 \\ 0 & \delta\ell' \end{bmatrix} \tag{25}$$

where

$$\delta\ell' = \delta\ell_I + a\gamma a^T \tag{26}$$

and that

$$|I - G\delta L'| = |I - g_I \delta\ell'| \tag{27}$$

where g_I is that partition of the perfect crystal Green matrix that refers to the same space as $\delta\ell$. Finally, from (6),(9),(15),(19) and (27), we have

$$s_I = -\tfrac{1}{2}\, k\, \ln[(\hbar\omega_I/kT)^6 \lim_{\omega \to 0} |I - g_I \delta\ell'|] + 3k. \tag{28}$$

The evaluation of formula (28) requires the computation of ω_I, the independent perfect crystal Green functions that make up g_I, and the force constant changes that make up $\delta\ell_I$ and a. Because of the partitioning introduced, g_I and $\delta\ell'$ are each of dimensions 42×42.

2.2 An anion vacancy in the fluorite structure

The dynamical matrix L_V of a crystal with a vacancy may be written as

$$L_V = L_t - \delta L \tag{29}$$

where L_t is L truncated by removing the 3 rows and columns belonging to the ion that is no longer present and δL represents the changes in L that comprise force constant changes due to (i) the relaxation of ions resulting from the presence of the vacancy and (ii) changes in the diagonal "self-term" sums because of the removal of a particular ion. Suppose that the ion to be removed to infinity when a vacancy is formed is located at $\ell'=0$, $\kappa'=1$. Then in a perfect crystal

$$\phi_{\alpha\beta}(\ell\kappa; \ell\kappa) = -\sum_{\substack{\ell',\kappa' \\ \neq \ell,\kappa}} \phi_{\alpha\beta}(\ell\kappa; \ell'\kappa') \tag{30}$$

but in a crystal with a vacancy

$$\phi'_{\alpha\beta}(\ell\kappa; \quad \ell\kappa) = -\sum_{\substack{\ell',\kappa' \\ \neq \ell,\kappa \\ \neq 0,1}} \phi'_{\alpha\beta}(\ell\kappa; \quad \ell'\kappa') \ . \tag{31}$$

Define

$$\underset{=}{G}_V = \underset{=}{L}_V^{-1} \tag{32}$$

and partition L_V as

$$\underset{=}{L}_V = \begin{bmatrix} \underset{=}{L}_0 & \underset{=}{B}_t \\ \underset{=}{B}_t^T & \underset{=}{\ell}_t + \delta\ell_V \end{bmatrix} \tag{33}$$

where $\underset{=}{L}_0$ is that part of $\underset{=}{L}_V$ which is unperturbed by the vacancy, $\underset{=}{\ell}_t$ is $\underset{=}{\ell}$, the dynamical matrix for the defect space, comprising the ion at the site of the vacancy-to-be and the ions that are its first and second neighbours, truncated as above. $\delta\ell_V$ is the matrix that describes the changes in the dynamical matrix $\underset{=}{L}$ that are induced by the vacancy, limited to the defect space, so that

$$\underset{=}{\delta L} = \begin{bmatrix} 0 & 0 \\ 0 & \delta\ell_V \end{bmatrix} \tag{34}$$

$\underset{=}{B}_t$ connects the unperturbed region with the defect space and $\underset{=}{B}_t^T$ is its transpose. Similarly

$$\underset{=}{L} = \begin{bmatrix} \underset{=}{L}_0 & \underset{=}{B} \\ \underset{=}{B}^T & \underset{=}{\ell} \end{bmatrix} \ . \tag{35}$$

Notice that $\underset{=}{B}_t$ in (33) is assumed to have the same elements as $\underset{=}{B}$ in (35) apart from truncation. Physically, this amounts to the assumption that the formation of a vacancy does not alter significantly the coupling between the remaining ions of the defect space and those in the rest of the crystal.

Solving (14) and (32) gives

$$\frac{|\underset{=}{G}|}{|\underset{=}{G}_V|} = \frac{|(\underset{=}{g}_V^{-1})_t - \delta\ell_V|}{|\underset{=}{g}_V^{-1}|} \tag{36}$$

where $\underset{=}{g}_V$ is that part of the perfect crystal Green matrix that refers to the defect space-to-be. (Notice that $\underset{=}{g}_V$ has three more rows and columns that $\delta\ell_V$).

It follows from (6),(9),(15) and (36) that the entropy change on removing an anion to infinity is

$$s_V = -\tfrac{1}{2} k \ln[m_-^3 (kT/\hbar)^6 \lim_{\omega \to 0} |\underline{g}_V| |(\underline{g}_V^{-1})_t - \delta\underline{\ell}_V|] - 3k \tag{37}$$

where m_- is the mass of the anion removed. The entropy of formation of a Frenkel defect in a fluorite crystal is therefore given by the sum of the expressions (28) and (37): the result is

$$s_F = s_I + s_V$$

$$= -\tfrac{1}{2} k \ln[m_-^3 \omega_I^6 \lim_{\omega \to 0} \{|\underline{g}_V| |(\underline{g}_V^{-1})_t - \delta\underline{\ell}_V| |\underline{I} - \underline{g}_I \delta\underline{\ell}'|\}]. \tag{38}$$

3.   The Green function method: calculation

The method used to find $\delta\underline{\ell}$ is to calculate the force constant matrices $\underline{\ell}$ and $\underline{\ell}'$ for the perfect and imperfect crystals and then compute $\delta\underline{\ell}$ from $\underline{\ell}'-\underline{\ell}$. Because of the inherent difficulties involved in including the long range Coulomb inter-action exactly in a lattice dynamical calculation on an imperfect crystal, in the computation of s_I the Coulomb interaction was included only between the interstitial and the remaining atoms of the defect space. Thus we are implicitly assuming that the changes in the Coulomb interactions between the ions of the defect space themselves and between these ions and those of the rest of the crystal are less important than the Coulomb contributions arising from the introduction of the charged interstitial. Furthermore, in the latter contribution only inter-actions with the ions of the defect space were included. Thus in computing the 42×42 matrix, $\underline{\ell}$, only the short-range terms in the elements, $\ell_{\alpha\beta}(ij), i,j=2,\dots,15$ were calculated. $i=1$ designates the interstitial. The matrix $\underline{\ell}'$ is of dimensions 45×45 and comprises the 3×3 partition $[\ell'_{\alpha\beta}(1,1)]$ which is Λ, the 42×42 partition $[\ell'_{\alpha\beta}(i,j)]$ $i,j=2,\dots,15$ which is $\underline{\underline{\ell}}+\delta\underline{\underline{\ell}}$, and the 42×3 matrix, $\underline{\underline{a}}$, and its transpose $\underline{\underline{a}}^T$. For the off-diagonal terms, $i\neq j$, $\alpha\neq\beta$, when $i,j\neq 1$ only short-range terms were calculated but when $i,j=1$ (elements of $\underline{\underline{a}}$ or $\underline{\underline{a}}^T$) the Coulomb interaction was included. For diagonal terms $i=j$, $\alpha=\beta$, for each $i\neq 1$ the term $j'=1$ is included in the summation

$$\phi_{\alpha\beta}(ii) = - \sum_{j'\neq i} \phi'_{\alpha\beta}(i,j') \tag{39}$$
$$(i \neq 1)$$

and so the 'self-terms' have both Coulomb and short-range interactions. Thus in $\delta\underline{\underline{\ell}}=\underline{\underline{\ell}}'-\underline{\underline{\ell}}$ both the off-diagonal terms and the diagonal terms for $i,j > 1$ include the Coulomb and short-range interactions with the interstitial.

For the vacancy the matrix $\underline{\underline{\ell}}$ is 33×33. For off-diagonal terms $i\neq j$, $\alpha\neq\beta$, the terms with i or $j=1$ included Coulomb and short-range interactions, but those with i and $j\neq 1$ the short-range interaction only. For the diagonal terms $\alpha=\beta$, $i=j$: when $j'\neq 1$ in the self-term sums

$$\phi'_{\alpha\beta}(ii) = - \sum_{j' \neq i} \phi'_{\alpha\beta}(i,j') \qquad (40)$$

the short-range interaction only was computed, but for j'=1, the Coulomb interaction was included as well in $\phi'_{\alpha\beta}(i,1)$. The vacancy defect space contains only 10 ions and so ℓ' is the 30×30 matrix $[\ell'_{\alpha\beta}(i,j)]$ i,j=2,...,11. Only the short-range interaction is involved in these terms so that $\delta\underline{\ell}=\underline{\ell}'-\underline{\ell}$ includes the Coulomb interaction only for the ion that is subsequently removed in forming the vacancy.

In the calculations of the self-terms in both $\delta\underline{\ell}_I$ and $\delta\underline{\ell}_V$, each ion of the defect space was given its full complement of first and second neighbours. The relaxations of these ions were calculated using the lattice simulation procedures, discussed in Chapter (1), and described in references (12) and (13). The crystal potentials of Catlow, Norgett and Ross[14] were used for the calculations of the ionic positions and for the calculation of the force constant matrices.

The Green matrices $\underline{g}_V$ and $\underline{g}_I$ needed for the evaluation of formulae (28) and (37) were computed from (16) using 262 independent wave vectors which generate by symmetry 8000 points in the Brillouin zone. 23 independent Green functions are needed for g_I and 21 independent Green functions for g_V. The eigenvectors were obtained from the shell model fitted to the phonon dispersion of SrF_2[12]. At first these were computed at one temperature only and the Green functions at other temperatures obtained by a scaling procedure. Latterly the Green functions at $\omega=0$ have been computed at each temperature using the best-available data on lattice constants. It might appear that there is a minor inconsistency here to use one potential model for the displacement and another for the eigenvectors but this inconsistency is more apparent than real since it was verified that the Catlow, Norgett and Ross potentials also give excellent representations of the phonon dispersion curves.

4.   The Green function method: results

So far the method has been applied only to the calculation of the entropy of formation of a Schottky defect in KCℓ and the entropy changes associated with the substitution of a cation[16] or an anion[17] in KCℓ, in addition to the entropy of formation of a Frenkel defect in CaF_2[9]. Results for the entropy of formation of a Frenkel defect in SrF_2 at three temperatures are given in Table 1. These calculations are based on the following approximations: the defect space was limited to the first and second neighbours of the interstitial or vacancy; the long-range Coulomb term was included only for the interaction of the defect with the ions in the defect space; the Green functions at the required T were obtained from those at the temperature at which the phonon dispersion had been measured, by a scaling procedure.

Table 1

Entropy of formation, at constant volume, s^V of a Frenkel defect in SrF_2, calculated by the Green function method. The corresponding entropy changes at constant pressure were obtained as described in the text.

SrF_2:	T/k	390	780	1040
	s^V/k	4.49	3.64	3.34
	s^P/k	5.74	5.68	5.28

The calculations described in section 3 yield the entropy change at constant volume, s^V. Comparison with experiment requires the entropy change at constant pressure. To first order in v^P, the volume change accompanying defect formation at constant pressure, the corresponding Helmholtz energy change at constant volume f^V is equal to the Gibbs energy change at constant pressure g^P, for the same process. Therefore, the entropy change at constant pressure s^P is related to that at constant volume s^V, by

$$s^P - s^V = (\partial f^V/\partial v)_T \, \beta^P \, v \qquad (41)$$

where β^P is the expansivity $(1/v)(\partial v/\partial T)_p$ and v is the volume of the crystal per molecule. If the lattice constant a, and therefore v, is known as a function of T, the right hand side of (41) can be evaluated in the quasiharmonic approximation by calculating u^V (using the lattice simulation methods) and s^V (by the Green function method) at each a(T). The results obtained are clearly of the correct order of magnitude. For SrF_2, the experimental value of s^P in the range 570-1100K is 4.1 k[18] and the mean calculated value in this range is approximately 5.4 k. Thus for SrF_2 the agreement is satisfactory.

Similar calculations are in progress for BaF_2, and we are also examining the seriousness of the scaling approximation by computing the Green functions at $\omega = 0$ at each of the required temperatures. The proper inclusion of long-range interactions is a formidable problem which is also receiving attention. The size of the defect space is a limitation but probably not a very serious one since the defect calculation shows that the displacements of the first and second neighbours are much larger than those of the third and subsequent shells; in computing the force constant changes each of the ions in the defect space had its full complement of (relaxed) first and second neighbours even though these ions lay outside the defect space. A point of crucial importance is the need for reliable experimental data on the temperature-dependence of the lattice constant. For SrF_2 only the relatively old X-ray data of Sirdeshmukh and Desphande[19] were available and a considerable (linear) extrapolation of a(T) was needed. Despite this drawback, the SrF_2 results seem physically reasonable.

Acknowledgements

This research is supported in part by the Natural Sciences and Engineering Research Council of Canada.

References

1. Theimer, O., Phys. Rev. 112, 1857 (1958).

2. Chandra, S., Pandey, G.K. and Agrawal, V.K., Phys. Rev. 144, 738 (1966).

3. Roy, D. and Ghosh, A.K., Phys. Rev. B3, 3510 (1971).

4. Agrawal, V.K. and Garg, H.C., Phys. Rev. B8, 843 (1973).

5. Govindarajan, J., Jacobs, P.W.M. and Nerenberg, M.A., J. Phys. C. 10, 1809 (1977).

6. Mahanty, J., Physics Lett. 29A, 583 (1959).

7. Mahanty, J. and Sachdev, M., J. Phys. C3, 773 (1970).

8. Maradudin, A.A., Montroll, E.W., Weiss, G.H. and Ipatova, I.P., Theory of Lattice Dynamics in the Harmonic Approximation (Academic Press, New York 1971).

9. Haridasan, T.M., Govindarajan, J., Nerenberg, M.A. and Jacobs, P.W.M., Phys. Rev. B20, 3481 (1979).

10. Nerenberg, M.A.H., Govindarajan, J., Jacobs, P.W.M. and Haridasan, T.M., Can. J. Phys. 58, 803 (1980).

11. Nerenberg, M.A.H., Haridasan, T.M., Govindarajan, J. and Jacobs, P.W.M., J. Phys. Chem. Solids, 41, 1217 (1980).

12. Norgett, M.J., AERE Harwell Report R.7650 (1974).

13. Catlow, C.R.A., Corish, J., Diller, K.M., Jacobs, P.W.M. and Norgett, M.J., J. Phys. C12, 451 (1979).

14. Catlow, C.R.A., Norgett, M.J. and Ross, T.A., J. Phys. C10, 1627 (1977).

15. Elcombe, M.M., J. Phys. C5, 2702 (1972).

16. Govindarajan, J., Jacobs, P.W.M. and Nerenberg, M.A., J. Phys. C9, 3911 (1976).

17. Govindarajan, J., Nerenberg, M.A. and Jacobs, P.W.M., J. Phys. Chem. Solids 39, 527 (1978).

18. Jacobs, P.W.M. and Ong, S.H. To be published.

19. Sirdeshmukh, D.B. and Desphande, V.T., Ind. J. Pure and Appl. Phys. 2, 405 (1964).

CHAPTER 3

CHARACTERISTIC VOLUMES OF POINT DEFECTS IN IONIC CRYSTALS

by

M.J. Gillan and A.B. Lidiard
Theoretical Physics Division, A.E.R.E. Harwell, Oxon OX11 ORA

1. Introduction

The thermodynamic parameters governing defect concentrations that are derived from experiment are almost universally those appropriate to conditions of constant pressure, P, i.e. the free energies are Gibbs free energies ΔG^P corresponding to defect formation, activation and interaction at constant pressure. These free energies may be decomposed in the usual way into internal energy, entropy and free volume terms, viz.

$$\Delta G^P = \Delta U^P - T \Delta S^P + P \Delta V^P \ . \tag{1}$$

In particular, the free volume ΔV^P is given by

$$\Delta V^P = \left(\frac{\partial \Delta G^P}{\partial P}\right)_T \ , \tag{2}$$

and it is this quantity which determines the effect of pressure upon the corresponding defect 'reaction'. In recent years there have been a number of determinations of ΔV^P for point defects in ionic crystals having the rocksalt[3-8], caesium chloride[9], fluorite[10-13] and other[14] structures. In general, the free volumes of formation of Schottky defects are greater than or roughly equal to the molecular volume while for Frenkel defects they are a fraction of the molecular volume. Volumes of defect activation and of association with impurities are generally small fractions of the molecular volumes.

It has long been one of the aims of theory to relate these quantities to the relaxation of the crystal lattice around the defect. However, as calculations of defect structure and energy are most conveniently done at constant crystal volume, it is not immediately obvious which is the most convenient way to calculate ΔV^P. Most considerations have, in fact, used the relationship between the volume of relaxation and the magnitude of the elastic strain field around the defect - an approach originally due to Kanzaki[15] and to Hardy[16]. This so-called "lattice statics" approach yields explicit expressions for the elastic strengths of defects in terms of the forces acting on the atoms of the lattice. Thus volume changes and, more generally, shape changes[17] can be obtained once these forces and the displaced positions of the atoms are known. In particular the elastic strength tensor is

$$G_{ij} = \sum_{\ell} R_{i\ell} F_{j\ell} (\underline{R}_\ell + \underline{\xi}_\ell) \tag{3}$$

while the volume change

$$\Delta V^P = \kappa_T \ \mathrm{Tr} \ \underline{\underline{G}} \tag{4}$$

where κ_T is the isothermal compressibility. In equation (3) $\underline{F}(\underline{R}_\ell + \underline{\xi}_\ell)$ is the force exerted by the defect on the atom at the displaced position $\underline{R}_\ell + \underline{\xi}_\ell$ which was initially at $\underline{R}_\ell$ while i,j signify Cartesian components. The summation is over all atoms ℓ. However although the results of this approach appear often to be satisfactory it is now quite clear that they are inadequate for defects bearing a net charge in ionic crystals; thus for Schottky defects the calculated relaxation volumes are negative[18,19], whereas experiment shows them to be positive[5,8]. A similar discrepancy arises with V_k-centres[20]. This has presented something of a puzzle. In this article we draw attention to one limitation of this approach, the removal of which allows us to obtain much better theoretical values which are of the correct sign.

At the same time we shall also discuss an alternative and more direct approach, which is especially convenient when one has access to modelling programs such as HADES and CASCADE which were discussed in Chapter (1). The approach is based upon the fact that a change in defect configuration of the crystal (formation, activation, etc.) at constant crystal volume leads to a change ΔF^V in the Helmholtz free energy and a corresponding change in external pressure of

$$\Delta P^V = - \left(\frac{\partial \Delta F^V}{\partial V}\right)_T . \tag{5}$$

Removal of this increment of pressure then gives an expansion

$$\Delta V^P = \kappa_T V \Delta P^V \tag{6}$$

and establishes the crystal in the same defect state as is obtained by carrying out the same change in the defect state at constant pressure. (N.B. in connection with the formation of Schottky defects ΔV^P refers to the relaxation volume alone and does not include the molecular volume associated with the replacement of the ions on the surface; this is because the defect calculations are done at constant lattice parameter). From (5) and (6)

$$\Delta V^P = - \kappa_T V \left(\frac{\partial \Delta F^V}{\partial V}\right)_T . \tag{7}$$

Calculations of ΔF^V are mostly carried out within the quasi-harmonic approximation in which, at temperatures high compared to the Debye temperature, the internal energy change ΔU^V is just the change in lattice potential energy while the entropy change ΔS^V results just from the change in lattice vibrational frequencies[21,22]. The result (7) (neglecting the entropy contribution) has been given before in several sources[22-25] but has hitherto only been used for detailed calculations in metals[23-25]. Here we shall report recent results for both Schottky and Frenkel defects in ionic crystals.

2. Calculations and Results

A careful examination of the derivation of the lattice statics expression for ΔV^P shows that it demands a strictly harmonic model for the lattice surrounding the defect region, i.e. it depends upon the assumption that the harmonic force constants are independent of crystal volume. Such a requirement is, we believe, too severe. If instead we employ a quasi-harmonic approximation then corrections to the usual expression can be evaluated explicitly for charged defects by using the Mott-Littleton description of the distant polarization and displacement field[26]. For the formation of Schottky defects these corrections prove to be large and positive so that, as shown in Table 1 (column 2), the net relaxation volumes are also positive.

Table 1

The volume of relaxation accompanying the formation of Schottky defects in units of the molecular volume. (N.B. the directly measured formation volume is the sum of the relaxation volume plus the molecular volume).

Substance	Calc[1]	Calc[2]	Expt.
NaCl	0.52	0.57	0.5 to 0.8
NaBr	0.43	0.47	0.2
KCl	0.50	0.57	0.5 to 0.6
KBr	0.48	0.54	0.1
MgO	0.7 to 0.1	-	-

[1] These values have been calculated by evaluating the correction to the lattice statics formula[26] and adding this to values previously calculated by this means[18,19].

[2] These values have been calculated directly[27] by means of the HADES program and equation (7). Both sets of calculated values are for low temperatures. They increase as the temperature is raised mainly due to the increase in κ_T.

At the same time we have also obtained the volumes of relaxation associated with the formation of Schottky defects by calculating their formation energies at several values of the lattice parameter and hence obtaining ΔV^P from equation (7) (to the neglect of the entropy contribution). These calculations have been carried out by means of the HADES program and are described more fully in reference 27. The results are also given in Table 1 (column 3) and although the two physical models differ in detail the two sets of results in columns 2 and 3 agree well. They also agree well with the experimental results; in particular the previous conflict over the sign of the relaxation volume no longer exists.

The evident success of these calculations encouraged us also to calculate the formation volumes of anion Frenkel defects in the alkaline earth fluorides[27].

The results are given in Table 2. These are again in good agreement with the experimental values. These results also show the expected variation with temperature; for these materials this variation results not only from the changes in κ_T but from the energy factor as well.

Table 2

The volumes of formation of anion Frenkel defects in the alkaline earth fluorides in units of the molecular volume.

Substance	Calculated		Expt.
	T = 0	T = $2T_m/3$	
CaF_2	0.39	0.28	0.3 to 0.4
SrF_2	0.19	0.25	0.2 to 0.3
BaF_2	0.12	0.19	0.1

These values have been calculated[27] by means of the HADES program and equation (7). Here T_m is the absolute melting temperature. The temperature variation for these materials appears more complex than for the alkali halides.

Lastly, Figure 1 shows the volume of formation of cation Frenkel defects in AgCl obtained in the same way, but at various temperatures[28]. Some of the temperature variation comes from the thermal expansion of the crystal, some from the temperature variation of κ_T. Unfortunately, this latter variation is not well known for AgCl above room temperature so that these predictions may not be reliable in detail. However, the experimental value of 16.7 cm^3/mole is a mean derived from experiments[4] in the region 450-600 K and is seen to lie close to the range of calculated values.

3. Conclusions

In summary it now seems clear that defect volumes can be calculated with much the same reliability as defect energies when we use equation (7) directly. So far it has generally been assumed that the dependence of ΔS^V upon V is much less important than that of ΔU^V, but additional calculations of the entropy contributions are now being made[29]. Preliminary indications are that they may amount to about 10% of the total volume of formation of Frenkel defects in the alkaline earth fluorides. Experimental results[30] on the volumes of activation for defect motion in the alkaline earth fluorides have been interpreted[31] as showing that the entropy term is as much as 25%. Clearly there is a need for further entropy calculations. But in approaching the calculations we can be confident that the models and methods which have proved reliable for the defect structures and energies will also yield good results here too. At the same time, as all these calculations are founded on the quasi-harmonic approximation, there is good reason to examine carefully the predictions of the models for bulk lattice

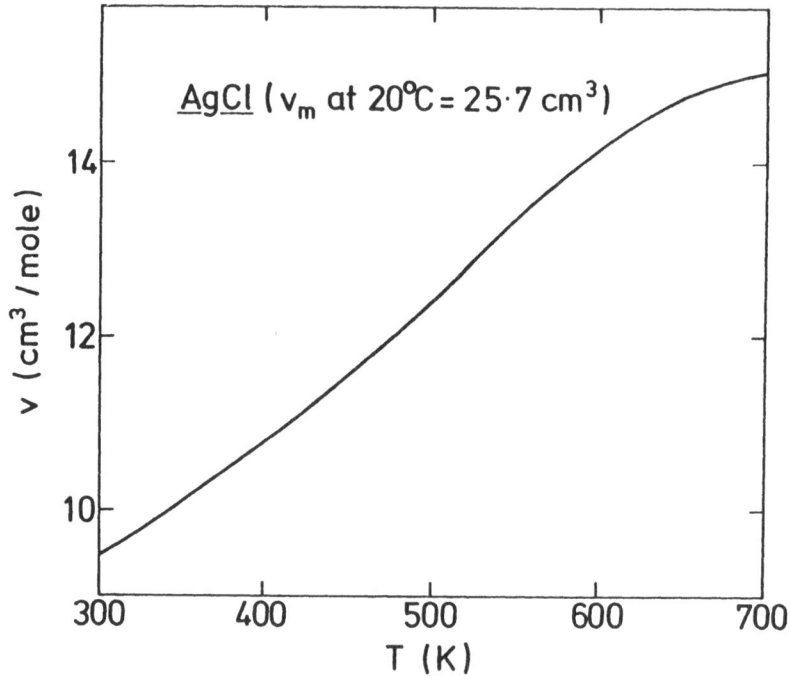

<u>Figure 1</u>   The volume of formation of cation Frenkel defects in AgCl as a
function of temperature calculated by means of the HADES program as
described in reference 28.

properties too (e.g. thermal expansion, dependence of elastic and dielectric
constants upon temperature).   Furthermore, as one may not wish simply to abandon
the more analytical treatments of the Kanzaki-Hardy lattice statics method some
additional analysis is required into the way the method fails for defect volumes.
We have argued that it does so because the analysis makes a too restrictive
assumption about the implications of the harmonic approximation and have evaluated
a correction to it within the framework of the <u>quasi</u>-harmonic approximation.   Further
examination of this correction term for other cases is desirable to determine its
magnitude and the conditions when it can be significant.

References

1.   Corish, J. and Jacobs, P.W.M., in Surface and Defect Properties of Solids,
Vol. 2 (Chemical Society Specialist Periodical Reports, 1973).

2.   Corish, J., Jacobs, P.W.M. and Radhakrishna, S., in Surface and Defect
Properties of Solids, Vol. 6 (Chemical Society Specialist Periodical
Reports, 1977), p.218.

3.   Kurnick, S.W., J. Phys. C. <u>20</u>, 218 (1952).

4.   Abey, A.E. and Tomizuka, C.T., J. Phys. Chem. Solids <u>27</u>, 1149 (1966)

5. Bierman, W., Z. Phys. Chem. (Frankfurt), $\underline{25}$, 90 and 253 (1960).

6. Taylor, W.H. II., Daniels, W.B., Royce, R.S.H. and Smoluchowski, R., J. Phys. Chem. Solids $\underline{27}$, 39 (1966).

7. Pierce, C.B., Phys. Rev. $\underline{123}$, 744 (1961).

8. Yoon, D.N. and Lazarus, D., Phys. Rev. $\underline{B5}$, 4935 (1972).

9. Samara, G.A., Phys. Rev. Letts. $\underline{44}$, 670 (1980).

10. Lallemand, M., C.R. Acad. Sci. Paris $\underline{B267}$, 715 (1968).

11. Lallemand, M., Thesis, Faculté des Sciences de l'Université de Paris VI (1972).

12. Samara, G.A., J. Phys. Chem. Solids $\underline{40}$, 509 (1979).

13. Oberschmidt, J. and Lazarus, D., Phys. Rev. $\underline{B21}$, 5813 and 5823 (1980).

14. Cleaver, B., Z. Phys. Chem. N.F. $\underline{45}$, 360 (1965).

15. Kanzaki, M., J. Phys. Chem. Solids $\underline{2}$, 24 and 107 (1957).

16. Hardy, J.R., J. Phys. Chem. Solids $\underline{15}$, 39 (1960) and $\underline{29}$, 2009 (1968).

17. Stoneham, A.M., J. Phys. $\underline{C6}$, 223 (1973).

18. Faux, I.D. and Lidiard, A.B., Z. Naturforsch $\underline{26a}$, 56 (1971).

19. Catlow, C.R.A., Faux, I.D. and Norgett, M.J., J. Phys. $\underline{C9}$, 419 (1976).

20. Mainwood, A. and Stoneham, A.M., J. Phys. $\underline{C8}$, 3059 (1975).

21. Vineyard, G.H. and Dienes, G.J., Phys. Rev. $\underline{93}$, 265 (1954).

22. Lidiard, A.B. and Norgett, M.J., in Computational Solid State Physics. Eds. F. Herman, N.W. Dalton and T.R. Koehler, p.385 (Plenum Press, New York 1972).

23. Hardy, J.R., J. Phys. Chem. Solids $\underline{29}$, 2009 (1968).

24. Finnis, M.W. and Sachdev, M., J. Phys. $\underline{F6}$, 965 (1976).

25. Jacucci, G. and Taylor, R., J. Phys. $\underline{F9}$, 1489 (1979).

26. Lidiard, A.B., Phil. Mag. $\underline{A43}$, 291 (1981).

27. Gillan, M.J., Phil. Mag. $\underline{A43}$, 301 (1981).

28. Catlow, C.R.A., Corish, J., Jacobs, P.W.M. and Lidiard, A.B., J. Phys. $\underline{C14}$, L121 (1981).

29. Harding, J. Private communication.

30. Andeen, C., Hayden, L.M. and Fontanella, J., Phys. Rev. $\underline{B21}$, 794 (1980).

31. Varotsos, P. and Alexopoulos, K., Phil. Mag. $\underline{A42}$, 13 (1980).

CHAPTER 4

FREE ENERGY CALCULATIONS FOR CRYSTALS

by

G. Jacucci
Dipartimento di Fisica
and
Instituto per la Ricerca,
Scientifica e Techologica,
Libera Universita di Trento,
38050 Povo, Italy.

N. Quirke
Department of Chemistry,
Royal Holloway College,
Egham, Surrey TW20 OEX, England.

1.   Introduction

Most of the chapters in this book are concerned with the calculation of internal energies or enthalpies. The present chapter will, in contrast, describe techniques for the direct calculation of free energies. Such calculations can in principle yield the stable structure of a crystal lattice, the concentrations of structural defects and their clusters, the orientation of grain boundaries, and the distribution of impurities; in addition the phenomenon of melting can be studied providing the relevant thermodynamic potentials can be calculated.

The simplest approach to free energy calculations is based on the quasi-harmonic approximation. This relies, first, on the truncation to second order of the Taylor expansion of the potential energy as a function of the atomic displacements from a given lattice configuration, and gives rise to a representation of the solid in terms of normal modes; this is, of course, the simple harmonic approximation.   The second assumption of the approach is that the effect of the higher order terms in the series, which lead to thermal expansion, can be taken into account by choosing the values of the lattice spacing, in the configuration about which the expansion is made, so that this matches the equilibrium value at desired temperature; it is this latter feature which leads to the term 'quasi-harmonic' approximation.   The treatment is adequate provided the n-body potential energy surface is harmonic within an energy range kT of the given equilibrium configuration. When this condition is not satisfied, the many-body correlations are no longer described by free quasi-particles as implied by the normal-mode analysis. However, given the validity first of classical mechanics, which is generally applicable in the temperature range where anharmonicity is important, and secondly of pair-wise additive interatomic potentials, machine calculations have proved to be a powerful tool in systems where the quasi-harmonic approximation breaks down; they can be used to obtain reliable equilibrium and dynamic properties of condensed systems containing a few hundred particles.   They are based on the calculations of thermal averages by numerical experiments using Monte-Carlo sampling or dynamical techniques either Newtonian or stochastic. The description of the Monte Carlo techniques is the main concern of this chapter; dynamical studies are considered in Chapters (5) and (18).

2. Quasi-harmonic studies and machine calculations

The standard method for calculating the free energy of crystals from a normal coordinate expansion of the atomic displacements, was used with great success by Born and collaborators[1]. An application of the general method to the phase transition from a cubic close packed to a body-centred cubic solid for inverse power potentials, can be found in a paper by Hoover et al[7]. The thermodynamic properties of non-ideal crystal lattices were studied in the quasi-harmonic approximation by Lifschitz[2] and independently by Montroll and Potts[3] and by Kanzaki[4]. These treatments have been extensively applied in the early sixties by Nardelli and co-workers[5] to a number of problems connected with local irregularities in the lattice. In particular, in an important paper, published in 1963, Nardelli and Terzi[6] evaluated the free energy of formation of a vacancy in fcc Lennard-Jones lattices.

The inadequacies of the harmonic approach became apparent, however, in the careful investigations of experimental data on alkali halide crystals by Tosi and Fumi[8] and on argon by Kuebler and Tosi[9]; this work indicated that the anharmonic contribution to the free energy of the lattice could not be neglected. New theoretical methods were clearly required in order to account for anharmonic effects. A number of theories, which nevertheless retained the simplicity of the single particle approach, have been proposed[10]. The development of more complex theories was greatly assisted by the availability of computer simulation data for the systems studied[11]. However, more recently, approximate theoretical calculations have been abandoned in favour of the exact approach provided by the simulations themselves. The computational studies yield thermal averages of functions of the particle coordinates. Properties such as the internal energy or pressure of the system can be written in terms of these coordinates. This is not, however, the case for the entropy. But the entropy difference between two different thermodynamic states can be calculated, if it is written as an integral of the differential quantity

$$ds = \frac{1}{T} dE + \frac{P}{T} dV \qquad (1)$$

along a thermodynamically reversible path linking the two states. This route was employed by Hoover and Ree[12] to locate the melting transition and to calculate the entropy of solids, and by Hansen and Verlet to study the liquid vapour transition[13]. The free energy of vacancy formation has also been calculated as the work necessary to uncouple a particle from its neighbours, by Squire and Hoover[14].

Following this work, more direct methods for the calculation of free energy differences have become available[15,16,17], based on the calculations of the energy distribution functions[18].

Although a number of intermediate ensembles may still be required to calculate the desired free energy difference, the procedure does not rely upon an approximate quadrature of the integral based on the evaluation of the integrand at intermediate points. For example, Jacucci and Ronchetti[19] using these methods have repeated the calculations of Squire and Hoover[14] to obtain the free energy of vacancy formation in a Lennard-Jones crystal, by a direct comparison of energy difference distributions between the perfect and defective crystal, without the need for intermediate ensembles; this topic is discussed further below.

If the systems for which the free energy differences are required are very similar, as is the case for a crystal and its quasi-harmonic description, the free energy difference can be obtained as a thermal average evaluated in one of the two ensembles. Pollock[20] used this direct method to calculate the anharmonic contribution to the free energy of a rare gas solid. However, this situation is rarely encountered. Later sections describe the energy distribution function method which can be employed to calculate free energy differences between a very wide range of systems. A comprehensive guide to past and present work in the field is not possible in this chapter. We concentrate therefore on current applications of computer simulations to free energy calculations for solid state problems, and we suggest possible directions of future work. Our approach is, first, to review, in section 3, the basic methodology of Monte-Carlo simulations; section 4 then describes the statistical mechanical formulation of the problem. Techniques and algorithms for Monte-Carlo simulations are discussed in the final sections of the chapter.

3. Monte Carlo simulation

This chapter is mainly but not exclusively concerned with the application of the Metropolis[32] Monte Carlo simulation technique to problems concerned with crystal lattices. In this section we give a brief outline of the Metropolis method.

Once the independent variables appropriate to a particular thermodynamic system have been chosen, statistical mechanics provides formal expressions for the dependent thermodynamic variables as averages in the appropriate ensemble. For example, if we keep N, the number of particles, V the volume and T the temperature fixed, then the configurational average of a function $f(q^{3N})$ of the particle coordinates (q^{3N}) is given in the canonical ensemble (fixed N,V,T) by

$$<f> = \int f(q^{3N}) \; e^{-\beta v(q^{3N})} \; d^{3N}q / \int e^{-\beta v(q^{3N})} \; d^{3N}q \qquad (1)$$

where $f(q^{3N})$ could be, for example, the potential energy function $v(q^{3N})$.

<f> would then correspond to the configurational part of the internal energy of the system. For problems of interest in this chapter, N is of the order of 10^{23} and $u(q^{3N})$ a complicated function of the particle coordinates. No analytical solution of equation (1) is therefore possibly, nor is a numerical solution practical unless the number of particles is reduced dramatically. This reduction of N is conventionally achieved by modelling a small number of particles (32-10,000) in the computer and imposing periodic boundary conditions to remove surface effects[41]. The limited number of significant figures which can be used in a computer to represent particle co-ordinates requires that we must rewrite equation (1) as a sum, not an integral, over all possible states (N_S)

$$\langle f\rangle \simeq \sum_{i=1}^{N_S} e^{-\beta u_i} f_i / \sum_{i=1}^{N_S} e^{-\beta u_i}$$

$$= \sum_{i=1}^{N_S} p_i f_i \qquad (2)$$

$$p_i = e^{-\beta u_i} / \sum_{i=1}^{N_S} e^{-\beta u_i}$$

However despite the approximation introduced above, N_S is still so large that a direct computation of all states is not possible. In order to make progress a way of selecting those states which are of high probability must be found. One possibility would be to sample configurations from the N_S available with a probability proportional to p_i of equation (2). An algorithm which does this can be derived from the theory of Markov chains[41], describing processes that can be characterised by a one step transition probability $P_{\ell,\ell+1}$ of going from a state ℓ to a successor state $\ell+1$, where $P_{\ell,\ell+1}$ does not depend on the states prior to ℓ. By imposing certain constraints on $P_{\ell,\ell+1}$ it is possible to generate succesive states such that in the limit of an infinite number of states in the chain, they are distributed according to a desired probability distribution function. In our case, the chain consists of sets of particle configurations and in the remainder of this section we describe the simple prescription due to Metropolis[32] for $P_{\ell,\ell+1}$ which ensures that in the limit of a very large number of configurations generated, using this one step transition probability, they are distributed according to the canonical ensemble distribution function p_i of equation (2). This allows the ensemble average <f> to be calculated from the unweighted average $\bar{f}$ of the function for the generated configurations N_{MC}

$$\langle f\rangle \sim \bar{f} = \sum_{j=1}^{N_{MC}} f_j \quad . \qquad (3)$$

Configurations are generated using the following procedure. Given an initial configuration of the N particles in a volume V, for example the lattice structure

of the system, a new trial configuration is obtained by a random displacement of
one of the particles. The potential energy of this new configuration (u') is
calculated and compared with the potential energy (u) of the old, starting
configuration. If $\Delta u = u' - u$ is negative then the trial configuration is accepted and
becomes the second configuration in the Markov chain. If it is positive,
the factor

$$B = \exp(-\Delta u / kT) \tag{5}$$

is calculated and compared to a random number R between 0 and 1. If $B \geq R$ then
the trial configuration is accepted as before. If B<R it is rejected and the old,
initial configuration is used again. It is therefore possible for successive
configurations to be identical. The second configuration in the chain now
becomes the old configuration and a new trial configuration is obtained as before
by a random displacement of one of the particles. This procedure is repeated
until N_{MC} configurations are obtained.

Unless the starting configuration is an equilibrium configuration of the
system studied, the first part of the simulation is used to bring the system to
equilibrium. The equilibrium state is assumed to have been reached when the desired
average ($\bar{f}$) or averages are observed not to drift on increasing the number of
configurations but to fluctuate about some mean value. The number of equilibrium
configurations required for convergence to the ensemble average depends upon the
quantity being averaged and the system studied, but is usually in the range of 10^5
to 10^7 configurations.

Our discussion now continues with an account of the various methods for
calculating free energies and the problems to which they have been applied.

4.   Energy distribution methods

"The introduction by Valleau and colleagues and by Bennett, building upon earlier
work of McDonald and Singer, of methods for the calculation of free energy
differences constitutes in our opinion the most important recent innovation in
applying the Monte Carlo method".

The remark, made in Wood and Erpenbeck's[21] review of molecular dynamics and
Monte Carlo calculations in statistical mechanics indicates the impact of the new
methods of calculating free energy differences. In this section we shall introduce
the equations upon which these methods are based and discuss the practical problems
which may arise in their application. Equations will be derived in the canonical
ensemble at constant number of particles N, volume V and temperature T. The
thermodynamic potential in this ensemble is the Helmholtz free energy A, related
to the configurational integral Q by the equations

$$\beta A = -\ln Q \tag{6}$$

$$Q = \int d^{3N}q \ \exp[-\beta u'(q^{3N})] \tag{7}$$

in which $\beta=1/kT$, and where the 3N q's are the particle coordinates and u'(q) is the internal energy.

4.1 Comparison of a system at two temperatures

By rewriting the configurational integral over particle coordinates as a one dimensional integral of the internal energy of the system of interest, McDonald and Singer[18] introduced the normalised energy probability

$$f_\beta(u) = \langle\delta(u'-u)\rangle$$

$$= \exp(-\beta u) \int \delta(u'-u) \ d^{3N} q/Q$$

$$= \exp(-\beta u) \ g(u)/Q \tag{8}$$

where $\langle...\rangle$ is a canonical average and g(u) is the density of states of the system. The free energy difference between two states of a system at different temperatures T_0, T_1, which is given by

$$\beta_1 A_1 - \beta_0 A_0 = \ln \frac{Q_0}{Q_1} \tag{9}$$

may be written using equation (8) as

$$\Delta(\beta A) = +\ln \int_{-\infty}^{\infty} f_{\beta_0}(u) \ \exp(-u(\beta_1-\beta_0)) \ du. \tag{10}$$

This is an exact expression, allowing the calculation of $\Delta(\beta A)$ if $f_\beta(u)$ is known at one temperature $\beta_0 = \frac{1}{kT_0}$. McDonald and Singer[18] pointed out that the function $f_{\beta_0}(u)$ can be estimated in a Monte Carlo simulation by compiling a histogram $f^*_{\beta_0}(u)$ of the probability of finding a configuration in the simulation at β_0, with internal energy between u and u+δu. However since the most commonly used Monte Carlo algorithm (the Metropolis algorithm discussed in section 5), is designed to avoid configurations of low probability, $f^*_{\beta_0}(u)$ will be zero in those regions of u where $f_{\beta_0}(u)$ is very small. If the integrand of equation (8) is significant in these regions the use of $f^*_{\beta_0}(u)$ will lead to very large errors. McDonald and Singer[17] found that reasonable results could be obtained for temperature differences of approximately 15%. For larger temperature differences, the high and low temperatures states favour regions of configurational space which are too widely separated to be explored by the Metrolopis algorithm at one temperature only. This is a general problem with such one sided (here one temperature) calculations. A two sided evaluation of $\Delta\beta A$ is provided by the ratio of the two distributions

$$f_{\beta_0}(u) = e^{-\beta_0 u} \ g(u)/Q_{\beta_0} \tag{11}$$

$$f_{\beta_1}(u) = e^{-\beta_1 u} \ g(u)/Q_{\beta_1} \tag{12}$$

$$\frac{Q_{\beta_0}}{Q_{\beta_1}} = e^{-u(\beta_0-\beta_1)} \frac{f_{\beta_1}(u)}{f_{\beta_0}(u)} . \tag{13}$$

Using this relationship we require only a partial overlap between the regions of u in which the two distributions have substantial values. This point will be discussed in detail later.

4.2 Comparison of two systems with different Hamiltonians

We now turn to the more general problem of the calculation of the free energy difference between two systems with different Hamiltonians and consider in more detail the question of the application of energy difference distribution functions obtained in both systems of interest - a procedure which we will refer to as 'two-sided' evaluations. For two systems with Hamiltonians $H_0=k+u_0$ and $H_1=k+u_1$, acting on the same configuration space (q), we have

$$e^{-\beta(A_0-A_1)} = \frac{Q_0}{Q_1} = \frac{\int e^{-\beta u_0} d^{3N}q}{\int e^{-\beta u_1} d^{3N}q} = \frac{\int e^{-\beta(u_0-u_1)} e^{-\beta u_1} d^{3N}q}{\int e^{-\beta u_1} d^{3N}q}$$

$$= <e^{-\beta(u_0-u_1)}>_1$$

$$\frac{\int e^{-\beta u_0} d^{3N}q}{\int e^{-\beta(u_1-u_0)} e^{-\beta u_0} d^{3N}q} = \frac{1}{<e^{\beta(u_0-u_1)}>_0} \tag{14}$$

where $<...>_0$ and $<...>_1$ are canonical averages with Hamiltonian H_0 and H_1 respectively. The calculation of either average would correspond to a one-sided evaluation of the free energy difference. Note that if $u_1=u_0+\varepsilon V$ then these equations are the starting point for perturbation theory calculations

$$A_1-A_0=-\ln<e^{-\beta\varepsilon V'}>_0 = \beta\varepsilon<v'>_0 + \frac{\beta^2\varepsilon^2}{2} (<v'>_0^2 - <v'^2>_0) +... \tag{15}$$

If we introduce energy distribution functions of the potential energy difference $\Delta=u_0-u_1$

$$f_0(\Delta) = <\delta(\Delta-(u_0-u_1))>_0$$

$$f_1(\Delta) = <\delta(\Delta-(u_0-u_1))>_1 \tag{16}$$

then, for example

$$<e^{-\beta(u_0-u_1)}>_0 = \int_{-\infty}^{\infty} f_0(\Delta) e^{-\beta\Delta}d\Delta . \tag{17}$$

As in the case of the temperature difference calculation discussed previously, if $f_0(\Delta)$ is obtained from a Monte Carlo simulation then unless $H_0 \sim H_1$, the function $f_0^*(\Delta)$ is unlikely to cover a sufficiently wide range of Δ to allow the integral to be accurately computed. The usefulness of equation (17) can be extended using the Umbrella sampling technique of Valleau and colleagues[15] in which further biased Monte Carlo simulations are performed to extend the range of Δ for which $f_0(\Delta)$ is known. The relationship between distributions of Δ in the two systems 0 and 1 is given by;

$$
\begin{aligned}
f_0(\Delta) &= \int \delta(\Delta-(u_0-u_1))\ e^{-\beta u_0}\ d^{3N}q/Q_0 \\
&= \int \delta(\Delta-(u_0-u_1))\ e^{-\beta(u_0-u_1)}\ e^{-\beta u_1}\ d^{3N}q/Q_0 \\
&= e^{-\beta\Delta} \int \delta(\Delta-(u_0-u_1))\ e^{-\beta u_1}\ d^{3N}\ q/Q_0 \\
&= e^{-\beta\Delta}\ f_1(\Delta)\ Q_1/Q_0 \ .
\end{aligned}
\tag{18}
$$

The free energy difference can therefore be written as

$$
\exp[-(\beta A_0 - \beta A_1)] = \frac{Q_0}{Q_1} = e^{-\beta\Delta}\ \frac{f_1(\Delta)}{f_0(\Delta)} \ .
\tag{19}
$$

In contrast to equations (16), this equation contains distribution functions from both systems 0 and 1 and corresponds to a two-sided evaluation of the free energy difference.

An alternative equation for the two-sided evaluation was proposed by Bennett[16] and called the acceptance ratio method. This and other methods, including equation (16), are discussed and compared in an illuminating article[16] of Bennett's, in which a complete treatment of the relative statistical errors can be found. Bennett's equation can be obtained from equation (18) by writing

$$
\omega(\Delta)\ Q_0\ f_0(\Delta) = \omega(\Delta)\ Q_1\ e^{-\beta\Delta}\ f_1(\Delta)
\tag{20}
$$

where we have introduced an arbitrary function $\omega(\Delta)$.

Integrating over all Δ we obtain

$$
\frac{Q_0}{Q_1} = \frac{\langle \omega\ e^{-\beta\Delta} \rangle_1}{\langle \omega \rangle_0}
\tag{21}
$$

The function $\omega(\Delta)$ was introduced by Bennett[17] in order to optimise the estimation of the free energy difference $\ln Q_0/Q_1$. From statistical arguments Bennett found that the choice

$$
\omega(\Delta) = \text{const}\ [\ 1 + (Q_1 n_0/Q_0 n_1)\ \exp(-\beta\Delta)]^{-1}
\tag{22}
$$

where n_0, n_1 are the number of statistically independent configurations from each

Monte Carlo simulation, minimised the expectation value of $(\Delta A_{est} - \Delta A)^2$; substituting equation (20) into equation (19) gives

$$\frac{Q_0}{Q_1} = \frac{\langle I(\Delta+C)\rangle_1}{\langle I(\Delta-C)\rangle_0} \ \exp(+C)$$

where

$$C = \ln \frac{Q_0 n_1}{Q_1 n_0} \ , \qquad I(x) = \frac{1}{1+\exp(x)} \qquad (23)$$

4.2.1. The problem of insufficient overlap

The use of Monte Carlo simulation to calculate a free energy difference using either equation (18) or equation (22) depends upon there being sufficient overlap between $f_0^*(\Delta)$ and $f_1^*(\Delta)$. This is clearly seen in equation (18) since, if there is no overlap, the ratio $f_0^*(\Delta)/f_1^*(\Delta)$ cannot be estimated. Where no overlap occurs several techniques are available which still allow a calculation of the free energy difference. The first is the use of bridging distributions. By constructing Hamiltonians which are intermediate between H_0 and H_1, energy distribution functions can be obtained which overlap with each other and those obtained from Monte Carlo simulations of H_0 and H_1. Repeated applications of either equation (18) or equation (22) enables the required free energy differences to be calculated. For the cases of hard-core potentials, Valleau and colleagues[15] called the repeated application of equation (16) "multistage sampling". A second, graphical method, was suggested by Bennett[7]. Taking the logarithm of both sides of equation (16) we obtain

$$\ln f_0(\Delta) + \tfrac{1}{2}\beta\Delta = \ln f_1(\Delta) - \tfrac{1}{2}\beta\Delta + \beta[A_1-A_0] \ . \qquad (24)$$

From this equation we see that a plot of the terms $\ln f_0(\Delta) + \tfrac{1}{2}\beta\Delta$, and $\ln f_1(\Delta) - \tfrac{1}{2}\beta\Delta$, against Δ produces two parallel curves separated by the constant term $\beta(A_1-A_0)$. For the case where there is no common region of Δ, both curves can be extrapolated into the gap as shown in Figure 1. The free energy difference can then be read from the graph as the vertical shift needed to superimpose the two curves in the gap region. The graphical method is limited to cases where the gap is not much wider than the broadness of the two distributions.

4.2.2 Single particle distributions

A major drawback of the approaches described above is that as the number of particles (N) used in the Monte Carlo simulation is increased the width of the energy distribution function and hence the amount of overlap decreases approximately as $1/\sqrt{N}$. We have seen that the amount of overlap between energy distributions is all important in applying the free energy difference methods, so that this decreasing overlap tends to limit free energy calculations to small systems. For

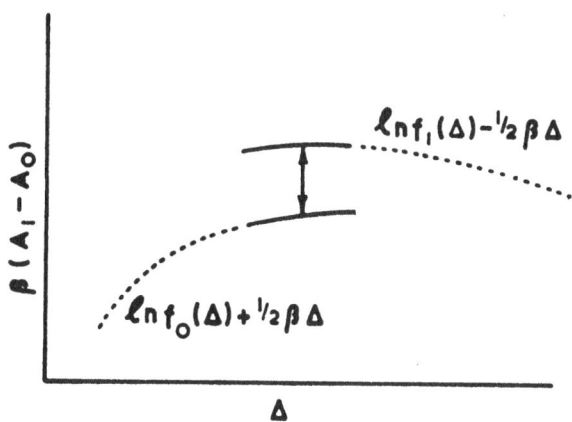

Figure 1   The graphical method of Bennett (7) (see text)

simple liquids this does not appear to be a severe problem since a number of studies[15,22,23] have shown that the free energy difference reaches its limiting value rather quickly.

This is probably not the case for crystals, where somewhat larger systems have been found to be necessary to reach a limiting value of ΔA[35].   The origin of the $1/\sqrt{N}$ behaviour is of a quite general nature, that is in the vanishing magnitude of relative fluctuations of extensive properties as the thermodynamic limit is approached. For homogeneous systems, this problem could in principle be avoided by dealing with distributions of energy and energy differences of individual particles.   These distributions might have $1/N$ type corrections to an otherwise N-independent average and width.   Two sided expressions for the difference between systems with two Hamiltonians, based on individual particle properties. are easily obtained as shown below.

First, let $u = \sum_i \phi_i$; $\phi_i = \frac{1}{2} \sum_j V_{ij}$ and let $<\delta(\phi-\phi_i)>$ be the individual energy distribution.   Then for the case of different temperatures we can write

$$\langle\delta(\phi-\phi_1)\rangle_0 = \int e^{-\beta_0 u} \delta(\phi-\phi_i) \, d^{3N}/Q_0$$

$$= \int e^{-(\beta_0-\beta_1)u} \delta(\phi-\phi_i) \, e^{-\beta_1 u} \, d^{3N}q/Q_0$$

$$= \frac{Q_1}{Q_0} \langle\delta(\phi-\phi_i) \, e^{-(\beta_0-\beta_1)u}\rangle_1 \tag{25}$$

For the case of different Hamiltonians, dropping the subscript i of ϕ_i and denoting by ϕ_0 and ϕ_1, the values corresponding to u_0 and u_1, we have

$$\langle\delta(\Delta-\phi_0+\phi_1)\rangle_0 = \int \delta(\Delta-\phi_0+\phi_1) \, e^{-\beta u_0} \, d^{3N} \, q/Q_0$$

$$= \int e^{-\beta(u_0-u_1)} \delta(\Delta-\phi_0+\phi_1) \, e^{-\beta u_1} \, d^{3N}/Q$$

$$= \frac{Q_1}{Q_0} \langle\delta(\Delta-\phi_0-\phi_1) \, e^{-\beta(u_0-u_1)}\rangle_1 \tag{26}$$

$$= \frac{Q_1}{Q_0} e^{-\beta\Delta} \langle\delta(\Delta-\phi_0+\phi_1) \, e^{-\beta(u_0^*-u_1^*)}\rangle_1$$

where $u^*=u-\phi_1$.

The presence of the exponential term inside the averages of equation (25) is bound to make it less reliable numerically then equation (19). The utility of these asymmetric two sided expressions is still to be tested.

4.3 Widoms method

Single particle energy distributions are also of interest because of the current applications of Widom's[27] equation for the free energy per particle in a homogeneous fluid[28]. The configurational chemical potential is

$$\beta \left(\frac{\partial A}{\partial N}\right)_{V,T} = -\ln \frac{Q(N+1,V,T)}{Q(N,V,T)} = -\ln \frac{V}{N} \langle\exp(-\beta\phi)\rangle \tag{27}$$

and the computational task consists in the evaluation of the mean Boltzmann factor of a test particle.

There has been some discussion as to the accuracy of such a method for fluid systems in various thermodynamic states. We point out that the problem just outlined can be interpreted as the problem of finding the free energy difference of two systems H_0 and H_1, differing only in that in system H_0 one of the N+1 particles in the volume V is uncoupled from the rest. Then $u_0=u_1$, $u_0-u_1=\phi$, and equation (11) reads

$$\exp(-\beta(A_0-A_1)) = \frac{Q_0}{Q_1} = \langle\exp(-\beta\phi)\rangle_1 \tag{28}$$

which corresponds to equation (27), once differences in the dimensions of the configurational integrals are taken into account. As a consequence we can regard current applications of Widom's equation as one sided evaluations of a free energy difference. A two sided evaluation can be obtained from equation (21), which reads in this case as

$$\exp(-\beta(A_0-A_1)) = \frac{Q_0}{Q_1} = \exp(-\beta\Delta) \frac{f_1(\phi)}{f_0(\phi)} \qquad (29)$$

where $f_1(\phi)$ is the distribution of the energy of the test particles in the system H_1 having only N interacting particles, and $f_0(\phi)$ is the distribution of the energy of any one of the N+1 interacting particles of the system H_0[29]. We have pointed out that one sided evaluations may be numerically inaccurate if the distributions do not overlap entirely. In general, a two sided evaluation does not suffer from this problem and should be either used in place of (27) or at least used to check its range of validity. We note in passing that Widom's equation cannot be applied in this fashion to obtain the free energy per particles of a perfect lattice; the problem is discussed below.

4.4 The N.P.T. Ensemble

The methods outlined in this section are not limited to the canonical (N,V,T) ensemble. Analogous expressions can be derived for other ensembles, for example the isothermal-isobaric (NPT) ensemble. In this ensemble the thermodynamic potential is the Gibbs free energy (G) and the required free energy difference is

$$\beta\Delta G = -\ln(\Phi_1/\Phi_0) \qquad (30)$$

where Φ_1 and Φ_0 are the isothermal-isobaric configurational integrals, which can be obtained from the distributions of the enthalpy difference $\Delta h=\beta h_0-\beta h_1\equiv\beta u_0-\beta u_1$ for the same configuration and volume

$$f_0(\Delta h) = \int dv' \int d^{3N}q \ e^{-\beta h_0} \ \delta(\Delta h - (h_1-h_0))/\Phi_0$$

$$f_1(\Delta h) = \int dv' \int d^{3N}q \ e^{-\beta h_1} \ \delta(\Delta h - (h_1-h_0))/\Phi_1 \ , \qquad (31)$$

as

$$\exp(-\beta\Delta G) = \frac{f_0(\Delta h)}{f_1(\Delta h)} \ \exp(\Delta h) \ . \qquad (32)$$

However, if the relative value of the equilibrium volume difference between the two systems at constant P and T does not exceed a few percent, then ΔG will be essentially the same as the constant volume result for the same system, i.e. ΔA[25]

$$\Delta G = G_0(P,T) - G_1(P,T) = U_0(P,T) - TS_0(P,T)$$
$$- U_1(P,T) + TS_1(P,T) + P(V_0(P,T) - V_1(P,T))$$
$$= A_0(V_0,T) - A_1(V_1,T) + P(V_0-V_1) \qquad (33)$$

to first order and remembering that $\left(\frac{\partial A}{\partial V}\right)_T = -P$

$$\Delta G - A_0(V_0,T) - A_1(V_0,T) - \frac{\partial A}{\partial V}(V,T)(V_1-V_0) + P(V_0-V_1) \tag{34}$$

$$= \Delta A(V_0,T) \, .$$

The N.P.T. ensemble will probably be of most use for the isobaric investigation of the variation of entropy and volume.

5. Remarks on sampling algorithms

In order to apply the ideas of the last section we must evaluate certain thermal averages. These can be calculated by sampling configuration space using statistical techniques which are either based on importance sampling or, using the ergodic theorem, on the integration of the equations, Newtonian or stochastic, of the system of interest. The former is the Monte-Carlo and the latter the Molecular Dynamics technique.

Since Metropolis et al[32] described their Monte Carlo method, relatively little has been done to study the properties of the algorithm or to find possible improvement.[39] Some outstanding exceptions are Fosdick's reduction of variance[33], the use of different acceptance recipes[40] and the so called "Smart Monte Carlo"[37,38]. Molecular dynamics has recently been extended to NVT and NPT canonical ensembles[34] and to cope with periodic cells of fluctuating shape[30]. Its sampling efficiency has undergone only minor improvements since its introduction. In the rest of this section we point to those methodological topics, relevant to the general problems tackled in this chapter, which we believe deserve further study.

5.1 Sampling efficiency

The Metropolis Monte Carlo algorithm accepts or rejects a new configuration whose internal energy (u') is more positive than the previous configuration (u) by comparing the Boltzmann factor of the difference $\beta(u'-u)$ with a random number between 0 and 1. Note that for new configurations to have a reasonable chance of being accepted during the run, this internal energy change must be of order kT. It has become almost an unwritten law that new configurations be generated by moving only one of the N particles to a new position. In principle, however, by moving all N particles a small distance to new positions, the internal energy change can be kept in the region of kT. It is of practical interest to ask whether there is an optimum number of particles (M≤N) which allows a more efficient sampling of configuration space than the standard, one particle move method. For 32 particle systems our experience is that one gains in moving all particles at once while for 108 it is better to move one at a time. At the present time we have no information for 1<M<N.

The estimator for the thermal average of the quantity X(q) is usually taken to be

$$<X> = \frac{1}{N} \sum_i X_i \qquad (35)$$

where X_i is the value the quantity takes in the ith configuration of the Metropolis chain. An estimate with a smaller variance is available however[39]. At the ith step in the Metropolis chain, two configurations and hence two values of X are available; the new trial configuration with value $X_i^!$ and the previous configuration, X_{i-1}. The probabilitities, before the Metropolis test, of accepting the new or the old configuration as the ith are P_{new} and P_{old} respectively. The new estimator is now

$$<X> = \frac{1}{N} \sum_i (P_{old} X_{i-1} + P_{new} X_i^!) . \qquad (36)$$

5.2 Sampling by molecular dynamics

The recent advances that have made it possible to study the canonical and isothermal-isobaric ensembles using the molecular dynamics[34] technique, provide another route to the energy distributions discussed above. Since force biased methods, when used in Monte Carlo simulations, lead to some improvement in the sampling efficiency[37,38] we can expect molecular dynamics methods to show a similar improvement; this has in fact been found to be the case[35]. Standard molecular dynamics can be considered to represent the microcanonical ensemble. There the fluctuations of internal energy are modified by the requirement that the total energy be constant. This is likely to affect the high and low u values in the distribution f(u) but has much less effect on the distribution of energy differences between two Hamiltonians since there may be no obvious correlation between the internal energies of the two systems. In this case one can use the f(Δ)'s from molecular dynamics without altering the canonical equations[35]. Another possibility is to rewrite the formalism of section 2 for the microcanonical ensemble[17].

5.3 Correlation times

The comparison of the sampling efficiency of various algorithms is perhaps best done in terms of the empirically estimated autocorrelation time τ introduced by Bennett[17] in the evaluation of the variance of the acceptance ratio estimate. Given the n subsequent configurations of the Markov chain of a Monte Carlo run, or of a molecular dynamics trajectory, we can count the number, n, of configurations that can be chosen so infrequently as to be uncorrelated; $\bar{n} = \tau^{-1} n$, where τ is an empirically estimated autocorrelation time with respect to the function X. This autocorrelation time can be defined as the large k limit of the quantity

$$k \, Var[X^{(k)}]/Var[X] \qquad (37)$$

where $X^{(k)}$ denotes the mean of k consecutive values of X. Of course τ can be accurately estimated only if n>>τ. The cost of a statistically independent data point can then be compared for different algorithms. For Molecular Dynamics and Monte Carlo for example, the cost of a single Monte Carlo move should be multiplied by τ_{MC} and the result compared to the product of τ_{MP} with the cost of a single molecular dynamics step.

5.4 Temperature fluctuations

Strictly speaking, averages obtained using the Metropolis, Monte Carlo or canonical molecular dynamics correspond to the required ensemble averages only in the limit of an infinite number of steps.   Therefore, all averages obtained from these algorithms are sub-averages of the infinitely long run.   Quirke and Jacucci[23] have shown that in the canonical version of the Metropolis procedure (where N,V,T are used as independent variables) these sub-averages can be associated with canonical ensemble averages at an effective temperature T.   In those applications of the free energy difference methods outlined in the last section, where the $\Delta\beta A$'s are small, the energy distribution functions must be corrected so as to represent truly the desired temperature of the systems studied.   To first order in the temperature difference $\delta\beta = \dfrac{1}{k\overline{T}} - \dfrac{1}{kT}$, the sub-average energy distribution function of the internal energy u, i.e. $f_{\overline{T}}(u)$, is related to the large (infinite) run function $f_T(u)$, via

$$f_{\overline{T}}(u) = f_T(u) + (\overline{u}-u)\ f_T(u)\ \delta\beta \qquad (38)$$

where T is the average internal energy at temperature T of the long run.   For free energy difference calculations between systems with different Hamiltonians, we require the energy distribution function of the quantity $\Delta=u_0-u_1$, in both systems. Since the simulations used to calculate them will be of finite length, each distribution will correspond to a different effective temperature $\beta_0=1/kT_0$, $\beta_1=\dfrac{1}{kT_1}$, not necessarily equal to the required temperature $\beta=1/kT$ inserted into the Metropolis algorithm.   In order to calculate the required free energy difference using equation (16) the distributions $f_{\beta0}(\Delta)$ and $f_{\beta1}(\Delta)$ must be corrected so as to represent the same temperature β for each Hamiltonian.   To first order in $\delta\beta$, the following equation holds

$$\frac{f^{\beta_1}(\Delta)[1+(<u_1(\Delta)>_1 - <u_1>_1)\ (\beta_1-\beta)]}{f_0^{\beta_0}(\Delta)\ 1+(<u_0(\Delta)>_0 - <u_0>_0)(\beta_0-\beta)]}\ e^{+\beta\Delta}\ =\ \frac{Q_1(\beta)}{Q_0(\beta)} \qquad (39)$$

here $<u_1(\Delta)>_1$ is the mean value of the internal energy of system 1, in the subset of configurations for which u_0-u_1 has the particular value Δ.   The left hand side of equation (36) contains functions of Δ whereas the right hand side is a constant. In the absence of a separate procedure for fixing the effective temperatures β_0

and β_1[23], they can be obtained from equation (33) by varying β_0 and β_1 until the left hand side is a constant in the overlap region of the two distributions $f_1^{\beta_1}(\Delta)$ and $f_0^{\beta_0}(\Delta)$. We note in passing that the distributions $<u_0(\Delta)>_0$ and $<u_1(\Delta)>_1$ are simply related by

$$\frac{Q_0}{Q_1} \, e^{+\beta\Delta} <u_0(\Delta)>_0 \; = \; <u_1(\Delta)>_1 + \Delta f_1(\Delta) \tag{40}$$

and by the analogous expression involving $f_0(\Delta)$.

The use of molecular dynamics to study the canonical ensemble enables a further check to be made on the idea of effective temperatures for subaveraging of a run. In this case an extra temperature can be defined in terms of the kinetic energy of the particles. It has been observed that the effective temperatures calculated from the potential energies agrees with those obtained from the kinetic energy of the particles[35]. In general for canonical simulations the internal energy will change due to two processes: a slow process, that for exchange with the bulk and a faster exchange between individual particles and between the total potential kinetic energies of the system. In finite runs the average of internal or potential energy will have a different value to that for the ensemble and hence the effective temperature of the calculation is not that of the bath but is that corresponding to this value of the internal energy. In the microcanonical ensemble the potential energy exchanges only with the kinetic energy. As a consequence, if we compare the microcanonical and canonical molecular dynamics results, we find a reduction in the correlation time τ for the potential energy in the microcanonical case[35]. We suggest that an indication of the different relaxation times involved in the two processes, that of correlation with the bath (E) and the internal correlations (F) could be obtained by measuring the correlation times τ of U and ϕ. The suggestion which is intuitive for molecular dynamics may, we believe, also hold for Monte Carlo. If $\tau_E = \tau_U$ and $\tau_I = \tau_\phi$, different algorithms will have a different ratio $R = \tau_E/\tau_I$. The smaller the value of R the better the definition of the temperature of the system with respect to the bath.

6. Current applications

The method described in section 2 are very flexible; we can study systems at different temperatures, densities and with different Hamiltonians. All that is required is that knowledge of the potential energy functions should be available for the appropriate particle configurations. They have been used to study a wide range of problems. From the point of view of thermodynamic perturbation theories, these methods provide an economical way of investigating the effect of various properties of adding extra terms to the intermolecular potential. For example, Jacucci and Quirke[22] used these methods to calculate directly the free energy difference between hard and soft core molecular fluids. More generally, these techniques allow the calculation of phase diagrams of pure liquids[15,23,24] and liquid mixtures[26]. In the solid state, vacancy formation entropies and volumes can be calculated by comparing two ensemble one of which contains an uncoupled particle[19,36]. In this section we describe some aspects of a few current applications of the methods.

6.1 Free energies of vacancy formation

Our first example is that of the defect calculation referred to above[19], where H_0 is the Hamiltonian of the perfect lattice and H_1 represents the Hamiltonian of the same lattice but with one uncoupled particle. All particles are constrained within single occupancy cells chosen to be a sphere within the Wigner seitz cell. This keeps the uncoupled particle from wandering through the lattice and prevents the neighbours of the vacancy from jumping into it. The free energy difference at constant volume between the two systems with Hamiltonian H_0 and H_1 can be calculated by applying equation (14) or (18) of section 2 using energy distribution functions obtained by Metropolis Monte Carlo simulation. In order to calculate the distribution functions $f(\Delta)$, where $\Delta = u_0 - u_1$, the following procedures are used. For the perfect lattice distribution $f_{LAT}(\Delta)$ a histogram of the potential energy difference Δ calculated in each perfect lattice configuration is constructed. For the lattice with a vacancy $f_{VAC}(\Delta)$ is calculated by constructing a histogram of Δ for a series of random insertions of the extra particle in the spheres contained within the Wigner Seitz cell.

It is interesting to note that in this case one of the two one sided evaluations of the free energy difference is found to be accurate, that is, the evaluation based on $f_{VAC}(\Delta)$

$$e^{-\beta \Delta A} = \int e^{-\beta \Delta} f_{VAC}(\Delta) \, d\Delta \tag{41}$$

It is unfortunate that the useful one sided evaluation is that determined by $f_{VAC}(\Delta)$; the distribution $f_{LAT}(\Delta)$ can be obtained with greater statistical accuracy since it can be calculated for each of the particles in the lattice and averaged over all of them. This is perhaps a consequence of the fact that the defective crystal can assume configurations which are appropriate to the perfect crystal with a smaller energy change than is the case for the perfect crystal when it assumes relaxed configurations. The one sided evaluation of equation (38) exactly corresponds to the procedure used when applying Widom's equation to calculate the free energy per particles in fluids; but it is conceptually different in that it does not yield the free energy per particles but rather a combination of the free energy per particle for the perfect solid plus the defect free energy. The equilibrium concentration of vacancies can then be found if one also has this free energy per particle of the perfect solid. These calculations have recently been extended to obtain the formation volume $\frac{dG}{dp})_{defect}$ for a Lennard-Jones crystal near melting where it is found to be 1.2 atomic volumes[36].

6.2 Polymorphic transformations

The second example is the hypothetical polymorphic transformation FCC to BCC of systems having various interatomic potentials. Such transformations have been observed to occur spontaneously in molecular dynamics calculations using a Lagrangian which allows for the variation of the shape and size of the system[30]. The direction of this spontaneous transformation was such that the FCC structure was obviously the stable structure for a Lennard-Jones type potential while the BCC structure was stable for a metallic potential appropriate to an alkali metal. To investigate the difference in free energy of the two structures for the same interaction potential, equation (17) has recently been used[35]. Fortunately, the sign of ΔA has been found to agree with the direction of spontaneous transition observed in the simulation. Two paths have been used to link the crystal structures. The first uses the existence of a physical path for the transformation. This is achieved by distorting the system in one direction and letting it shrink in the two normal directions but keeping the volume constant. The second uses as intermediate systems the respective harmonic crystal. These calculations have been carried out with both Monte Carlo and Molecular Dynamics methods. The Molecular Dynamics method has been found to be vastly superior in sampling efficiency and has been used to deal with systems containing a larger number of particles than would have otherwise been too costly to perform using Monte Carlo techniques.

6.3 Clusters

We conclude with an interesting application that has recently been made to the evaluation of the configurational free energy of a cluster of N_{spins} in the Ising model[31]. For this calculation one compares configurations of clusters differing by one spin. This problem has a peculiarity that the distribution of energy differences is a histogram containing only a few discrete values.

This calculation has provided a very accurate numerical test in a completely solveable case of the validity of classical capillarity theory, permitting amongst other things, the identification of a logarithmic term in the cluster free energy as a function of the number of particles.

References

1. Born, M. and Huang, K. Dynamical theory of crystal lattices (Oxford University Press) (1950).

2. Lifschitz, I.M. Suppl. Nuovo. Cim. 4, 716 (1956).
    Lifschitz, I.K. Kosevich, A.M. Reports on Proceedings in Physics XXXIX, 217 (1966).

3. Montroll, E.W. and Potts, R.B. Phys. Rev. 100, 525 (1955).

4. Kanzaki, H.J. Phys. Chem. Solids 2, 24 (1957).

5. Nardello, G.F. Rendiconti Della Scuola Internazionale di Fisica "E Fermi", XVIII Corso, Pergamon Press, Oxford (1963).
    Nardelli, G.F. Repanai-Chiarotti, A, Nuovo Cim 18, 1053 (1960).
    Nardelli, G.F. and Tettamanz, N. Phys. Rev. 126, 1283 (1962).

6. Nardelli, G.F. and Terzi, N. J. Phys. Chem. Solids 25, 815 (1964).

7. Hoover, W.G., Young, D.A. and Grover, R. J. Chem. Phys. 56, 2207 (1972).

8. Tosi, M.P. and Fumi, F.G. Phys. Rev. 131, 1458 (1963).

9. Kuebler, J. and Tosi, M.P. Phys. Rev. 137A, 1617 (1965).

10. For a comprehensive analysis of cell models see Barker, J.A. "Lattice theories of the liquid state" (MacMillan, New York 1963).

11. For example, Hoover, W.G., Ashurst, W.T. and Grover, R. J. Chem. Phys. 57, 1259 (1972), and Barker, J.A. and Gladney, H.M. J. Chem. Phys. 63, 3870 (1975).

12. Hoover, W.G. and Ree, F.H. J. Chem. Phys. 12, 4873 (1967).

13. Hansen, J.P. and Verlet, L. Phys. Rev. 184, 151 (1969).

14. Squire, D.R. and Hoover, W.G. J. Chem. Phys. 50, 701 (1969).

15. Valleau, J.P. and Torrie, G.M. "Modern theoretical chemistry", Vol. 5, edited by B.J. Berne (Plenum Press), chapter 5 and references (1977).

16. Bennett, C.H. "Exact calculations on point defects in model substances". IBM Research Report RC4648, December 1973. Published in Diffusion in Solids. Editors: Nowick, A.S. and Burton, J.J. (Academic Press) (1975).

17. Bennett, C.H. J. Comp. Phys. 22, 245 (1976).

18. McDonald, I.R. and Singer, K. Discuss Faraday Society 43, 40 (1967).

19. Jacucci, G. and Ronchetti, M. Solid State Comm. 33, 35 (1980). A similar calculation was independently underaken by Pollock, E.L. and Flynn, L.P. (unpublished).

20. Pollock, E.L. J. Phys. C. 9, 1129 (1976).

21. Wood, W.W. and Erpenbeck, J.J. Ann. Rev. Phys. Chem. 27, 319 (1976).

22. Jacucci, G. and Quirke, N. Molec. Phys. 40, 1005 (1980).

23. Quirke, N. and Jacucci, G. Molec. Phys. 45, 811 (1982).

24. Kohler, F. and Freasier, B.L. Proc. 6th Conf. on Thermo., Merseburg (Gdr), 33 (1980).

25.   Flynn, C.P. "Point defects and Diffusion" (Oxford, Clarendon Press), Chapter 7 (1972).

26.   Torrie, G.M. and Valleau, J.P. J. Chem. Phys. 66, 1402 (1977).

27.   Widom, B., J. Chem. Phys. 39, 2808 (1963).

28.   See for example: Adams, D.J. Molec. Phys. 28, 1241 (1974).
      Romano, S. and Singer, K. Molec. Phys. 37, 1765 (1979).
      Powles, J.G. Molec. Phys. 41, 715 (1980).
      Shing, K.S. and Gubbins, K.E., Molec. Phys. (in press) 1982.
      Powles, J.G., Evans, W.A.B. and Quirke, N., Molec. Phys. (in press) 1982.

29.   A similar equation has been used by Sing, C. and Gubbins, K.E. (preprint 1981).

30.   Parinello, M. and Rahman, A. Phys. Rev. Letts. 45, 1196 (1980).

31.   Jacucci, G., Martin, G. and Perini, A. Nucleation Workshop, Les Houches (1981).

32.   Metropolis, N., Rosenbluth, A.W., Rosenbluth, M.N., Teller, A.H. and Teller, E. J. Chem. Phys. 21, 1087 (1973).

33.   Fosdick, L.D. Methods in Computational Physics, 1, 245 (1963).

34.   Anderson, H.C. J. Chem. Phys. 72, 2384 (1980).

35.   Jacucci, G. and Rahman, A. Private communication.

36.   Jacucci, G. and Ronchetti, M. To be published.

37.   Rao, M., Pangali, C. and Berne, B.J. Molec. Phys. 37, 1773 (1979).

38.   Rossky, P.J., Poll, J.D. and Friedman, H.L. J. Chem. Phys. 69, 4628 (1978).

39.   Kalos, M.H. and Whitlock, P.A. Monte Carlo Methods, Vol.1 (Lecture Notes, to be published).

40.   Peskun, P.H. Biometrika 60, 607 (1973).

41.   Wood, W.W. (1968). in "Physics of Simple Liquids" (H.N.V. Temperley, J.S. Rowlinson and G.S. Rushbrooke, Eds.), North Holland, Amsterdam.

CHAPTER 5
MOLECULAR DYNAMICS SIMULATIONS OF
CRYSTALLINE IONIC MATERIALS

by

J.R. Walker
Department of Chemistry,
University College London,
20 Gordon Street, London WC1 OAJ

1. Introduction

As mentioned in the previous chapter, the majority of the simulation studies
discussed in this book use 'static' descriptions of the lattice in which no explicit
account of thermal motions is included. Such an approach is adequate for solids
at low and ambient temperature; and indeed for many solids the approach is still
useful at temperatures up to the melting point. However, in several materials of
technological importance, the operating temperatures of the system in which they are
used are such that the thermal energy of their constituent particles is of the same
order as the energy barrier opposing their motion. The prime examples are the
'superionic' conductors used in solid-state batteries; in addition, however, we
should note that the fuel materials in fission reactors under operating conditions may
fall into this class of materials. Simulation studies of these systems require a
different type of technique, which is discussed in this chapter; further discussion
of dynamic simulation methods is also presented in Chapter (18).

The phenomenon of superionic conduction has been extensively studed in recent
years[1]. These compounds are essentially normal ionic solids which however,
exhibit abnormally high ionic conductivity, which may approach the values
typical of molten salts at temperatures well below the melting point. Such
materials may be used to form the electrolytes of advanced battery systems. The
possibility of constructing such batteries has provided powerful motivation for
research into superionics, as the battery systems could provide the basis for
efficient automotive power units, and allow a greater utilisation of nuclear energy
by 'load-levelling'.

'Static' simulation methods[2,3,4,5] of the type discussed in Chapter (1) have
shed great light upon the atomistic properties of superionics. They are, however,
unable to deal in a straightforward manner with the correlations between
diffusing ions; this may be of considerable importance in superionics where
there are often large levels of disorder. Moreover, the use of static techniques
in studying particle transport is based on the assumption of hopping models for
transport, the validity of which requires that the activation barrier to migration
be considerably greater than KT. However, we know that this latter condition is
often not obeyed in superionics. For these reasons the molecular dynamics

technique, which has been extensively used in studies of liquids,including molten salts[6], has recently been applied to the study of superionic solids. The reader should consult reference (2) for a general account of the techniques and a discussion of the applications to molten salts which most closely resemble applications to superionics. The present chapter concentrates on the modifications necessary for the treatment of solids. The next section considers the question of interatomic potentials and of simulation techniques with special emphasis on the problems raised by ionic crystal simulations. The chapter ends with an account of the application of the technique to the case of high temperature UO_2. The Appendix then gives a description of a computer program that has recently been developed specifically for simulations of this class of material.

2. Techniques and Potentials

2.1 Techniques

In essence, the molecular dynamics technique takes an ensemble of particles (typically 100-1000 in number) to which periodic boundary conditions are applied and whose time evolution is followed by solving in an iterative manner the equations of motion of the system; potentials for the atomic interactions must, of course, be specified. The use of periodic boundary conditions means that each ensemble or 'box' of particles is surrounded by an infinite array of identical boxes; the simulated system therefore has no surface. Moreover, the imposition of an artificial periodicity may cause problems if a small number of particles is used.

In general, the iterative simulation proceeds as follows. A starting configuration for the ensemble is specified, i.e. the particles are assigned coordinates and velocities for time t=0. The velocities are chosen so that the mean kinetic energy accords with the temperature at which it is desired to perform the simulation; the coordinates are assigned in accordance with known structural properties, although care must be exercised here as will be discussed in greater detail below. The forces acting upon the particles are then calculated using the specified potentials. The velocities and coordinates of the particles may be determined at a time, Δt, later, using a suitable numerical algorithm. The process is then continued until a sufficient number of 'time steps', have been taken for the dynamical properties of the system to be adequately sampled.

Careful consideration must be given both to the choice of the numerical algorithm for updating the velocities and coordinates,and to the magnitude of Δt. Regarding the former, we have found that the algorithm due to Beeman[7] proves to be particularly satisfactory in solid state simulations. This gives the velocities $V(t+\Delta t)$ and coordinates $x(t+\Delta t)$ at time $(t+\Delta t)$ relative to their values $V(t)$ and $x(t)$ at time t, according to the expressions

$$V(t+\Delta t) = V(t) + [2A(t+\Delta t)+5A(t)-A(t-1)].\ \underline{\Delta t} \tag{1}$$

$$x(t+\Delta t) = x(t) + V(t).\Delta t + [4A(t)-A(t-1)].\ \frac{\Delta t^2}{6} \tag{2}$$

where A(t) is the acceleration of the particle at time t. The time step, Δt, clearly must be chosen so as to be less than the 'characteristic time' of any important type of motion (e.g. particle vibrations) in the material under investigation. Thus in simulating solids we must ensure that $\Delta t \ll v^{-1}$ where v is the highest phonon frequency of the solid.

The use of techniques, summarised above, for obtaining dynamical information on solids assumes of course that the system evolves ergodically. The initial time steps will almost certainly not represent the true system; and in general it is necessary to allow several hundred steps for the system to equilibrate. The special problems posed for solid state studies by this initial equilibration are discussed further below.

In general, several thousand time steps (corresponding to ~ 10 picoseconds 'real time' if Δt is set to 10^{-14} sec) are needed for an adequate simulation; as a result the large amounts of computer time are required for dynamical studies, in contrast to the static techniques for which, although large cpu memory may be needed, are modest as regards their demands on time. The 'expense' of dynamical simulations thus severely limits their application.

2.1.1 Special considerations for ionic crystal simulation

The first point we consider applies generally to solids, and concerns the 'box' used for the ensemble of particles in the simulation, which for solids must be a super-cell of the unit-cell of the crystal. Thus if the crystal is non-cubic, the box must be non-cubic, whereas for liquid simulations cubic 'boxes' are adequate. Adaptation of computer codes for non-cubic systems does not, however, present serious difficulties, and has been achieved in the FUNGUS program described in the Appendix.

The second problem which applies specifically to ionic crystals, concerns the electrostatic simulations which converge only slowly in real space; it is usual therefore to use the Ewald[8] technique which is also employed in the static simulationsdiscussed in Chapter (1) and which transforms the summations into reciprocal space in which the convergence is far more rapid. Despite this improved convergence, evaluation of the electrostatic term may take up to 30% of the cpu time required for the simulations, and is indeed responsible for the increased cpu requirements for simulations of ionic systems when compared with non-polar systems.

The next problem to be considered concerns the initiation of the simulation
to which reference has already been made.   The most obvious procedure is to
assign particle coordinates corresponding to the crystallographically determined
structure. A problem inherent in this approach is that on moving away from the
lattice sites the ions gain potential energy and lose kinetic energy. As a
result, an excessive number of time steps (possibly of the order of 1,000) are
taken in achieving the correct distribution of energy between potential and
kinetic terms; the problem is particularly severe in systems such as UO_2 where
the ionic charge is high and the forces correspondingly strong.   To reduce the
difficulties caused by this problem it is desirable to start the simulation with
an 'equipartitioned' distribution of kinetic and potential energy.   This may be
achieved by calculating force constants for displacements of the ions, from which
one is able to estimate an average displacement for each type of ion which
corresponds to an equipartitioned potential energy at the temperature of the
simulated system. We have found that this procedure very considerably reduces the
number of time steps required in the preliminary equilibration.

A general purpose computer program (FUNGUS) is now available for simulation
studies on ionic crystals.   It may be applied to crystal structures of any
symmetry, and incorporates the Ewald technique for Coulomb summations and the
equipartition procedure described above. A more detailed discussion of the program
is presented in the Appendix.

2.2 Generation of Potentials

Given the adequacy of the simulation techniques and the use of a sufficiently
large number of particles and of time steps, the reliability of the results of the
simulation depends on the quality of the interatomic potentials used in the study.
Potentials for ionic crystals are reviewed in Chapter (10), which also describes
how potential models may be parameterised either by fitting to bulk crystal data or
by electron gas calculation of the short range interionic repulsions.

A special problem arises with dynamical simulations in the description of
ionic polarisation. As noted in Chapter (10), polarisation is normally described
in the static simulations by the shell model of Dick and Overhauser[9].   In
molecular dynamics studies it has, with few exceptions, not been possible to
include polarisability, since, for example, if the shell model were used,
it would be necessary to determine the equilibrium configuration of the shells
with respect to the cores after each iteration, since shell (i.e. electron)
configurations respond instantaneously to changes in nuclear coordinates.
Shell equilibration each time step increases the total cpu time required for the
calculation by a factor of between five and ten.   It seems likely therefore that
with presently available computer power, dynamical simulations will be limited

to rigid ion potentials. In simulations of superionic systems, the deficiencies of such potentials may be partially overcome by ensuring that the potential models correctly generate the static dielectric constant of the crystal. For, as discussed in Chapter (10), correct reproduction of this quantity is the most important requirement for a lattice potential which is used in simulating defect properties; and in most solids, defect mechanisms are responsible for ionic transport processes.

Many of the points made in this section are illustrated by the simulation of high temperature UO_2 presented below. For further illustrations of the application of dynamical techniques to solids we refer to the Chapter (18), and to the recently published work of Rahman and co-workers[10,11].

3.   Simulation of stoichiometric UO_2

The study of high temperature UO_2 has many important applications in the field of reactor technology.   In particular, accurate knowledge of transport and thermodynamic properties of the system is necessary for prediction of, for example, the behaviour of oxide fuels during abnormal operation of the reactor. Experimental studies of high temperature UO_2 are, however, very difficult, and guidance from simulation studies is   therefore of great value.

The experimental Arrhenius energy for oxygen diffusion in stoichiometric UO_2 at low temperatures is ~ 3 eV[5,12].   The Arrhenius energy for the stoichiometric material for a conventional defect model based on anion Frenkel disorder is equal to half the anion Frenkel energy plus the anion vacancy activation energy[5]. It is clearly desirable that the potentials used in the dynamical simulation study should yield defect energies which reproduce this value.   This had been successfully achieved by Catlow[5] who developed shell model potentials of UO_2. Walker and Catlow[13], however, showed that, by ensuring that the static dielectric constant was correctly reproduced, an accurate value of the Arrhenius energy could still be calculated for a rigid ion potential.

With the aid of the FUNGUS program discussed in the Appendix it has been possible to simulate UO_2 at several temperatures. Detailed results have been obtained for three temperatures $1101^\circ K$, $2193^\circ K$ and $4201^\circ K$. At the lowest temperature the material is a 'normal' solid, while at the highest it is a molten salt.   The simulation studies showed, however, that at $2193^\circ K$ the system was 'superionic' with an oxygen diffusion coefficient of ~ 2×10^{-6} cm^2 sec^{-1}.   This result demonstrates that UO_2 resembles other fluorite structured solids (e.g. $SrCl_2$, BaF_2) in becoming superionic of elevated temperatures;   further discussion of these latter systems are presented in Chapter (18).

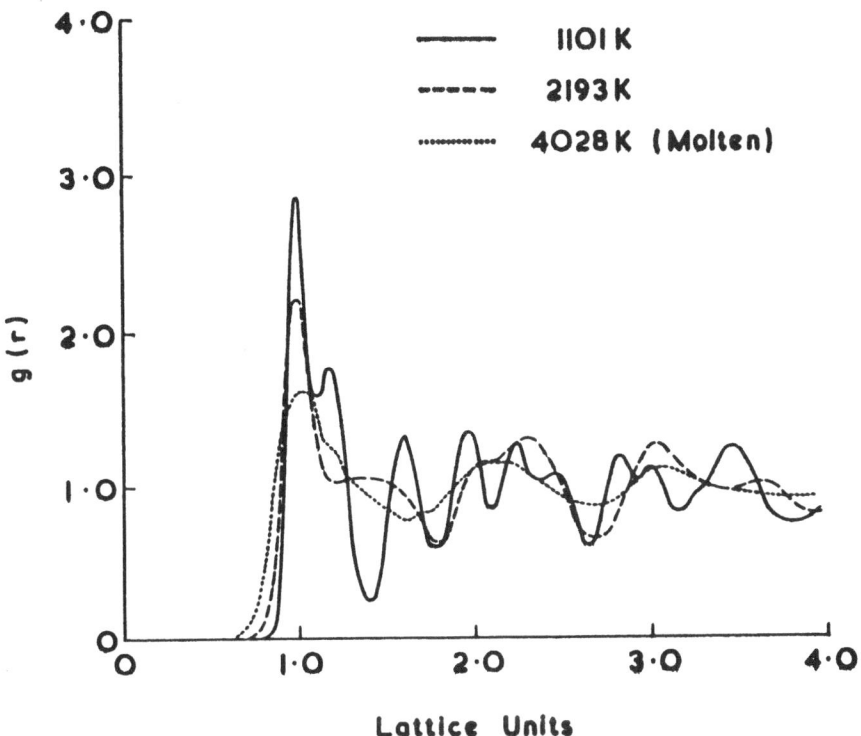

Figure 1 $O^{2-} - O^{2-}$ radial distribution functions for UO_2 at 1101 °K, 2193 °K and 4028 °K.

For the other fluorite structured crystals, it is clear that superionic properties are associated with the generation of anion disorder although the nature and extent of this is controversial. The most important features of the structural properties of high temperature UO_2, as revealed by simulations, are summarised by the radial distribution functions shown in Figure 1. At $1101^\circ K$ we note the Gaussian shaped peaks characteristic of a normal solid. At $2193^\circ K$, however, the correlation function has lost a great deal of structure, and more closely resembles that for the liquid than for the low temperature solid.

It is tempting to attribute the loss of structure to disorder on the anion sublattice. However, a more careful examination of the radial distribution function shows that the peaks which have disappeared are those corresponding to correlations in the <111> direction. The reason for this behaviour is seen by examination of the density map for UO_2 - a projection of which on the (100) plane for the system at $2193^\circ K$ is shown in Figure 2. The latter clearly shows that the distribution of oxygen about the regular anion sites is no longer spherical. The oxygen density now has tetrahedral symmetry with lobes pointing towards the four interstitial sites which surround the regular fluorite positions.

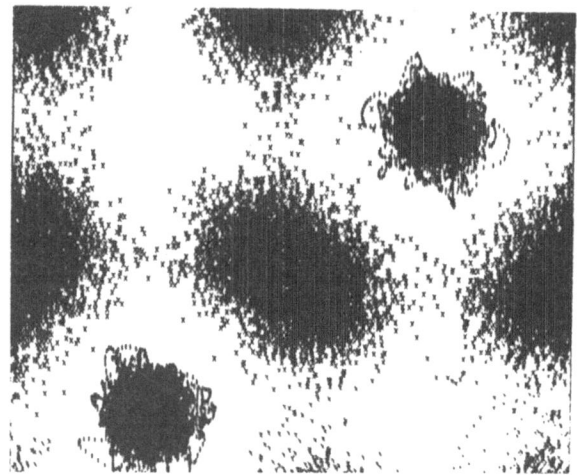

<u>Figure 2</u> Density plot for UO_2 at 2193 °K projected
onto [100] plane.

Figure 2, however, shows that there is no evidence for the occupation of the
cube centre interstitial sites. Similar results were found for other fluorites as
discussed in Chapter (18). The structural properties of high temperature UO_2
cannot therefore be simply interpreted in terms of the generation of interstitial
disorder. Indeed it is unlikely that any conventional defect description is
appropriate.

Finally, we note that by performing simulations at several temperatures it is
possible to calculate the activation energy for anion diffusion; a value of 0.67 eV
was obtained for UO_2 in the superionic region. This is slightly higher than the
calculated vacancy activation energy of 0.5 eV[13], but far less than the energy
that would be predicted from an analysis of a normal 'intrinsic' diffusion
mechanism, for which, we we recall, an Arrhenius energy of ~ 3 eV is obtained at
low temperatures. Our result suggests that in the high temperature region, defect
formation is making little contribution to the Arrhenius energy, a result which
is in line with experimental data on other high temperature fluorite structured
materials[14]. This and other features of those simulations are discussed in greater
detail in reference (13).

Conclusions

Molecular dynamics simulations are clearly capable of yielding useful structural
and dynamical information on high temperature solids. The main problems with the
technique are the excessive cpu times that are required for detailed studies. As
we have seen, this factor may seriously limit the nature of the potentials used
in the simulations. Extensive use of the techniques for solids will therefore
probably need considerable further expansion in computer power.

The Molecular Dynamics program FUNGUS

The FUNGUS computer code is a flexible, molecular dynamics package intended for the simulation of superionic conductors.

The code was originally intended for use on the CRAY-1s series of computers and contains many features which enable the code to take advantage of the vector processing facilities of these machines. The program is also available with CRAY-1s simulation routines which enable the code to run with the CDC FTN5 (A.N.S.I. FORTRAN-77 superset). Minor modifications, however, enable the program to run on any computer system.

The program has the essential features of a standard molecular dynamics program in that it employs periodic boundary conditions, a Born-Mayer-Huggins form for the short range potentials* and the Ewald method to facilitate the evaluation of the Madelung sums. It does, however, contain a number of novel features which greatly extend its usefulness for the study of superionic conductors and solids in general.

First, the program is not restricted to a cubic box for the simulation. The FUNGUS computer code is constructed so that any box shape may be used, without any increase in computational requirement. Many important superionic conductions have hexagonal space groups (e.g. β-Al_2O_3, Li_3N) and cannot easily be studied with conventional molecular dynamics packages. If a supercell of true crystalline unit cell can be provided which will fill three-dimensional space then the program is able to use this supercell as the basis for the molecular dynamics simulation.

Secondly, the program incorporates the kinetic and potential energy equipartitioning routine discussed in section 2. The use of this equipartitioning process has been found greatly to reduce the number of iterations needed to reach an equilibrium distribution of velocities. It is possible, therefore, to reduce significantly the amount of computer time required to perform a molecular dynamics simulation, and hence allow more complicated materials to be modelled.

Finally, the program is constructed so that the entire package may be controlled from one driver routine. Each basic function of the code (e.g. the calculation of forces, and the displacement of ions) is assigned to a different subroutine. This design of program allows a great deal of flexibility to the parameters used in the simulation and allow modifications (for example to the

---

*Alternative forms in the short range potential may, however, readily be included.

updating algorithms) to be made with the minimum of difficulty.

We have employed the FUNGUS code for the simulations of UO_2 reported in this chapter.

References

1. Hayes, W. 'Contemporary Physics", 19, 469 (1978).

2. Catlow, C.R.A., Corish, J., Diller, K.M., Jacobs, P.W.M. and Norgett, M.J., J. Phys. C 12, 451 (1978).

3. Walker, J.R. and Catlow, C.R.A., Nature, 286, 473 (1980).

4. Walker, J.R. and Catlow, C.R.A., Phil. Mag. 43, 265 (1980).

5. Catlow, C.R.A., Proc. Roy. Soc. A353, 533 (1977).

6. Sangster, M.J. and Dixon, M., Adv. Phys. 25, 247 (1976).

7. Beeman, D., J. Comput. Phys. 20, 130 (1976).

8. Ewald, P.P., Ann. der Physik., 64, 253 (1921).

9. Dick, B.G. and Overhauser, A.W., Phys. Rev. B112, 90 (1958).

10. Jacucci, G. and Rahman, A., J. Chem. Phys. 69, 4117 (1978).

11. Vashishta , P. and Rahman, A., in 'Fast Ion Transport in Solids'. (Eds. Vashishtra, P., Mundy, J.N. and Shenoy, G.K.), North Holland, p.527 (1979).

12. Matthews, J. UKAEA Report AERE-M.2643(1974).

13. Catlow, C.R.A. and Walker, J.R., J. Phys. C14, L979 (1981).

14. Carr, V.M., Chadwick, A.V. and Saghafian, R., J. Phys. C11, 55 (1978).

15. Walker, J.R. To be published.

CHAPTER 6
AB INITIO CLUSTER CALCULATIONS FOR DEFECTS IN THE SOLID STATE

by

E.A. Colbourn and J. Kendrick

I.C.I., PLC, The Heath, Runcorn, Cheshire WA7 4QE

1. Introduction

The defective solid state, because of its complexity, poses a number of interesting challenges to the theoretician. Ideally, to model both the perfect and the imperfect lattice, one would choose to work with a very large number of atoms whose positions could be specified (and varied) explicitly, in order to investigate, in detail, the electronic structure. In practice, of course, such an approach is seldom feasible. In the case of a perfect lattice, where there is translational symmetry, the use of reciprocal space rather than real space allows a treatment based on energy band theory[1]. However, as soon as an imperfection is introduced into the lattice, translational symmetry is lost and energy band theory can neither be used nor easily generalised for the study of the defective solid. Alternative approaches must therefore be adopted in this case.

One of the most fruitful procedures, particularly for the investigation of localised phenomena, has been to study a small number of atoms representing an explicit region of the solid in real space. Molecular orbital or valence bond methods can then be applied within this cluster to determine its electronic structure. This approach has proved very satisfactory for studying a variety of effects such as chemisorption at surfaces, point defects within a solid, and the determination of many spectroscopic properties including valence electron excitation and ionisation spectra. These so-called "cluster methods" are also helpful in the investigation of ligand field effects[2] and for studying essentially electronic phenomena such as excitons[3]. Molecular orbital calculations, either in some suitably modified form or in combination with lattice defect relaxation procedures, seem to be of some promise for the study of defects in solids. This chapter then attempts to summarise, if only briefly, some of the technical points associated with such calculations.

One of the most attractive features of cluster methods is their versatility. Not only can they be applied to a range of problems, some of which have been mentioned above, but they can be used for a considerable variety of solids. For example ionic solids, which can also be studied using lattice relaxation techniques such as the HADES procedures outlined elsewhere in this book, have been treated using cluster methods, and an interfacing of the two techniques, which treat different fearures of the solid, has proved useful in a number of cases. Cluster methods can also be used to examine metals and covalent semiconductors. Each type of material necessitates the use of special boundary conditions, and this will be discussed in section 5 of this chapter.

An additional feature of the cluster approach is that the electronic structure can be examined with a variety of molecular oribital procedures. In some cases empirical pseudopotential techniques have been used, while other studies have involved semiempirical molecular orbital methods such as extended Hückel theory[4] or complete neglect of differential overlap, CNDO[5]. As computer technology becomes more sophisticated, however, it becomes possible to work at increasingly accurate levels of approximation, so that for the present our discussion is concerned exclusively with first-principles methods. These involve no parameterisation based on experimental data. We include among these, all-electron ab initio molecular orbital treatments and those pseudopotential methods in which the core-valence interactions are represented by an effective potential based entirely on ab initio calculation, but in which the valence electrons are treated accurately and explicitly at a minimum level of Hartree-Fock theory. Alternative cluster methods that have been used particularly for colour centre problems are dealt with in Chapter (7).

Our aim will be to discuss some of the considerations necessary to perform cluster calculations; we will deal especially with the choice of basis set, with the inclusion of the correct boundary conditions and with the necessity in some cases of including lattice relaxation effects. We will not attempt a complete review of the published literature but will illustrate instead the type of problems which have been, or can be, tackled by cluster methods.

Of necessity we are unable to describe in detail the fundamental principles of the molecular orbital methods considered here. Further information can be obtained from the book by McWeeny and Sutcliffe[6], or, in a simpler form, in that by Richards and Horsley[7].

2. Ab Initio Methods

Ab initio methods are potentially the most powerful and flexible techniques available for studying small clusters. All integrals are calculated rigorously, and no specific form of the potential (such as the muffin-tin potential used in the X_α method[8]) needs to be assumed. The only approximations which remain are the truncation of the basis set (discussed below) and the orbital approximation. This assumes that each electron moves in the averaged (rather than the instantaneous) field of the other electrons, so that the problem reduces to that of solving a coupled set of one-electron (Hartree-Fock) equations rather than a single, more complicated many-electron equation. However, electron correlation, which is neglected in such an approach, can be included subsequently by methods such as configuration interaction, the generalised valence bond approach, or many-body perturbation theory.

Cluster molecular orbitals are generally expressed as linear combinations of atomic-orbital-like basis functions which are most often centred on the nuclei. The usual forms for the atomic basis functions are either Slater-type orbitals of the form $Nr^{n-1} e^{-\zeta r} Y_\ell^m(\theta,\phi)$ (where n is the principal quantum number and $Y_\ell^m(\theta,\phi)$ is a spherical harmonic) or Cartesian Gaussian-type oribitals of the form $N e^{-\zeta r^2} x^p y^q z^t$. In both cases, N is a normalisation constant for the functions. Gaussian orbitals do not in general give as good a representation of atomic orbitals as do Slater-type functions. However, the evaluation of multicentre integrals, particularly those of the inter-electronic repulsion which are inevitably required in cluster calculations, is considerably easier for Gaussian functions and this usually more than compensates for the extra number of functions required. A complete basis set would contain an infinite number of functions. It is therefore necessary to truncate the basis while retaining functions which span the important regions of space. This truncation must be done with some care, however, since the accuracy of an ab initio calculation is generally influenced primarily by the quality of the basis set.

A major advantage of any ab initio method over semi-empirical procedures is that the total energy of the system, calculated using a technique such as the molecular orbital method, represents an upper bound to the exact energy. In other words, a more accurate calculation (for instance one with a more aptly chosen basis set) must lead to a lowering of the total energy of the system. This means that we can, if necessary, improve the accuracy of the calculations in a systematic fashion. This feature is not readily available for semiempirical calculations.

As the number of basis functions n increases, the number of integrals requiring evaluation increases as n^4 due to the interelectronic exchange and repulsion term in the Hartree-Fock Hamiltonian. For many systems of potential interest, therefore, the amount of computer time needed escalates rapidly. However, only the valence electrons are important in many localised phenomena. In view of this, a number of pseudopotential techniques have been developed to allow the replacement of the core electrons by an effective potential.

The pseudopotential formalism discussed here is the one most commonly used in cluster calculations, and is described in detail by Kahn, Baybutt and Truhlar[9]. We will sketch only briefly the most important concepts of the method. The effect of the core electrons is included by modifying the one-electron Hartree-Fock operators with the addition of Coulomb and exchange potentials due to the core electrons. These effective core potentials can be made to depend on the angular momentum. They are independent of the basis set employed for a particular atom and are most accurately determined from neutral-atom wavefunctions. This point deserves some comment, since clearly the potentials are more easily determined when there is no valence-valence interaction, i.e. when there is only one electron outside the core. The single-valence-electron ion has therefore frequently been used to determine pseudopotentials[10] but this has proved unsatisfactory for ions in highly charged states because of the contraction of the electron density. Melius, Olafson and Goddard[11] have considered effective potentials for iron based on a neutral atom and an Fe^{7+} (one valence electron) ion, and found that those from the Fe^{7+} ion lead to errors in the atomic excitation and ionization energies of approximately 1-2 eV, which is over ten times the error obtained when the effective core potential was based on the neutral atom.

It is important to emphasize that pseudopotentials such as these are ab initio in form; no parametric functional forms are assumed, as they often are in model pseudopotentials, and no experimental data is needed to construct them. For computational convenience the pseudopotentials are frequently expanded in Gaussian form, and this reduces the computer time required by several orders of magnitude compared to the use of numerical potentials. A sufficient number of Gaussian functions is used to describe the pseudopotential so that the fit is accurate; this procedure therefore does nct lead to a model potential.

The chief source of error arising from the use of pseudopotentials lies in the assumption that the core is frozen; that is to say, it is completely insensitive to changes in the valence electron structure. For the transition metals, where the 3 s, 3p and 3d orbitals are rather close in energy, the problem would seem to be at its most acute when the 3s and 3p orbitals are included in the effective core, since rearrangement of the d-electrons is known to influence the core. Surprisingly, however, the errors arising from such a treatment amount to no more than a fraction of an electron volt[11]. Finally, it is worthwhile noting that correlation effects can be included in the pseudopotential[10] so energies and wave-functions obtained in this way can be of very high quality.

3. Basis set choice

Since the choice of basis set is so important a factor in ab initio calculations, accurate reliable basis sets, suitable for molecular calculations, have been tabulated for most atoms up to and including the transition metals. Some of the more popular bases are those of Dunning[12,13], Veillard[14] and Wachters[15], although numerous other tabulations exist. Basis sets such as these are normally optimised for the free atom, and in some cases, for a few ionic states; they are of sufficient flexibility to allow for the changes in electronic distribution which occur when the atom is incorporated into a molecule. However, in some cases it is necessary to add functions of higher angular momentum, commonly known as polarization functions, to describe bond formation accurately. Whether or not tabulated basis sets such as those mentioned above, which have been optimized for molecular systems, are suitable for use in solid state cluster calculations remains largely uninvestigated, although it is an important point for the future.   An extreme example of this is the O^{2-} ion.   In the free state it does not exist, since it is unstable with respect to the dissociation, $O^{2-} \rightarrow O^- + e$.   In oxides such as MgO, however, the Madelung potential stabilizes the 2p electrons by about 10-11 eV, so that atomic wavefunctions are likely to require some modification in the solid state.   In Table 1 we illustrate the sort of changes that result from basis set and exponent variation for O^{2-} in a point-ion lattice.   The Gaussian basis

Table 1

Basis set improvements of O^{2-} and Mg^{2+} in MgO.   The electronic energy E_e is given in atomic units and is lowest for the optimum choice of basis set

O^{2-} in MgO lattice	$E_e(a.u)$
Dunning basis for oxygen atom	-82.983595
Dunning basis plus diffuse p function (unscaled)	-83.025112
Dunning basis plus diffuse p function (scaled by $\zeta=0.9$)	-83.018145
Dunning basis plus diffuse p function (scaled by $\zeta=1.1$)	-83.031467
Mg^{2+} in MgO lattice	
Veillard basis for magnesium atom	-190.098757
Veillard basis with $\zeta=0.06$ for diffuse s function	-190.098800
Veillard basis with $\zeta=0.03$ for diffuse s function	-190.098729
Veillard basis with most diffuse p function uncontracted	-190.099170

set proposed by Dunning[12] is used throughout.   As shown, there is a decrease in energy of approximately 1 eV upon addition of a diffuse p function to the oxygen atom basis, but thereafter little change as the diffuse function orbital exponent, ζ, is varied.   For Mg^{2+} in the same point-ion lattice, however, the situation is quite different.   Here we have used the Gaussian basis of Veillard[14] to represent the lattice Mg^{2+} ion, and have investigated the effects of varying the most diffuse s function.   Table 1 shows that this has a negligible effect on the energy, which supports the view that the 3s level in MgO is essentially unoccupied.   Even uncontracting the most diffuse p function, which should allow greater flexibility in the basis, does not give a signifiant lowering of the energy (less than 0.03 eV). On the other hand, for Fe^{2+} in a point-ion FeO lattice, the addition of a diffuse s and p function to the iron atom basis set of Wachters[15], and the uncontraction of the most diffuse d function, do lead to appreciable changes[16].   These subsequently affect a number of calculated defect properties such as the polaron formation energy[17].

Generally speaking, some modification of free atom basis set appears necessary for a realistic modelling of ions in solids. For lattice anions, the use of a free ion basis set, or one including diffuse functions for the valence electrons, would appear to be an essential starting point. For cations with closed valence shells, such as Mg^{2+} or Al^{3+}, an adequate description seems to be given by a basis set for the neutral free atom. However, for those lattice cations which have an unfilled d shell, the change in electronic structure of this outer shell appears to require at minimum the uncontracting of the most diffuse d function, or alternatively the addition of another diffuse d function. It is perhaps worthwhile cautioning against the use of minimal basis sets, those in which each electron pair is represented by only one Slater-type function. These are often barely adequate for molecular calculations, and when such potentially drastic changes in electronic structure as those caused by the lattice are introduced, their limitations are likely to be serious unless very extensive exponent optimisation is undertaken.

At this point it is of interest to digress briefly to discuss the question of ionicity found in oxide and halide clusters, in part because it emphasizes the importance of the effect of the lattice on the electronic structure. For example, Murrell, Tennyson and Kamel[18] have found that free LiF is 85% ionic, whereas in the lattice it is 99% ionic. For MgO, the free molecule (at the experimental lattice internuclear separation of 2.106Å) is calculated to be about 40% ionic, while in the lattice it too is nearly 99% ionic[19]. Molecular FeO (at an internuclear separation of 2.16Å) is estimated to be 45% ionic, and in the lattice is 85% ionic. This concurs with the results of Bagus et al[2] who studied the FeO_6^{10-} cluster as well as several states of FeO_6^{9-}; they also find little evidence for covalency. In all these systems the presence of the lattice greatly enhances the ionicity of the atomic species involved. Even for O^- in the {001} surface of MgO, where an Mg_5O^{9+} cluster (with 59 electrons) embedded in a point-ion field is considered, the net charges are found to be very close to -1 on the oxygen and +2 on each magnesium ion[19].

The previous discussion has illustrated that the atomic components of ionic lattices are sufficiently different from the free atoms that a certain degree of basis set optimization, along the lines sketched previously, may be warranted. It is not so clear whether polarization functions, i.e. d orbitals on magnesium or oxygen, or p orbitals on hydrogen, need also to be included in the solid state. In recent cluster calculations of hydrogen chemisorption on oxide surfaces Colbourn and Mackrodt[20] have found that polarization functions are important for both the oxygen and the hydrogen atoms. For the OH^- group on the {001} surface of MgO the calculated bond strength is increased by about 10 kcal/mole, while the corresponding equilibrium internuclear separation is decreases by 0.05Å upon the addition of polarization functions. However, in cases where a weaker bond is formed, such as MgH at the {001} surface of MgO, polarization appears not to be so important. Murrell et al[18] have included polarization functions for the fluoride ion in LiF, where the polarizability is associated with p-d excitations, and have also included them on the lithium ions to allow for charge transfer from F 2p orbitals to the Li 2p. By comparing these extended basis set calculations with those involving minimum basis sets, they concluded that polarization functions, by permitting the charge transfer, were necessary for an accurate calculation of the three-body energy. Bounds and Hinchcliffe[21] have also included polarization functions in their cluster calculations on LiCl, and like Murrell et al[18] find them to be important in calculating the contribution of three-body terms to the energy. As yet there appears to be no systematic study of the role or polarization functions in the solid state, but as the above examples show, they are important whenever a chemical bond is formed (as in many chemisorption studies) and also may be necessary to describe charge-transfer processes accurately. Obviously their role may also be significant in covalently-bonded clusters, such as those found in silicon, in order that the bond formation be described correctly.

4. Spin states

For ionic solids in particular, the spin states of the constituent atomic species are influenced by the local lattice environment, and so deserve special attention. The Hartree-Fock molecular orbital theory in its restricted form (RHF, in which α-spin orbitals are constrained to be spatially identical to those with β-spin) is easily applied to closed-shell systems, where the Hamiltonian can be completely defined in terms of doubly occupied and virtual (unoccupied) orbitals[6]. When there are open shells, it is possible to use unrestricted Hartree-Fock theory, where α-spin orbitals may be spatially different from those with β spin. However, this can produce wavefunctions which are not eigenfunctions of the spin operator S^2 and which, therefore, cannot give a fully accurate representation of the physical system. For this reason it can be more satisfactory to modify the restricted Hartee Fock procedure to extend its range of application, and a necessary first step in this is the partitioning of the orbitals into sets of doubly occupied, singly occupied and virtual orbitals. When the set of singly-occupied orbitals contains only one member, the Hamiltonian describing the system is generally straightforward except when the singly occupied orbital is equivalent by symmetry to some of the doubly occupied orbitals. For example, in gas-phase transition metal atoms and ions there are five equivalent d orbitals (usually designated $d_{xy}, d_{xz}, d_{yz}, d_{x^2-y^2}$ and d_{z^2} and, when there are fewer than ten electrons in these, some symmetry equivalencing in the Hamiltonian will be necessary. This requires the specification of some parameters in the Hamiltonian to define it correctly for the system, as discussed by Guest and Saunders[22]; they also tabulate the expressions needed to study d^n systems.

Problems of greater magnitude arise when there are two open (partially occupied) shells. Generalised restricted Hartree-Fock procedures have been developed for such cases[23] but the energy expression must be evaluated for each individual case and to date very incomplete tabulations of them exist. For solid state cluster calculations the problem is particularly acute for transition metals in a cubic environment as, for example, Fe^{2+} in FeO. In such cases the d orbitals are split into t_{2g} (three-fold degenerate) and e_g (double degenerate) sets. To return to our example, Fe^{2+}, with a d^6 configuration, this may be either in a low spin (t_{2g}^6) or high spin ($t_{2g}^4 e_g^2$) state. Similarly for Fe^+, which has seven d electrons, the two possibilities are $t_{2g}^6 e_g$ and $t_{2g}^5 e_g^2$. In general, since it is not intuitively obvious which state will be the more stable in any given environment, both cases must be considered.

Our discussion so far has concerned mainly the determination of spin states in octahedral clusters modelling face-centred cubic solids, since the orbital degeneracies can lead to greater complexities in this case. When the symmetry of the system is lower than this, as in, for example, hexagonal systems like $\alpha-Al_2O_3$, symmetry equivalencing of the orbitals may no longer be necessary so the calculation is simplified.

5. Boundary conditions for the cluster

One of the main problems with the use of clusters to study solid state phenomena is the simulation of the rest of the solid, and its effect upon the cluster itself. This has sometimes been neglected entirely in the literature, as for example in the investigation by McOmber et al[24] of the mobility paths in fast-ion conductors. They have used pseudopotential techniques to study MI_4^{3-} (M = Na,K,Ag and Cu) clusters, but have not included the effects of the remaining lattice. It would be of interest to see whether its inclusion would affect the nature of their results. Kunz and Guse[25] have considered an $(MgO)_3$ cluster as a model of the catalytically-active V_I centre in MgO. This consists of a micro {111} face of three oxygen ions in a triangular array, but they too appear to have neglected the effects of the lattice exterior to their cluster. They find that $(Mg^+O^-)_3$ is a good representation of the charge state of the cluster, rather than $(Mg^{2+}O^{2-})_3$, but again it is quite possible that the influence of the rest of the lattice could modify this.

Since in quasi-ionic solids the energy level ordering of the orbitals is strongly affected by the Madelung potential, it is important that the correct potential be used. For ab initio calculations, the most suitable method, both for the accuracy it provides and for its ease of application, is to embed the explicit cluster in a larger array of point charges. A Watson sphere[26] surrounding the whole cluster could be used, and indeed frequently is for X_α cluster calculations. However, as Kunz and Klein[27] have pointed out, for accurate values of energy differences for orbitals on different ions within the cluster, a simple Watson sphere is inadequate because the potential inside such a sphere is constant. Hence although a Watson sphere may give the correct Madelung potential at the centre of the cluster, the potential at adjacent ions may be radically in error. Kunz and Klein propose an alternative formalism, valid for either X_α or ab initio methods, which avoids this difficulty by modifying the one-electron Hartree-Fock equations[27]. They have compared their method both with the Watson sphere and with total neglect of the external lattice for $(Li_6F)^{5+}$ clusters and find that, for orbitals on different centres, the energy level ordering may be in error by as much as 1.5 eV when either the Watson sphere or free-space boundary conditions are used.

By embedding the cluster in an array of point ions one effectively modifies the one-electron Hartree-Fock equations. Provided a sufficient number of point charges can be included, the Madelung potentials at different centres can be specified accurately, giving a realistic representation of the effect of the rest of the lattice. This approach has been widely adopted. It has been used by Stoneham[3] in studying the self-trapped exciton in NaCl: he used a 4x4x3 block of ions of which four (two Na and two Cl) were treated explicitly. Stoneham and Sharma[28] have also used point ions to give the correct potential for the F^+ centre in the surface of MgO, with either zero, one or five Mg ions included explicitly. They find that the geometry differs slightly when zero or one Mg ion beneath the plane is taken. When the Mg is explicitly specified, rather than taken as a point charge the stable position of the F centre is 0.32 a_o below the surface rather than 0.256 a_o as in the point-ion case. Surratt and Kunz[29] have also embedded an $(NiO_6)^{10-}$ cluster in a 5x5x4 ion array to study the structure of NiO, and find good agreement with experimental photoemission spectra. They have used a similar model to investigate hydrogen atom chemisorption at defect sites on the surface of NiO[30,31]. Their block of point ions is sufficiently large that the Madelung potential is given correctly when charges of ± 2 are taken, but it is possible to use smaller arrays of point ions and to adjust the charges of the outermost ones to give the Madelung potential accurately.

We emphasize here a point which may be obvious to quantum chemists familiar with ab initio molecular orbital techniques but which may not be so clear to others. It concerns the size of the point array which can be used. By definition there are no basis functions centred on the point ions, so the point ions do not affect in any way the calculation of two-electron integrals of the interelectronic repulsion operator. At most it is necessary to calculate only some additional one-electron nuclear attraction integrals when the size of the point ion field is increased, and these integrals are very quickly evaluated compared to the two-electron integrals. Inclusion of a large cluster of point ions should not therefore greatly increase the amount of computer time required. Furthermore, regardless of the number or even of the positions of the point ions, the two-electron integrals need not be recalculated unless the geometry of the explicitly-treated cluster, where there are basis functions on each centre, is changed. This can result in substantial savings in computer time when studying the effect of the environment on the cluster, for example. Most ab initio programs usually set a maximum number of centres which are allowed, and this provides a constraint on the size of the point ion block unless the program can be appropriately modified.

So far we have discussed the boundary conditions suitable for ionic solids. However, cluster methods are also applicable to metals and to covalent solids, and these require a different treatment of the lattice outside the explicit cluster. For metals, the necessary boundary conditions are not well defined, and the usual approach has been to take larger and larger clusters until quantitative agreement with experiment has been found. Thus for example, Melius, Upton and Goddard[32] have studied clusters containing from 13 to 87 nickel atoms, replacing all but one electron on each centre by a pseudopotential. They find that the ionization potential converges by the time the cluster reaches 43 atoms in size, but that even with the largest cluster (87 atoms) the electron affinity is still in error by 2.5 eV; the calculated value of 2.75 eV agrees very poorly with experiment, 5.2 eV. For chemisorption on metals, smaller clusters have been used. Bagus and Seel[33] have taken only five copper atoms in their study of the binding of CO to copper, but this is adequate to give both equilibrium bond lengths and ionization potentials which agree well with experiment.

For covalent materials, the atoms at the boundary of the cluster would normally be bonded to other atoms. To simulate these bonds while still retaining the possibility of working with a finite cluster, the usual approach is to saturate the 'dangling bonds' of the cluster with hydrogen atoms. Relaxation of the {111} surface of silicon has been studied by Redondo et al[34] using Si_4H_9 clusters, in which one silicon atom is a 'surface' atom, while the other three correspond to the second layer of the solid; each of these three is bonded to three hydrogen atoms. The vertical ionization potential, calculated to be 5.78 eV, is within the experimentally measured range of 5.6-5.9 eV, which suggests that the model is a good one even though the cluster is small. In this case an effective core potential has been used for the ten core electrons on each Si atom, and this has been shown to give properties accurately to 0.1 eV[35]. The neutral vacancy in silicon has been studied by Surratt and Goddard[36] using a cluster Si_4H_{12}, and they have been able to determine the electronic symmetry of the ground state (^1E) as well as to find the excitation energies to the $^3T_1$ and $^5A_2$ states.

In the above-mentioned investigations the Si-H 'saturator' bond length has been fixed, usually to a value near that found in silane. Kenton and Ribarsky[37] have considered the effect of varying Si-H bond lengths on such properties as the cluster band gap, the total energy and orbital symmetry order, and the charge densities, in an Si_5H_{12} cluster. The orbital energy ordering in particular was found to be sensitive to the Si-H bond length, affecting the calculated value of both the cluster band gap and other energetic properties.

The E_4 centre in α-quartz has also been treated using a model Si_2O_6H cluster, with the addition of six hydrogen atoms to represent the six next-nearest silicon atoms external to the explicit cluster[38]. The E_4' centre involves an oxygen vacancy between the two silicon atoms of the cluster, and this is replaced by a hydrogen atom bound to one of the silicons. Calculations for this centre[38] have allowed the structure within the cluster to relax, although some errors have inevitably been introduced by the neglect of long-range interactions. Although the agreement between the calculated values and the measured EPR spectrum is not quantitative, the validity of this model for the E_4' centre is supported by the overall agreement. No investigation of the effect of saturator bond length has been made, the O-H bond length being simply fixed at $0.96\overset{o}{A}$. The findings of Kenton and Ribarsky[37], however, suggest that more studies of the influence of the saturator bond length on cluster properties would be valuable in this case.

6. Lattice relaxation effects

Within a cluster, particularly for defected systems, lattice relaxation is known to be important in determining the properties of point defects, particularly for the case of ionic materials. This is especially true for clusters in which there is a lattice vacancy, or an interstitial ion. One of the principal virtues of ab initio methods is that lattice relaxation effects can be included explicitly by specifying the atom positions within the cluster. Ideally one would choose to vary the positions of all the atoms in the cluster to determine the geometry corresponding to the minimum energy. Tennyson and Murrell[39] have adopted such an approach in calculating the (symmetric) outward relaxation for the next-nearest neighbours about the F centre in LiF, and found that it amounted to 2.5% of the lattice parameter. Isoya et al[38] have also allowed full geometry optimisation of their ($Si_2O_6H/6H$) cluster model for the E_4' centre in a α-SiO_2, taking advantage of the automatic equilibrium geometry optimization features of the Gaussian 70 molecular orbital package. However, inadequacies in representing the boundary conditions could possibly affect their results. Using an Si_4H_9 cluster, Redondo et al[35] have calculated the lattice relaxation in the {111} surface of silicon, finding that the one 'surface' silicon atom is displaced by $0.08\overset{o}{A}$.

In general this method is not feasible, particularly for less symmetric defects, because of the computational effort required to calculate the energies at a number of different geometries. However, even in these cases some attempt can be made to include lattice relaxation. Several approaches to this problem have been suggested. Surratt and Goddard[36] have included lattice relaxation about a vacancy in silicon using the displacement of 0.08Å calculated by Redondo et al[34] for the silicon {111} surface. Stoneham[3] included lattice relaxation for the self-trapped exciton in NaCl by moving the two chloride ions, since all the properties of the relaxed exciton could be described by a model involving the Cl_2^- ion. The Cl_2^- separation was obtained empirically by matching the optical transition energies and hyperfine constants, calculated as a function of internuclear separation, with the experimental values.

We suggest the further possibility of calculating relaxed positions for ionic solids using static lattice relaxation methods incorporated in defect lattice programs such as HADES[40] and CASCADE[41]. These calculate relaxed geometries which can then be used in a single ab initio calculation. The necessity of including lattice relaxation in cluster calculations is demonstrated by the results shown in Table 2, for the Cu-CO bond strength of CO adsorbed on Cu^+ and Cu^{2+} in the

Table 2

Effect of lattice relaxation on bond strength for CO chemisorption on copper in the {001} surface of MgO.

	Cu-CO bond strength (kcal/mole)	
	Unrelaxed lattice	Relaxed lattice
Cu^+ in {001} MgO	19.0	11.8
Cu^{2+} in {001} MgO	14.0	6.2
$Cu_{Mg}^+ - O_O^-$ in {001} MgO	10.4	12.2

MgO {001} surface[20]. In these calculations only a linear Cu-C-O cluster was used, however, and lattice effects (including relaxation) were modelled by representing the surface by point ions. The relaxed geometries, calculated using defect lattice procedures for surfaces[42] are shown in Figure 1. The movement of Cu^{2+}, and especially of Cu^+, out of the plane is particularly marked and must be taken into account in any cluster calculations with this surface.

These results indicate that in cases where lattice relaxation is expected to be substantial an explicit inclusion of these effects is necessary to obtain quantitative agreement between theory and experiment. At present, however, few studies are reported in the literature both with and without lattice relaxation to assess its influence; such investigations are currently being undertaken[20].

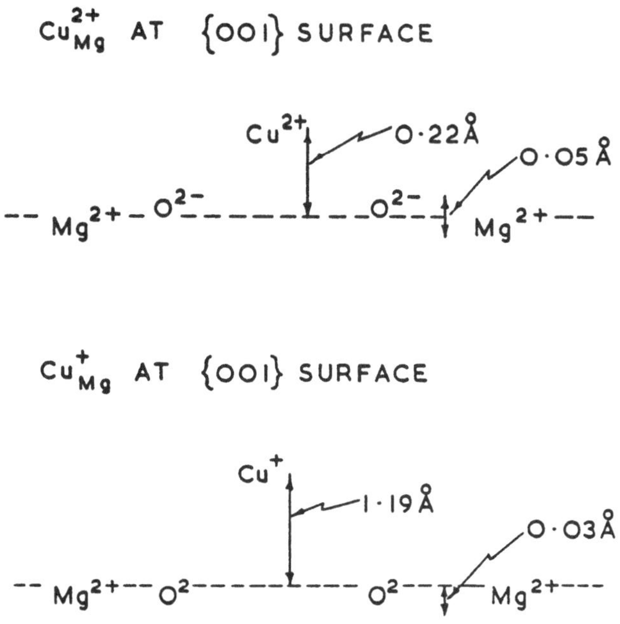

$$Cu_{Mg}^{2+} \ AT \ \{OOI\} \ SURFACE$$

$$Cu_{Mg}^{+} \ AT \ \{OOI\} \ SURFACE$$

Figure 1   The relaxed {001} surface of MgO containing Cu^+ and Cu^{2+} impurity ions

7.   Conclusions

It is clear that ab initio cluster calculations can provide significant information on the electronic structure of a wide variety of localized solid state phenomena, even from the limited number of calculations reported to date.   In this chapter we have emphasized some of the important considerations such as basis set choice, representation of the boundary conditions and inclusion of lattice relaxation effects, which have on occasion been neglected in the published literature. Although ab initio methods provide a powerful technique for describing the localized electronic structure of the solid state, their use is not yet a matter of routine. It is therefore especially important that the above-mentioned features receive further attention both to prevent any misuse of this technique arising from a lack of awareness of their importance and to build up experience of their effects on the quality of a calculation.

In a sense the present situation for solid state cluster calculations parallels that of fifteen to twenty years ago for molecular calculations, when basis sets for most atoms were not well documented and when the effect of changing the basis set was still largely only poorly understood. With current developments in computer technology, we foresee the same advances in the field of solid state chemistry and physics as have already been made for molecular quantum chemistry, as more calculations elucidate the importance of such fundamental effects as basis set choice and provision of adequate boundary conditions.   This can only be enhanced as pseudopotential techniques become more widely available and as advances in computer hardware and software permit calculations on larger clusters.

In conclusion we wish to emphasize the virtues of these ab initio techniques compared to less sophisticated semiempirical methods, or to the X_α method outlined elsewhere in this book by Harker. Aside from its high degree of accuracy and its independence of parameterization from experiment, the most attractive feature of the method is its flexibility. Lattice geometries, including any relaxation effects, are easily specified. Except for metals, where the problem of representing the solid external to the cluster remains intractable, it is also possible to model the boundary conditions, and we have outlined methods applicable both to ionic and quasi-ionic solids as well as to more covalent semiconductor materials.   In ionic solids this is significantly more satisfactory than the use of a Watson sphere about the cluster.   The absence of a muffin-tin form of the potential for the cluster also favours ab initio methods over the scattered-wave X_α approach because it provides more realistic approximation to the actual potential within the solid. For these reasons ab initio cluster techniques will play an increasingly important role in the understanding of localized solid state phenomena.

References

1.   Kittel, C., Quantum Theory of Solids (Wiley, 1962).

2.   Bagus, P.S., Brundle, C.R., Chuang, T.J. and Wandelt, K., Phys. Rev. Lett. 39, 1229 (1977).

3.   Stoneham, A.M., J. Phys. C. 7, 2476 (1974).

4.   Messmer, R.P. and Watkins, G.D., Phys. Rev. Lett. 25, 656 (1970).

5.   Haynes, M.R., Phys. Rev. B. 5, 697 (1972).

6.   McWeeny, R. and Sutcliffe, B.T., Methods of Molecular Quantum Mechanics (Academic Press, 1969).

7.   Richards, W.G. and Horsley, J.A., Ab Initio Molecular Orbital Calculations for Chemists (Clarendon Press, Oxford, 1970).

8.   Weinberger, P. and Schwarz, K., in Int. Rev. Sci. Phys. Chem. Series 2, Vol.1 (1975).

9.   Kahn, L.R., Baybutt, P. and Truhlar, D.G., J. Chem. Phys. 65, 3826 (1976).

10.   Melius, C.F., Goddard, W.A. and Kahn, L.R., J. Chem. Phys. 56, 3342 (1972).

11.   Melius, C.F., Olafson, B.D. and Goddard, W.A., Chem. Phys. Lett. 28, 457 (1974).

12.   Dunning, T.H., J. Chem. Phys. 53, 2823 (1970).

13.   Dunning, T.H. and Hay, P.J., in Methods of Electronic Structure Determination. Ed. H.F. Schaefer (Plenum, 1977).

14. Veillard, A., Theor. Chim. Acta. $\underline{12}$, 405 (1968).

15. Wachters, A.J.H., J. Chem. Phys. $\underline{52}$, 1033 (1970).

16. Kendrick, J., unpublished results.

17. Colbourn, E.A., Kendrick, J. and Mackrodt, W.C., unpublished results.

18. Murrell, J.N., Tennyson, J. and Kamel, M.A., Molec. Phys. $\underline{42}$, 747 (1981).

19. Colbourn, E.A., unpublished results.

20. Colbourn, E.A. and Mackrodt, W.C., unpublished results.

21. Bounds, D.G. and Hinchcliffe., Molec. Phys. $\underline{40}$, 989 (1980).

22. Guest, M.F. and Saunders, V.R., Molec. Phys. $\underline{28}$, 819 (1974).

23. Saunders, V.R. and Guest, M.F., ATMOL3, Part 9, The SCF Programs, Rutherford Laboratory Report (1976).

24. McOmber, J.I., Topiol, S., Ratner, M.A., Shriver, D.F. and Moskowitz, J.W., J. Phys. Chem. Solids $\underline{41}$, 447 (1980).

25. Kunz, A.B. and Guse, M.P., Chem. Phys. Lett. $\underline{45}$, 18 (1977).

26. Watson, R.E., Phys. Rev. $\underline{111}$, 1108 (1958).

27. Kunz, A.B. and Klein, D.L., Phys. Rev. B$\underline{17}$, 4614 (1978).

28. Sharma, R.R. and Stoneham, A.M., J. Chem. Soc. Faraday II, 913 (1976).

29. Surratt, G.T. and Kunz, A.B., Solid State Commun. $\underline{23}$, 555 (1977).

30. Surratt, G.T. and Kunz, A.B., Phys. Rev. Lett. $\underline{40}$, 347 (1978).

31. Wepfer, G.G., Surratt, G.T., Weidman, R.S. and Kunz, A.B., Phys. Rev. B$\underline{21}$, 2596 (1980).

32. Melius, C.F., Upton, T.H. and Goddard, W.A., Solid State Commun. $\underline{28}$, 501 (1978).

33. Bagus, P.S. and Seel, M., Phys. Rev. B$\underline{23}$, 2065 (1981).

34. Redondo, A., Goddard, W.A., McGill, T.C. and Surratt, G.T., Solid State Commun. $\underline{20}$, 733 (1976).

35. Redondo, A., Goddard, W.A. and McGill, T.C., Phys. Rev. B$\underline{15}$, 5038 (1977).

36. Surratt, G.T. and Goddard, W.A., Phys. Rev. B$\underline{18}$, 2831 (1978).

37. Kenton, A.C. and Ribarsky, M.W., Phys. Rev. B$\underline{23}$, 2897 (1981).

38. Isoya, J., Weil, J.A. and Halliburton, L.E., J. Chem. Phys. $\underline{74}$, 5436 (1981).

39. Tennyson, J. and Murrell, J.N., Molec. Phys. $\underline{42}$, 297 (1981).

40. Norgett, M.J., UKAEA Harwell Report AERE-R.7650 (1974).

41. Smith, W., Daresbury Laboratory Technical Memorandum DL/SCI/TM25T (1981).

42. Mackrodt, W.C. and Stewart, R.F., J. Phys. C$\underline{10}$, 1431 (1977).

CHAPTER 7   COMPUTATIONAL METHODS FOR THE ELECTRONIC
                 STRUCTURE OF DEFECTS IN INSULATORS

by

A.H. Harker
Theoretical Physics Division, A.E.R.E. Harwell, Oxon OX11 ORA

1.   Introduction

It is now more than 30 years since the first attempts were made to calculate
the electronic structures of F centres by variational methods in a manner which
included the discrete nature of the ionic lattice. Inui and Uemura[1] and
Kojima[2] used a linear combination of atomic orbitals on the nearest neighbours
to the F centre to accommodate the defect electron.   Kojima[3] and Gourary and
Adrian[4] used vacancy-centred envelope functions, and that approach then became
the more popular one.   In recent years computational techniques have advanced to
such an extent that one can contemplate self-consistent field calculations on the
defect and the surrounding lattice, and the linear combinations of atomic orbitals
method is receiving renewed attention.

The aim of this chapter is to describe the main features of point defect
electronic structure calculations as they have developed over the past ten years.
One of the reasons for the continued interest in simple defects in simple crystals
is that they provide extremely detailed tests of theory against experiment. Optical
transitions, spin resonance parameters and magnetic circular dichroism measurements
are fine probes of energy levels and wavefunctions.   Of necessity detailed
comparisons of theory with experiment have had to be omitted, and although the
important theoretical techniques are described the examples cited are far from
exhaustive.

2.   Non-self-consistent calculations

Calculations of this type must assume an electronic structure for the lattice.
The point ion model is the simplest representation of the lattice; the dominant
effect binding the defect electron is the Madelung potential, so point-ion results
are good first approximations.   The non-self-consistent calculations attempt to
improve the model of the lattice by providing some description of the electronic
structure of the ions, and possibly by incorporating polarisation and distortion.

2.1 Single-centre defect wavefunctions

The electronic structure of ions has three main effects on defect states:
first the Coulomb interaction with the defect electrons is altered;
secondly exchange interactions are introduced; thirdly, the defect wavefunction
must be orthogonal to the occupied ionic orbitals.   The last effect may be
incorporated by carrying out an explicit Schmidt orthogonalisation[6,7,8] or by

casting the Hamiltonian in pseudopotential form[9,10,11,12,13,14]. If an exact calculation is required, there is no difference in the effort required for the two approaches; the pseudopotential form can be more suitable for approximation. One of the greatest problems with such approximations is that they impose conditions on the defect wavefunction, such as smoothness, which may be violated when more flexible functions are used in variational calculations.

If the ionic wavefunctions are assumed not to overlap, an exact calculation involves only one and two centre integrals. These may be evaluated exactly by Löwdin's α-function technique[16] for any functional form, but if Slater functions or Gaussians are chosen standard methods of molecular quantum mechanics may be used[17]. Wood and Opik[7,8] devised their own method which involves an energy- and angular momentum-dependent effective potential for each ion; Gash[12] used α-functions; Harker[14] worked with Slater functions and Leung and Song[15] opted for Gaussians. There were other differences in detail. However, a fair measure of agreement between the different methods obtained as shown by the results in Table 1. Some care is needed in interpreting the absolute energy levels, but the energy differences may be compared validly.

Table 1

Energy levels of the 1s and 2p shells of F centre in NaCl with lattice of free ions (Hartree units). ΔE refers to the s-p excitation energy.

	Reference	E_{1s}	E_{2p}	ΔE
Opik and Wood	7	-.172	-.043	.128
Gash	12	-.256	-.156	.100
Harker	14	-.241	-.154	.087

We have mentioned the requirement that the defect wave function be orthogonal to those of the ions, but of course the ionic functions should themselves be orthogonal[16,19]. If Löwdin's symmetrical orthogonalisation method is applied then three and four centre integrals are introduced. Gash[12] and Harker[14] have included overlaps in F centre calculations, as shown in Table 2.

Table 2

Energy levels of F centre in NaCl with lattice of free or orthogonalised ions

Lattice	Author	Reference	E_{1s}	E_{2p}	ΔE
free	Gash	12	-.256	-.156	0.100
orthogonal			-.260	-.169	0.091
free	Harker	14	-.243	-.156	0.087
orthogonal			-.263	-.189	0.075

These calculations differed in the way in which the multicentre integrals were approximated and in the accuracy to which the ionic functions were orthogonalised. The energy changes on orthogonalising the ions are comparable with those involved in polarisation and distortion, as discussed below.

So far we have assumed that a superposition of free ions is an adequate representation of the solid. In fact the ionic functions will be modified[20] because of the electrostatic potential of their crystalline environment. Bauer[21] has discussed the balance between anion contraction in the Madelung potential and the consequent increase in kinetic energy. Kunz[22] has applied the local orbitals method to the calculation of ionic functions in crystals. In this procedure the usual requirement of orthogonality of one-electron functions is abandoned in favour of a requirement that the wavefunctions should be localised in space. This localisation allows the resulting equations to be expanded in powers of overlap and truncated at low order. Simmons et al[23] calculate optimized orbitals for a crystal by solving a self-consistent field problem for the perfect crystal at the centre of the Brillouin zone using atomic orbitals centred in the lattice sites. Then, by taking the relative weights of basis functions in each Bloch function, they define the coefficients of ionic wavefunctions.

The dependence on ionic wavefunctions has been studied by Leung and Song[15]. Typical results are shown in Table 3 for ionic wavefunctions of Tubis[24] and Clementi and Roetti[25].

Table 3

Dependence of F centre energy levels in LiF on lattice ion wavefunctions[15]

Source of ionic wavefunctions	Reference	Number of shells of ions included	E_{1s}	E_{2p}	ΔE
Tubis	24	2	-.221	-.040	0.181
Clementi and Roetti	25	5	-.153	+.075	0.228
Tubis	24	2	-.220	-.003	0.217
Clementi and Roetti	25	5	-.152	+.096	0.248

Clearly the results are very sensitive to the details assumed for the lattice electronic structure, although a full inclusion of overlap between ions will probably reduce the effect.

2.2 Linear combinations of atomic orbitals

In calculations by the linear combination of atomic orbitals (LCAO) method, the bottleneck is the calculation of multicentre integrals. For this reason Wood[26] had to resort to numerous approximations in LCAO calculations on the F centre in LiCl. By using Gaussian functions Chaney and Lin[17] were able to calculate all the integrals exactly. Their method was to take optimised crystal orbitals[23] for LiF, and assume that the crystal could be represented as an assembly of closed-shell ions with those wavefunctions. The crystal potential is formed under that assumption, and the secular equation solved for eigenfunctions which are linear combinations of atomic orbitals centred on the vacancy and on lattice sites up to sixth neighbour. The eigenfunctions will correspond to the filled states (Li^+1s; F^-1s, 2s, 2p) and to the F centre electron. However, a problem arises in that there will be filled crystal levels corresponding to ions beyond sixth neighbour, and these will contaminate the F centre function unless the basis functions are chosen carefully to have small overlaps with more distant sites. Song[28] discusses this difficulty, and suggests that it may be overcome by orthogonalising to these core states as is done in the one-centre expansion method.

2.3 Polarisation and distortion

Chaney and Lin[27] made what amounted to one iteration towards self-consistency, in an attempt to estimate the polarising effect of the F centre electron on the lattice. Their result was anomalous, however, in that it predicted larger polarisation effects in the ground state than in the excited state. The charge density in the excited state should be more diffuse than in the ground state, so the polarisation should be greater.

Wood and Opik[7,8] have made detailed studies of the polarisation and distortion problem. Their method involves the following features:

1. electronic structure of nearest-neighbour ions is included in detail;
2. an effective mass model is used beyond nearest neighbours;
3. dielectric polarisation is included in a modification of the Toyozawa-Haken-Schottky method[29a,b];
4. the nearest neighbours are allowed to relax, with interionic forces of the Born-Mayer type using effective radii from Tosi and Fumi[30].

The integration of all these methods requires great care to avoid double counting, and introduces a large number of parameters. Some of the parameters can be estimated from perfect crystal data, but others have to be selected to give

reasonable F centre results.   There are also difficulties of principle.   The
Toyozawa-Haken-Schottky model uses a dielectric constant, which will be dominated
for ionic crystals by the anion polarisability.   In F centres, the exponential
decay of the F centre wavefunction results in a far larger polarising field on the
nearest neighbour cations than on next neighbour ions.   Furthermore the
polarisation of the ions will distort their wavefunctions which were so carefully
included in detail.   Other shortcomings of the Toyozawa-Haken-Schottky model are
discussed by Stoneham[5].

The lattice statics model used by Stoneham and Bartram[31] avoids the continuum
approximation and uses a shell model for the ions in the crystal.   The lattice
distortion and polarisation are treated by representing each atom by a charged
shell and a nucleus.   The force constant between the shell and the nucleus
controls the ionic polarisability, and the shells interact with one another
electrostatically and through Born-Mayer repulsive forces.   This method is equally
applicable to defects of low symmetry.

3.   Self-consistent methods: cluster calculations

The difficulties in extending single-centre calculations to include
polarisation and distortion, and the fact that results are sensitive to the details
of the structure assumed for the ions, suggest that one should attempt to calculate
the structure of the lattice and the defect simultaneously, using the same
approximations for both. We enter the field of cluster calculations, in which
electronic structures are acknowledged to be determined primarily by local properties,
with little or no influence from possible periodic structures of solids[32].

The basic approach of a cluster calculation is to take a section of a solid
and calculate its structure as though it were a large molecule. If the cluster is
large enough its electron density near the middle should resemble that in the
solid; if it contains a defect the surrounding cluster should respond in the same
way as the solid would. With cluster calculations, the defect energy levels can be
related to the levels of the perfect solid. Lattice distortion is quite
straightforward within the cluster, but if the cluster is small careful thought may
be necessary if it is to be matched to a model of the surrounding lattice for
longer-range distortion and polarisation.

For F centres in ionic crystals extended-ion calculations[7,15] show that
it is only the first two shells of ions that have a large effect on the energy of
the ground state, but ions out to the third or fourth shell affect the excited
state. As some of the important questions in this subject involve small energy
differences between excited states it seems that a large cluster may be necessary.
This is even more so where defect formation processes, involving more than one
defect in a cluster, are concerned. To keep the calculation within bounds, one is
forced to use a semiempirical method. To fix the parameters of the semiempirical
scheme, band structure calculations on perfect solids may be used.

3.1 Perfect solids and boundary conditions

One of the worst problems in cluster calculations is that about half the atoms in the cluster are necessarily surface atoms, instead of one in 10^8 in a real crystal. Unless something is done to adjust the environment of these atoms on the outside of the cluster, any attempt to relate the electronic structure of the cluster to a band structure is likely to be frustrated by large numbers of surface states, and unrealistic charge inhomogeneities are likely to build up as electrons settle on the surface or in the centre of the cluster.

Fairly simple ways of dealing with this problem are available for ionic and covalent crystals. For ionic crystals, the surrounding matrix from which the cluster has been excised may be represented as an array of point ions[33]. For a covalent crystal such as diamond the "dangling bonds" at the surface may be saturated with hydrogen atoms[34] or with pseudoatoms which mimic appropriately hybridized crystal atoms. Both these approaches require some prior knowledge of the perfect crystal structure: for ionic materials, one has to choose an effective charge to represent the ions, and for covalent clusters the positions of the hydrogens or hybridizations of pseudoatoms have to be chosen appropriately. In magnesium oxide, for example, Hayns and Dissado[33] found that varying the effective charge of a matrix surrounding a 27-atom cluster between zero and fully ionic caused the band gap to vary between 7.3 and 6.5 eV, and broadened the valence band from 13.4 to 14.3 eV. The experimental band gap is 7.8 eV, and the valence band width 13 to 17 eV. For a cluster of 35 carbon atoms in a diamond lattice, Larkins[34] found that the band gap decreased from 9.5 to 8.5 eV when the surface of the cluster was saturated with hydrogen atoms. More importantly, though, whereas the 35-atom cluster of diamond with dangling bonds had a charge imbalance, leading to a departure from uniformity of charge of about 0.15 e, when the bonds at the surface were saturated with hydrogens this imbalance was removed. Müller and Scherz[35] adjusted the potentials seen by different atoms within the cluster to achieve charge uniformity, and Brescansin and Ferreira[36] adjusted the potential to ensure periodicity. Of course more elaborate ways of terminating the cluster are possible, by connecting each surface atom to a Bethe lattice[37], or by embedding the cluster in a matrix, with simplifying assumptions about the matrix-cluster coupling[38]. The single-atom or point-ion terminations are probably adequate, in view of the additional approximations that are usually made in calculating the electronic structure of the cluster.

There is probably little point in moving to large clusters in an attempt to reduce the surface effects without resorting to one of the above schemes. Kadura and Kühne[39] used a tight binding model with two parameters (a Coulomb interaction between sites and a resonance integral) and worked with clusters of up to 21^3 atoms. They showed in their model that the binding energy per atom and the valence band width were monotonic functions of cluster size. Messmer[40] used a similar model for a cluster of 9^3 atoms, and showed that the local density of states for the central atom was almost identical with that calculated from a semi-infinite model. Calculations with more complicated models show that monotonic variations cannot be guaranteed, and in any case convergence to bulk values is fairly slow. For example, Baetzold et al[41] compared cluster and band models of Palladium using Extended Hückel theory, and Messmer and Watkins[42] studied diamond clusters. The rate of convergence with cluster size appears to be much faster with the Self-Consistent Field Xα Scattered Wave method[43] for metal clusters, although other ab initio methods do not exhibit rapid convergence[41].

Overall, the conclusion is that once one has reached a cluster of about 30 atoms, corresponding roughly to a 3x3x3 cube, one has to work very hard for little reward if one increases the cluster size, and one ought to work with a fairly small cluster and treat the surface in some special way. If we are to devise an appropriate termination for the surface of a cluster, we need to model the bulk crystal using the same computational scheme as we use for the cluster. This means extending the cluster calculation to include the translational symmetry; and to monitor the success of the method it is convenient to relate it to the Bloch Functions for a periodic lattice. A self-consistent formulation of an LCAO method with periodic boundary conditions[44,45] involves a sum over wave-vectors. Such a formulation may be reduced to a conventional tight-binding form if the Fock matrix elements are formed from a superposition of free-atom or free-ion potentials. There would be no allowance for changes in the potential in the crystalline environment - that is, the calculation would not be self-consistent. This is a particular problem in partly covalent solids, where some charge transfer takes place, and in materials where the point symmetry of the atomic sites is so low that spherical atomic potentials are a bad approximation.

What is needed for the self-consistent scheme is a way of choosing a representative set of wave-vectors to include in the summation so that instead of using the full number, equal to the number of unit cells, we use an approximation to the average over wave-vectors. Cohen and co-workers have addressed the problem of generating sets of special points in the Brillouin zone[46] from which average quantities such as charge density or energy may be calculated. Another way of achieving a suitable average is to use a unit cell which is larger than the primitive cell[44]. By doing this we reflect non-zero wave-vectors back onto the $\underline{k}=0$ point. Provided that we choose the large unit cell so that the whole star of wave-vectors of equal magnitude and a particular symmetry reflect back to $\underline{k}=0$, the set of wave-vectors is adequate, but the calculation becomes simpler because we only need to consider the $\underline{k}=0$ point. The symmetries of the eigenfunctions calculated in the large unit cell allow us to relate them to the symmetry in $\underline{k}$-space, and the equations simplify in such a way that standard molecular orbital programs may be used for the calculation with straightforward modifications to the integrals.

3.2 Semi-empirical methods

There is a long history of attempts to simplify the full self-consistent field equations by omitting some of the two-electron integrals, approximating others, and introducing experimental electron affinities and ionisation potentials. Probably the most widely used self-consistent scheme is CNDO: the acronym stands for complete neglect of different overlap, and the approximations are fairly well known[47]. In this valence-only method, overlap integrals are neglected, all three- and four-centre integrals and many one- and two-centre ones are neglected, and the remaining terms are parameterised in terms of ionisation potentials, a resonance integral and simplified Coulomb integrals. The large unit cell calculation described above takes a particularly simple form in this approximation[45,48]. The parameters are essentially arbitrary, and available for fitting to experimental data, but physical intuition can be used to pick initial values and limit the range of values used. In the past, users of the CNDO method have tended to take the original parameters[47] which were obtained by a mixture of fitting to diatomic molecular calculations and interpolation, and apply them directly to the solid. Parameters seem not to be transferable between such different situations, and the method that Harker and Larkins have adopted[45] is to use the large unit cell method to refine the CNDO parameters, fitting to the lattice spacing, cohesive energy, and valence band-width, and trying to ensure that the ordering of various levels in the valence band is correct. The parameters are then applied to defect calculations. This approach is similar to that of, for example, the shell model for lattice calculations discussed in Chapter (10). There a functional form is chosen for interatomic potentials, based on physical arguments, and the parameters refined by fitting perfect crystal properties before the method is applied to defect problems. Important points for defect calculations are whether the results are largely independent of the size of cluster used in the large unit cell method, and whether reasonable values of bulk properties are obtained when parameters are transferred from large unit cell calculations to large clusters without period boundary conditions. Successful calculations[45] confirmed both these points.

One of the most extensive applications of these methods has been the study of radiation damage in alkali halides, involving the interaction of an F centre and an H centre. The evolution of the defect system in KCl has been modelled using both 42- and 57-atom clusters[49], embedded in a point-charge matrix, using CNDO with parameters which have good perfect crystal properties. In both 42 and 57-atom clusters the F and H centre were found to repel each other in the nearest-neighbour position but attract at the next-nearest separation. A wide range of geometries was studied, corresponding to the evolution of the self-trapped exciton to an F-H pair. From these it seemed that F-H formation from the ground state involved a considerable potential energy barrier, whereas when the hole was in an excited state the curve was essentially flat. As wavefunctions were available from the cluster calculation it was possible[50] to estimate the rates of the non-radiative transitions among the excited states of the self-trapped exciton. It appears that the decay of the lowest electron-excited state is more rapid into the lowest hole-excited state than into the ground triplet state. Thus the cluster calculation has allowed us to understand the whole sequence of radiation damage in KCl, from initial production of highly excited self-trapped excitons followed by rapid decay to F and H centres.

3.2 The X-α method

The approximation which is central to the Xα method is the replacement of the true exchange and correlation potentials by a single potential dependent solely on the local electron density. Electron gas theory shows[51] that a function dependent on the cube root of the density is appropriate, but the numerical constant is kept as a parameter (α) which may be determined by atomic structure calculations[52]. The second approximation which is usually made (and has been in defect calculations) is that of a "muffin-tin" potential. Space is divided into regions of three types: spheres centred on each nucleus, within which the potential is spherically symmetrical; intersphere regions with a constant potential; a region outside the cluster in which the potential is spherically symmetrical. This sort of model fits intuitive pictures of ionic lattices, but breaks down in some cases: for example, the C_2 molecule is not predicted to bind[53]. Computer programs are readily available[54] which calculate numerical wavefunctions. One important point is that Koopman's theorem does not hold for the Xα Hamiltonian, so that one-electron eigenvalues no longer relate simply to transition energies. If a calculation is made on a "transition state" with occupation numbers of the relevant spin-orbitals halfway between those of the ground and excited states[55,56] the relationship may, however, be used.

Brescansin and Ferreira[35] studied simple clusters of 27 atoms of NaCl, representing a central Na or Cl and three shells of neighbours. The small differences in charge between similar atoms in different shells shifted the relevant energy levels considerably, so that it was hard to identify the band structure. Yu et al[57] found that their calculated transition for defects oscillated as the size of their cluster increased. They suggested that this could be overcome by including the appropriate correction in each region of the cluster to account for the electrostatic potential of the surrounding lattice. A particular problem is that a large amount of the electron density of the excited states is in the intersphere regions where the potential is least well determined. Tang Kai et al[58] have shown that the choice of sphere radii is important, and that agreement with experiment is only achieved if radii are used which reflect the ionic radii. In a sense the radii are empirical parameters.

When Oliveira et al[59] attempted to treat F centres in alkaline-earth fluorides with the Xα method the agreement they obtained with experiment was far worse than a simple point-ion calculation would achieve. They found that different sphere radii, different α parameters, distortions or different crystal potentials in each sphere offered little improvement, and concluded that the problem lay in the muffin-tin potential. The Xα method can be cast in LCAO form and the muffin-tin approximation removed, for example in the discrete variational method[60]. To date, such schemes have not been applied to colour centres.

Finally, there is a growing application of ab initio LCAO method to solid state problems. A recent review is given by Bullett[32]; and a more detailed discussion is available in Chapter (6) of the present volume. Of particular relevance to the present chapter is the work of Murrell et al [61,62] who reported calculations on the F centre in LiF. We should note that problems were encountered in the calculations on the excited state which converged relatively slowly.

4. Conclusions

It is clear that one of the weakest points of current theories of point defects is in the treatment of the electronic structure of the host lattice. One of the advantages of cluster calculations is that the defect and lattice can be treated together. Great care is needed to incorporate the electrostatic field of the lattice outside the cluster correctly. Polarisation and distortion can be handled by cluster models where they are of short range, that is, for neutral defects. For charged defects hybrid models similar to those used by Wood and Opik[7,8] will have to be developed. The final method may well involve three regions: an innermost region in which the electronic structure is calculated in detail and self-consistently; a second region with model ions, interacting with each other through potentials of Born-Mayer type and with the inner region through pseudopotentials; and an outer continuum. The distortion of the whole would be controlled by some efficient algorithm similar to that used in the lattice simulation methods discussed in Chapter (1).

References

1. Inui, T. and Uemura, Y., Prog. Theor, Phys. 5, 252 (1950).

2. Kojima, T., J. Phys. Soc. Japan 12, 908 (1957).

3. Kojima, T., J. Phys. Soc. Japan 12, 918 (1957).

4. Gourary, B.S. and Adrian, F.J., Phys. Rev. 105, 1180 (1957).

5. Stoneham, A.M., Phys. Stat. Sol.(b), 52, 9 (1972).

6. Martino, F., Internat. J. Quantum Chem. 2, 217 (1968); 2, 233 (1968).

7. Opik, U. and Wood, R.F., Phys. Rev. 179, 772 (1969).

8. Wood, R.F. and Opik, U., Phys. Rev. 179, 783 (1969).

9. Gourary, B.S. and Fein, A.E., J. Appl. Phys. Suppl. 33, 331 (1962).

10. Kübler, J.K. and Friauf, R.J., Phys. Rev. 140, A1742 (1965).

11. Bartram, R.H., Stoneham, A.M. and Gash, P., Phys. Rev. 176, 1014 (1968).

12. Gash, P.W., Ph.D. Thesis, University of Connecticut (1970).

13. Wood, C.H. and Wang, S., Phys. Rev. B7, 2810 (1973); B7, 2827 (1973).

14. Harker, A.H., D.Phil. Thesis, University of Oxford (1973).

15. Leung, C.H. and Song, K.S., Can. J. Phys. 58, 412 (1980).

16. Löwdin, P.O., Adv. Phys. 5, 1 (1956).

17. Shavitt, I. and Karplus, M., J. Chem. Phys. 36, 550 (1962).

18. Shavitt, I., Methods in Computational Physics 2, 1 (1963).

19. Landshoff, R., Z. Phys. 102, 201 (1936).

20. Paschalis, E. and Weiss, A., Theoret. Chim. Acta. 13, 381 (1969).

21. Bauer, R., Phys. Stat. Sol.(b), 50, 225 (1972); 50, 491 (1972).

22. Kunz, A.B., Phys. Stat. Sol. 36, 301 (1969).

23. Simmons, J.E., Lin, C.C., Lafon, E.E., Fouquet, D.F. and Chaney, R.C.,
J. Phys. C8, 1549 (1975).

24. Tubis, A., Phys. Rev. 102, 1049 (1956).

25. Clementi, E. and Roetti, C., At. and Nucl. Data Tables 14 (1974).

26. Wood, R.F., Phys. Rev. Lett. 11, 202 (1963).

27. Chaney, R.C. and Lin, C.C., Phys. Rev. B13, 848 (1976).

28. Song, K.S., Solid State Commun. 21, 335 (1977).

29a. Toyozawa, Y., Progr. Theor. Phys. (Kyoto), 12, 422 (1954).

29b. Haken, H. and Schottky, W., Z. Phys. Chem. 16, 218 (1958).

30. Tosi, M.P. and Fumi, F.G., J. Phys. Chem. Solids 25, 45 (1964).

31. Stoneham, A.M. and Bartram, R.H., Phys. Rev. B2, 3403 (1970).

32. Bullett, D.W., Solid State Phys. 35, 129 (1980).

33. Hayns, M.R. and Dissado, L., Theoret. Chim. Acta. (Berlin), 37, 147 (1975).

34. Larkins, F.P., J. Phys. C. Solid State Phys. 4, 3065 (1971).

35. Müller, C.M. and Scherz, U., Phys. Rev. B21, 717 (1980).

36. Brescansin, L.M. and Ferreira, L.G., Phys. Rev. B20, 3415 (1979).

37. Koiller, B. and Brandi, H.S., Phys. Stat. Sol.(b), 92, 279 (1979).

38. Pisani, C., Dovesi, R. and Carosso, P., Phys. Rev. B20, 5345 (1979).

39. Kadura, P. and Kühne, L., Phys. Stat. Sol.(b), 88, 537 (1978).

40. Messmer, R.P., Phys. Rev. B15, 1811 (1977).

41. Baetzold, R.C., Mason, M.G. and Hamilton, J.F., J. Chem. Phys. 72, 366 (1980).

42. Messmer, R.P. and Watkins, G.D., Phys. Rev. 7, 2568 (1973).

43. Messmer, R.P., in Semiempirical Methods of Electronic Structure Calculation. Edited by G.A. Segal (Plenum Press, New York 1977).

44. Evarestov, R.A., Phys. Stat. Sol.(b), 72, 569 (1975).

45. Harker, A.H. and Larkins, F.P., J. Phys. C12, 2487, 2497, 2509 (1979).

46. Chadi, D.J. and Cohen, M.L., Phys. Rev. B8, 5747 (1973).

47. Pople, J.A. and Beveridge, D.L., Approximate Molecular Orbital Theory (McGraw-Hill, New York 1970),

48. Evarestov, R.A. and Lovchikov, V.A., Phys. Stat. Sol.(b), 79, 743 (1977).

49. Itoh, N., Stoneham, A.M. and Harker, A.H., J. Phys. C10, 4197 (1977).

50. Itoh, N., Stoneham, A.M. and Harker, A.H., J. Phys. Soc. Japan 49, 1364 (1980).

51. Slater, J.C., Phys. Rev. 81, 385 (1951); 82, 538 (1951).

52. Schwartz, K., Phys. Rev. B5, 2468 (1972).

53. Slater, J.C., Int. J. Quantum Chem. Symposium 8, 81 (1974).

54. Katsuki, S., Palting, P. and Huzinaga, S., Comp. Phys. Commun. 14, 13 (1978).

55. Slater, J.C. and Johnson, K.H., Phys. Rev. B5, 844 (1972).

56. Slater, J.C., Adv. Quantum Chem. 6, 1 (1972).

57. Yu, H.L., Siqueira, de M.L. and Connolly, J.W.D., Phys. Rev. B14, 772 (1972).

58. Tang Kai, A.H., Calais, J-L and Hassib, A., J. Phys. Chem. Solids 40, 803 (1979).

59. Oliveira, L.E., Oliveira, P.M. and Maffeo, B., Phys. Stat. Sol.(b), 87, 25 (1978).

60. Ellis, D.E. and Parameswaran, T., Int. J. Quantum Chem. 5, 443 (1971).

61. Murrell, J.N. and Tennyson, J., Chem. Phys. Lett. 69, 212 (1980).

62. Tennyson, J. and Murrell, J.N., Mol. Phys. 42, 297 (1981).

SECTION B

POTENTIALS

CHAPTER 8

INTERATOMIC POTENTIALS IN SOLIDS

by

N.H. March
Theoretical Chemistry Department, South Parks Road, Oxford

1. Introduction

The objective of the study of interatomic potentials is to represent the ground-state energy E of a solid as a function of nuclear positions $\{\underline{\ell}\}$. The Born-Oppenheimer approximation underlies the whole of the discussion given below.

Ideally one would like interatomic force fields for:

(a) Different structures (fcc, bcc, hcp)

(b) Perturbed lattice configurations (phonons, surfaces, vacancies, self-interstitials, line defects).

(c) Amorphous materials and for liquids (the latter to include discussion of the melting of solids).

Even if it should prove possible to calculate the ground-state energy, it would be impracticable to handle the output as $E(\{\underline{\ell}\})$. Whenever possible therefore, one wants to decompose this into pairwise additive, three-body and higher order terms. Subsequent chapters in this section of the book will deal with the extent to which such decompositions can be achieved for different classes of material. The present chapter concentrates on fundamental aspects which are applicable to solids in general.

2. Force fields and electron density

We wish to emphasise here the intimate relationship between force fields and the electron density $\rho(\underline{r}) \equiv \rho(r\{\underline{\ell}\})$ in the solid. Below, an approximate expression for $E\{\underline{\ell}\}$ will be written in terms of $\rho(\underline{r})$. But even if we solve for $\rho(\underline{r})$ in a perfect crystal, as we can by energy band theory, the Schrödinger equation will inevitably yield a $\rho(\underline{r})$ which belongs to the crystal as a whole.

Only in exceptional cases will it be possible, by examining contours of constant electron density, to assert that $\rho(\underline{r})$ can be built from localized electron distributions centred on specific nuclei. One such exceptional case, however is the NaCl crystal, where Castman et al[1] have shown that the electron density measured by X-ray scattering can be used to demonstrate that, to high accuracy, $\rho(\underline{r})$ can be regarded as built from free Na^{+} and Cl^{-} ions. This is possible because the electron density has dropped practically to zero at the boundary separating Na and Cl cells. We shall discuss the use of X-ray (and electron) scattering data in other cases below. But we must turn to the question of why we can use $\rho(r\{\underline{\ell}\})$ to describe the ground-state energy $E(\{\underline{\ell}\})$.

2.1 Density description of ground state

It has been known since Schrödinger's work that the ground-state energy E is a functional of the many-electron wave function Ψ. Shortly after, density descriptions were used, by Thomas and Fermi independently, to treat atomic ground states[2]. Any doubts which may have existed about the fundamental basis of the density description were resolved by the theorem of Hohenberg and Kohn[3] which demonstrated that the ground-state energy is a unique functional of the electron density.

Thus, it is quite natural in any study of $E(\{\underline{\ell}\})$ to focus attention on $\rho(\underline{r}\{\underline{\ell}\})$. But, as already pointed out above, to obtain a useful working form to represent the ground-state energy, we wish to decompose into pair, three-body and other many body terms. To do so, we wish to form $\rho(\underline{r})$, albeit approximately, from localised building blocks associated with individual nuclei, or with groups of nuclei (i.e. bonds or molecules).

A purely quantum mechanical analysis of $\rho(\underline{r})$ without appeal to chemical or physical instruction, shows that the construction of such localized building blocks, or 'blobs', of electron density is subject to arbitrariness. Let us return to the case of X-ray scattering from a perfect crystal, with say one atom per unit cell. Here crystallographers use the reciprocal lattice vectors $\underline{G}$ to Fourier analyze $\rho(\underline{r})$, which is then fully specified by its Fourier components $\rho_{\underline{G}}$. These Fourier components, we emphasize, are accessible to experiment via Bragg reflection intensities. But in the force field analysis, one would like to extract a localized picture by writing

$$\rho(\underline{r}\{\underline{\ell}\}) = \sum_{\underline{\ell}} \sigma(\underline{r}-\underline{\ell}): \quad \sigma(\underline{r}) \rightarrow 0, \quad \underline{r} \rightarrow \infty . \tag{1}$$

Any localized distribution $\sigma(\underline{r})$ will reproduce $\rho(\underline{r})$ if its Fourier transform $\sigma(k)$ is such that, at the discrete values $\underline{k} = \underline{G}$, $\sigma(\underline{G}) = \rho_{\underline{G}}$. Evidently then, since $\sigma(\underline{k})$ is fixed only at discrete values $\underline{G}$ one has many possible decompositions of the periodic density $\rho(\underline{r})$ into localized blobs $\sigma(\underline{r})$.

Nevertheless, important qualitative information about directionality is obtained in the X-ray scattering. Thus in bcc Fe, to be discussed further below, one has 'overlapping' reflections (411) and (330), corresponding to the same value of $\sin \theta/\lambda$, θ being the scattering angle and λ the X-ray wavelength. Using a single crystal, the intensity I(411) is found to be different from I(330) and no picture of spherical localized blobs $\sigma(|\underline{r}|)$ on the nuclei can explain the X-ray scattering from bcc Fe. In terms of interatomic potentials this tells us that central pair forces cannot be adequate for accurate studies. A more striking example, in a case where we relax the requirement of one atom per unit cell is that of the so-called 'forbidden' (222) reflection in crystalline Si. This is treated in some detail below, by taking the localized building block as one chemical bond embracing two Si nuclei.

Having stressed that it is $\rho(\underline{r})$ and hence $E(\{\ell\})$ that follow from a full quantum-mechanical analysis and not $\sigma(\underline{r})$ or the pair potential $\phi(\underline{\ell}_i - \underline{\ell}_j)$, let us nevertheless briefly summarize how one may obtain an approximate ground-state energy $E[\rho]$ from $\rho(\underline{r})$.

2.2 Energy as function(al) of density

One can write the ground-state energy E in the form

$$E[\rho] = \int t_r[\rho] \, d\underline{r} + \int \rho \, V_N \, d\underline{r} + \tfrac{1}{2} \int \rho \, V_e \, d\underline{r} +$$
$$+ \int \epsilon_{xc}[\rho] \, d\underline{r} + V_{nn} . \tag{2}$$

Here, one has written in the first term on the right-hand side the single-particle kinetic energy density $t_r[\rho]$, the correlation kinetic energy being incorporated in $\epsilon_{xc}[\rho]$ which is the energy density of exchange (x) and correlation (c) interactions. The other terms are classical electrostatic energies, V_N being the potential energy of the nuclei, V_e the potential energy of the electron cloud $\rho(\underline{r})$ and V_{nn} the nuclear-nuclear interaction energy.

Minimizing the energy (2) with respect to ρ, subject to the normalization requirement $\int \rho \, d\tau = N$, where N is the total number of electrons, amounts to imposing

$$\frac{\delta(E - \mu N)}{\delta \rho} = 0 \tag{3}$$

where μ is a Lagrange multiplier which has the interpretation from (3) of the chemical potential dE/dN. The Euler equation of the above variation problem is then:

$$\mu = \frac{\delta t_r[\rho]}{\delta \rho} + V_N + V_e + \frac{\delta \epsilon_{xc}[\rho]}{\delta \rho} \equiv \frac{\delta t_r[\rho]}{\delta \rho} + V(\underline{r}) . \tag{4}$$

This equation (4) expresses the chemical potential μ as a sum of a kinetic term, and a potential term, both of which vary with position $\underline{r}$ in the inhomogeneous electron cloud of the solid. The fact that μ is independent of $\underline{r}$ is a statement of the equilibrium condition, namely that the electrons have redistributed themselves in space so that no further charge flow is possible.

For present purposes, two points need emphasis: (i) since $t_r[\rho]$, by definition, is a single-particle kinetic energy, equation (4) shows that $\rho(\underline{r})$ can be calculated from a one-body potential $V(\underline{r})$, into which exchange and correlation is subsumed. Of course, exact knowledge of $\delta \epsilon_{xc}[\rho]/\delta \rho$ in equation (4) would require exact solution of the many-electron problem. But in numerous examples, useful approximations can be made to add this term to the Hartree potential $V_N + V_e$.

In a perfect crystalline solid, equation (4) shows us that band theory can allow
the calculation of $\rho(\underline{r})$ in the many-electron problem, in principle exactly, from
the one-potential $V(\underline{r})$ defined by (4). (ii) Though to solve (4) to high accuracy
is important, the variational principle (3) ensures that one can use a model of
$\rho(\underline{r}\{\underline{\ell}\})$ to calculate E to useful accuracy, a procedure adopted by Jensen, Gombás
and Wedepohl, and most recently by Gordon and Kim[5].

We turn immediately to an example of (ii) above in the case of ion-ion
interactions.

3.   Ion-ion interactions from superposition density

Gordon and Kim[5] studied ion pair interactions using $\rho(\underline{r})$ as the superposition
of Hartree-Fock densities for free ions. The usefulness of such calculations
for solids rests on the following assumptions.

(i) That formation of an ionic solid such as NaCl can be described as bringing
free atoms from infinity into an assembly of Na^+ and Cl^- ions at the spacings
relevant in the crystal.

(ii) That one can represent the force field by pair-wise additive potentials.

(iii) That one can usefully approximate the electron density of the two-centre ion
problems by the superposition density.

Of course, even accepting these assumptions (i) to (iii), which are
consistent for NaCl with the X-ray data[1], one has to calculate $E[\rho]$ from equation
(2) by assuming forms for $t_{\underline{r}}$ and ε_{xc}. Following earlier workers, Gordon and Kim[5]
chose the forms

$$t_{\underline{r}}[\rho] = c_k\, \rho^{\frac{5}{3}} \;\; ; \;\; \varepsilon_{xc}[\rho] = \varepsilon_{xc}^0[\rho] \, , \tag{5}$$

where both forms are taken from the theory of the homogeneous electron gas[4]. With
these assumptions Gordon and Kim studied:

(a) Ion pairs made of an alkali ion and a halide ion.

(b) Two alkali ions or two halide ions.

(c) Combinations of an alkaline earth ion and a halide ion.

It is now clear that such procedures can generate satisfactory potentials for other
ionic and semi-ionic solids as well, as discussed in greater detail in Chapter (10).
One important refinement of the method emphasised in this latter chapter is the use
of electron densities for the interacting ions that have been calculated for ions
in the crystal environment rather than in the free state. This feature turns out
to be of crucial importance for oxides.

Additional examples of the use of such potentials are to be found in the recent
molecular dynamics studies by Rahman[6] of the superionic phase of CaF_2. Chapter 10

takes up this topic in more detail.

Of course, it is clear that the above potentials contain relatively severe approximations, and must have limited applicability. One limitation, as Kim and Gordon clearly point out, is that the method fails to include forces arising from the induction by the charge of an ion of a dipole moment on a second ion. This, in turn, interacts with the charge producing a charge-dipole interaction that varies as r^{-4} at large ionic separation r. The induced dipole moment is due to the distortion of the electron clouds with respect to the nuclei. One would clearly have to transcend the superposition (iii) of free ion densities to include this as discussed further in Chapter (10). In addition to this neglect of the induction force, part of the dispersion interaction is also omitted by the above procedure. The magnitudes of dispersion energies are again taken up in Chapter (10).

3.1 Other consequences of ionic model in solids

It can be argued, of course, that one wishes to treat an ionic solid, whereas the above discussion is tantamount to discussing ion-ion interactions in free space. Therefore, we wish to draw attention to the important work of Ruffa[7] who argued that, consistently, with writing the density $\rho(\underline{r})$ as a superposition of 'localized blobs' representing ions, one could write the total wave function of the crystal as a product of ionic wave functions. Then he showed that the off-diagonal density, or first-order density matrix $\rho(\underline{r},\underline{r}')$ could be written also as a superposition. For M nuclear sites, and with the ith ion having n_i electrons, he obtained

$$\rho(\underline{r},\underline{r}') = \sum_{i=1}^{M} \rho_i^{\{n_i\}}(\underline{r},\underline{r}') \quad . \tag{6}$$

Furthermore, an explicit, but slightly more complicated expression for the two-particle density matrix can also be calculated from the product wave function of the ionic model.

The additive property of $\rho(\underline{r},\underline{r}')$ was utilized by Ruffa to show that the electron polarization $\underline{P}$ is a sum of the polarizations of the ionic constituents, namely

$$\underline{P} = \frac{e^2 h^2}{4\pi^2 m} \sum \epsilon_i (n_i/(\overline{E}_i)^2) \tag{7}$$

where ϵ_i is the local field at the ith ion while $\overline{E}_i$ is a mean excitation energy introduced by Ruffa[7] via the Thomas-Kuhn sum rule. This additive formula (7) is in accord with the important empirical analysis of Tessman et al[8]; moreover in the case of cations it allows one to go from a free space excitation energy, $\overline{E}_{if}$, to a value appropriate to the crystal environment, through the replacement $\overline{E}_{if} \rightarrow \overline{E}_{if} - eV_m$, V_m being the appropriate Madelung potential. The result thus obtained for the crystal polarizability, has thereby incorporated a considerable proportion of the environmental effects. The anion case is also discussed by Ruffa[7], to whose work the interested reader should refer. Use of Ruffa's analysis has been made recently by Shanker and Verma[9].

From this discussion in which the localized building blocks are ions, we turn to covalent solids where the localized densities are basically described by $\sigma(\underline{r})$ for one chemical bond. This leads us then to the valence force field in covalent materials.

4. Valence force fields with special reference to solid silicon

Valence force fields have been widely and successfully used to analyse the vibrational spectra of molecules. The concept on which this force field description rests is that of electrons localized in chemical bonds with highly directional electron densities.

Distortion of this directional electronic charge cloud gives rise to many-body forces. These are described by empirical parameters, for bond stretching and bond angle changes. The merits of such an approach are first that it often provides an accurate and intuitive phenomenological description, with a small number of force constants, and secondly that it is frequently possible to transfer to the crystal, values of force constants extracted from normal-mode analyses of suitable molecular models.

Below, we shall focus on such a description of the tetrahedrally bonded group IV semiconductors, and in particular on silicon. We shall consider experimental evidence, in the form of X-ray and electron scattering from both crystalline and amorphous Si, and of phonon frequencies and elastic constants in the crystalline material. Also, there is support from first principles for the approach, starting with the work of Adams[10] and Gilbert[11]. These workers showed how to write down defining equations for a localized atomic-like basis set, in such a way that the influence of the distant environment was only a weak perturbation. The approach was taken forward by Anderson[12] and especially by Bullett[13] who uses a chemical pseudo-potential approach to evaluate a valence force field in C and Si. Very briefly, to each bond, a bond equation of the type

$$H\phi_i - \sum_j \phi_j V_{ji} = \varepsilon_i \phi_i \qquad (8)$$

is assigned. Here, the suffices i,j label adjacent bond orbitals, while the V_{ji}'s are interaction parameters. Bullett demonstrates that the familiar sp^3 hybrid orbitals represent, in the group IV materials, approximate solutions of such localized orbital equations.

4.1 X-ray and electron scattering from crystalline and amorphous silicon

Clear support for the picture of an electron density $\rho(\underline{r})$ in Si, that is built from chemical bonds based on the sp^3 hybrids, is afforded by the work of Stenhouse et al[14]. By simple linear combination of these sp^3 hybrids, they constructed the chemical bond density embracing two Si nuclei, as shown in Figure 1a. Figure 1b shows the result that would have been obtained by superposing free Si atoms, that is without bonding effects.

Figure 2 shows the scattering factor f versus sin θ/λ for crystalline Si, obtained by superposition of these localised bonds. The most important point to emphasize is that the forbidden (222) reflection is immediately reproduced to excellent accuracy by the chemical picture. Band theory, based on the one-body potential $V(\underline{r})$ of equation (4), can also be used for crystalline Si. The results of these substantial calculations yield X-ray intensities of comparable quality with the very much simpler chemical bond calculations.

However, it is in the calculation of the X-ray and electron scattering from amorphous Si that the chemical bond description really comes into its own. Because of the lack of long-range order, band theory becomes very difficult to apply, as Bloch's theorem and the concept of Brillouin zones are no longer applicable, whereas simple superposition of chemical bond densities is straightforward, for a given, tetrahedrally bonded, local structure. Using a random network model of the structure, with inclusion of odd membered rings, Stenhouse et al[14] obtain excellent agreement with the X-ray and electron scattering results, as shown in Figures 3a and 3b.

We have, therefore, as would be anticipated for such a classical covalent system as Si, a direct demonstration of the validity of the picture of highly directional chemical bonds in which electrons are localized, as is required to validate the valence force field. Further discussion of bonding in semi-covalent and covalent semiconductors is given in Chapter (11).

4.2 Phonons and elastic constants in group IV semiconductors

The use of the valence force field in diamond was first proposed by Musgrave and Pople[15]. Later important work is that of McMurray et al[16], and effective use of such a force field in defect calculations has been made by Stoneham and Larkins[17]. We report briefly below the results of the very recent work of Tubino and Piseri[18], who calculated phonon frequencies (in 10_m^{2-1}) at points Γ, X and L for the group IV crystalline semiconductors, a sample of their results being shown in Table 1 below.

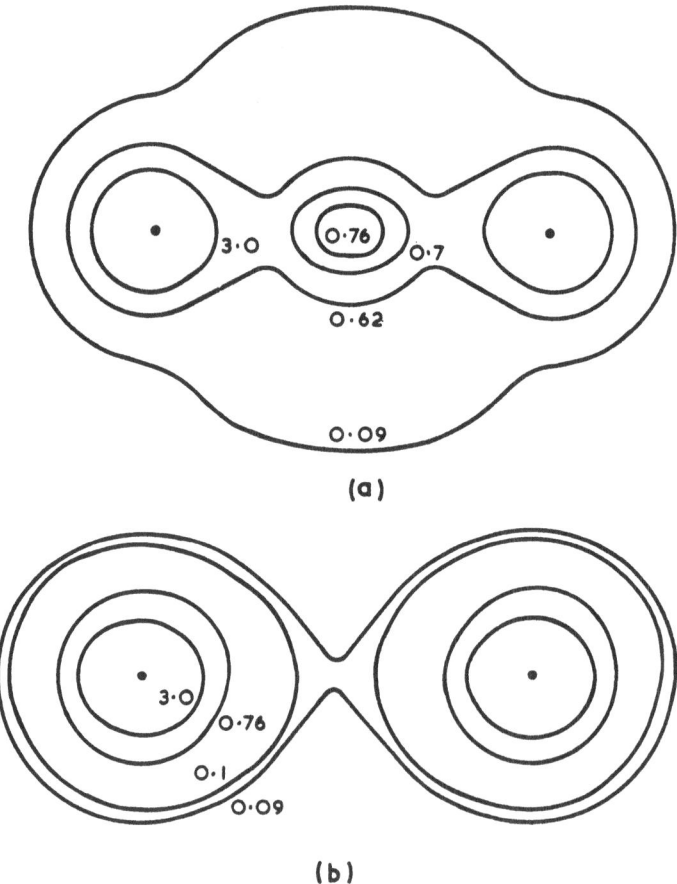

(a)

(b)

Figure 1   Contours of equal electron density for   (a) Si-Si bond,   (b) Superposition of isolated atom densities.

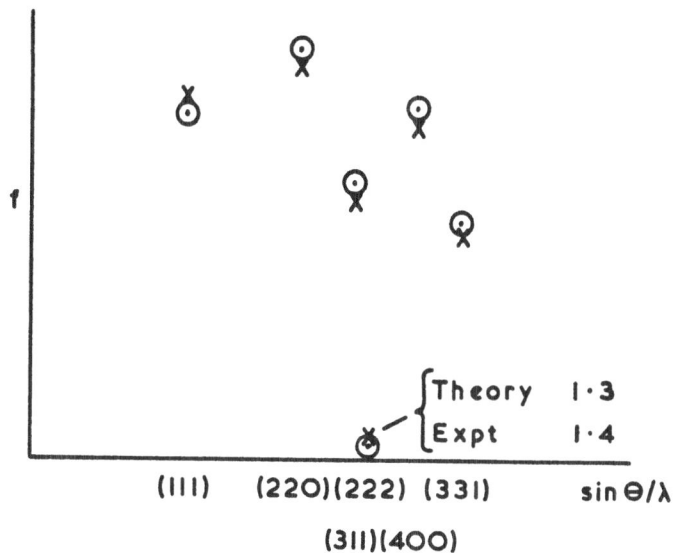

(111) (220)(222) (331)   sin Θ/λ

(311)(400)

<u>Figure 2</u>   X-ray scattering amplitude at Bragg reflections for crystalline Si.
Crosses - experiment, circles - chemical bond theory.

<u>Table 1</u>

	C		Si		Ge		Sn	
	Th	Expt	Th	Expt	Th	Expt	Th	Expt
$\omega_{TO}(\Gamma)=\omega_{LO}(\Gamma)$	1336	1333	514	518±8	293	300±10	198	200
$\omega_{TO}(X)$	1058	1077	457	463±10	270	275±10	182	184±3
$\omega_{LA}(L)$	1041	1034	350	378±10	197	215±10	125	138±1

The agreement with experiment is seen to be generally good, most frequencies
being within the experimental errors.   Only for the $\omega_{LA}(L)$ phonons of Si, Ge
and Sn are the calculated values systematically slightly lower than the observed
results: this suggests that the frequencies of these phonons are influenced, to
a small extent, by factors not incorporated in the valence force field used, for
example, anharmonicity and long-range forces.

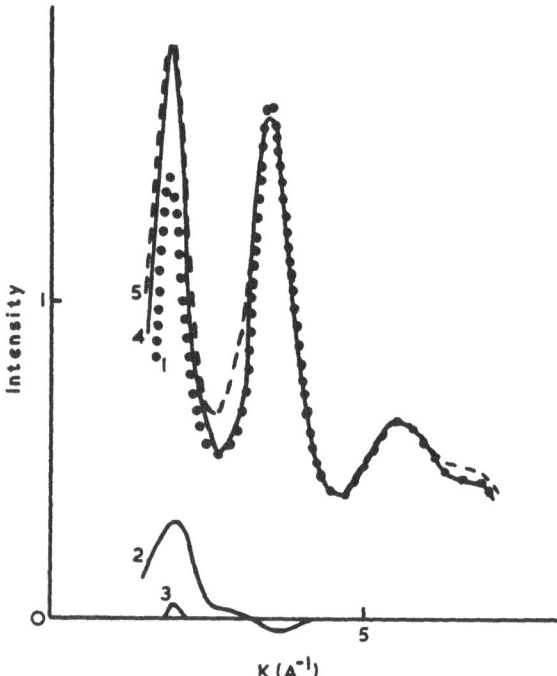

<u>Figure 3(a)</u> X-ray scattering intensity for amorphous Si. Curves (2) and (3), correspond to theory and experiment respectively. Curve (1) corresponds to no charge in the middle of Si-Si bond.

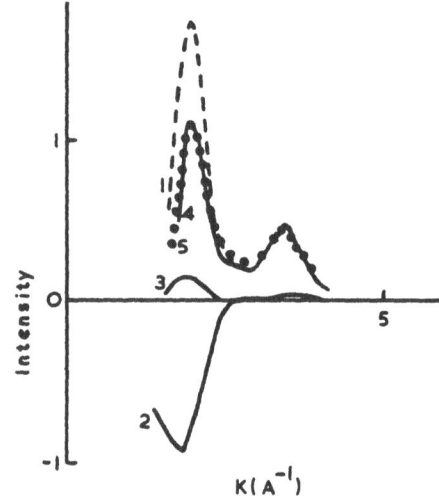

<u>Figure 3(b)</u> Electron scattering intensity for amorphous Si. Curves (4) and (5) correspond to theory and experiment. Curve (1) again corresponds to no charge in the middle of Si-Si bonds.

4.3 Related bond charge models

Though bond charge models mean quantitatively different things to different people, all such models describe electrons shared between atoms, i.e. covalency. In the case of the Si-Si density contours in Figure 1, Stenhouse et al[14] showed that these could be modelled, very precisely, by spherical blobs of charge centred on the two Si nuclei plus a spherical distribution of electrons centred at the mid-points of the Si-Si bond. This latter charge, however, was spread out in configuration space. Its profile is of course, important in the X-ray and electron scattering calculations.

It is relevant here to refer to two other examples: first the work of Rustagi and Weber[19] on the partially ionic 3-5 semiconductors and secondly the force fields for Fe set up by Matthai et al[20] and used to construct the phonon frequencies. With regard to the first of these examples, in heteropolar semiconductors, the bond charge is believed to shift from the mid-point of the bond towards the group $\underline{V}$ ion. Valence charge density calculations[21] find such shifts of the charge density maximum which Rustagi and Weber[19] identify as the position of the bond charge. The bond charge centre divides the bond length roughly in the ratio 5:3. The magnitude of the bond charge, which measures covalency, decreases with increasing ionicity. In this work, a six-parameter model provides a good description of the experimental data for GaAs, GaSb and InSb. A small seventh parameter is found to be needed in GaP; this describes weak second-neighbour interactions. Trends in the force constant parameters are evident in this type of description.

In the second example mentioned above, Matthai et al[20] derived central two-body forces which fitted the liquid structure data. Angularity, which is known from chemical hybridization arguments[22] to be more important in the bcc structure than in the close-packed structures was incorporated by placing charge at the centre of near neighbour and next near-neighbour bonds. After parameterisation of the model, the phonons were calculated; the results along [100], [111] and [110] are shown in Figure 4.

Comparison with the results of purely central pair force calculations shows that the quantitative agreement with experiment in the [100] direction plotted in Figure 4 requires the inclusion of angularity. Secondly, the theory, which still exaggerates the minimum in the [111] direction, is poorer in this respect without inclusion of bond charges. One defect of the theory, which is not removed by the inclusion of directionality, is the crossover shown in the [100] direction.

Following this discussion of the ways valence electron density, concentrated in rather localized and well directed bonds, can be incorporated in force field models, we turn to the fundamental electron theory of interatomic forces for small

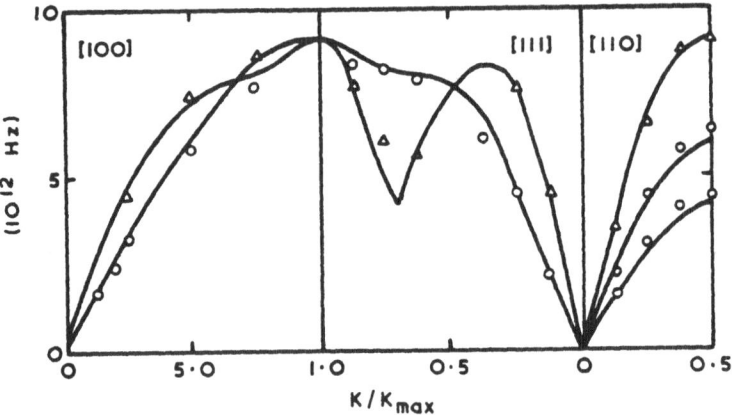

Figure 4   Phonons in bcc Fe calculated from parameterized bond charge model[20]

displacements from lattice equilibrium positions.   Here, we shall show that a first principle theory exists[23] though considerable progress will be needed before it may be applied in a fully quantitative manner.

5.   Electron theory of force fields for small displacements

We shall restrict the discussion below to a monoatomic lattice, with one atom per unit cell.   Furthermore we work within the adiabatic and harmonic approximations.

The results of the electron theory are most usefully stated by generalizing the old, rigid-ion model.   This can be done by appeal to equation (1).   In the rigid-ion model, it is assumed that the localized blob $\sigma(r)$ is fixed relative to the nucleus, and that as the ions vibrate, all the configurational dependence is in the nuclear positions $\{\underline{\ell}\}$, to first order in the displacements from the equilibrium crystal lattice positions.

Electron theory shows, as we sketch below, that, in fact, the localized 'blobs' σ deform as the ions move.   Though a discussion of this deformability can be useful, as shown by March and Wilkins[24], a fundamental decomposition of the periodic electron density requires one to focus on the gradient of the electron density $\nabla_r \rho(\underline{r})$. Thus, in the periodic case, one writes a generalization of equation (1) as:

$$\nabla_r \rho(\vec{r}\{\ell\}) = \sum_\ell \underline{R}(\underline{r}-\underline{\ell}) \quad . \tag{9}$$

Then it can be shown that the density change from the perfect lattice density, say $\Delta\rho(\underline{r})$, which is induced by phonon displacements $\underline{\mu}_\ell$, can be written exactly to first order in the displacements in terms of the localized vector field $\underline{R}(\underline{r})$ in equation (9). The theory shows that it is generally the case that $\underline{R}(\underline{r}) \neq \text{grad } \sigma(\underline{r})$. When $\underline{R}$ can be written as the gradient of a scalar, then a pairwise additive description is possible. In $\underline{k}$ space, and at the reciprocal lattice vectors $\underline{G}$, one can still write:

$$\underline{R}(\underline{G}) = i \ \underline{G} \ \rho_{\underline{G}} \tag{10}$$

where $\rho_{\underline{G}}$ are the Fourier components of the charge density discussed above. But for an arbitrary vector $\underline{k} \neq \underline{G}, \underline{R}(\underline{k})$ (apart from the imaginary i) is no longer parallel to $\underline{k}$ and one needs the concept of a tensor of charge density. The angle, $\theta_{\underline{k}}$ say, between $\underline{R}(\underline{k})$ and $\underline{k}$ for $\underline{k} \neq \underline{G}$ is a measure of the many-body forces.

5.1 Density change due to change in one-body potential

In equation (4), we focussed on the fact that the exact electron density $\rho(\underline{r})$ in the ground state of a solid could be obtained from a one-body potential energy $V(\underline{r})$, into which, however, are built many-electron effects through the exchange and correlation energy density $\varepsilon_{xc}[\rho]$. Therefore, in calculating the 'vector rigid ion' $\underline{R}(\underline{r})$ discussed above, the way forward is to calculate the density change $\Delta\rho(\underline{r})$ from the change $\Delta V(\underline{r})$ in the one-body potential from its periodic value. We stress that this can be done by linear response theory as

$$\Delta\rho(\underline{r}) = \int \ F(\underline{r},\underline{r}') \ \Delta V(\underline{r}') \ d\underline{r}' \quad . \tag{11}$$

The important point here is that the linear response function F is calculable by energy band theory from the periodic one-body potential $V(\underline{r})$. Of course, many-electron effects, as we have stressed, are built into both the periodic potential $V(\underline{r})$ and the change in it, $\Delta V(\underline{r})$. Concerning the latter, we may write, following Jones and March[23]

$$\Delta V(\underline{r}) = \Delta V_{Hartree}(\underline{r}) + \int \ U(\underline{r},\underline{r}') \ \Delta\rho(\underline{r}') \ d\underline{r}' \quad . \tag{12}$$

Here $\Delta V_{Hartree}$ is calculated merely from electrostatics while U incorporates electron-electron non-classical effects, i.e. exchange and correlation. One can now write $\Delta\rho(\underline{r})$ explicitly in terms of $\underline{R}(\underline{r})$ which, as set out fully by Jones and March[23], can be calculated from a knowledge of F and U.

5.2 Example of nearly free electron metals

Metal potentials are discussed in detail in Chapter (9). Suffice it to say here that, for nearly free electron metals, the functions F and U above reduce to

$$F \equiv F(\underline{r}-\underline{r}'): \quad U \equiv U(\underline{r}-\underline{r}') \tag{13}$$

which together determine the $\underline{k}$ dependent dielectric function $\epsilon(\underline{k})$. One then obtains the effective metal pair potential from a 'localized blob' of electron density, given in $\underline{k}$ space by

$$\sigma(k) = k^2 V_b(k) [\epsilon^{-1}(k)-1] \tag{14}$$

where $V_b(k)$ is the bare-ion potential. It can then be shown that the pair potential $\phi(r)$ is given by

$$\phi(r) = \int \frac{k^2 V_b^2(k)}{\epsilon(k)} e^{i\underline{k}\cdot\underline{r}} d\underline{k} \tag{15}$$

with long range oscillations of the form $\phi(r) \sim \dfrac{\cos 2k_f r}{r^3}$ where $2k_f$ is the diameter of the Fermi sphere.

In the example of bcc metallic Fe treated by Matthai et al[20] and discussed above. the central pair potential had similar oscillations even though, of course, Fe is not in any sense a simple, or nearly-free electron metal, because of the presence of d electrons. In the work of Matthai et al[20], the central pair potential was designed to describe the liquid phase, in which the Fermi surface is isotropic. The screening of an ion by electrons within an anisotropic Fermi surface has been discussed very recently by Flores et al[25]. The distance dependence of this screening, depending on Fermi surface shape, will eventually have to be incorporated into any force field for solid transition metals which is to explain quantitatively the Kohn anomaly in lattice vibrations. Put in terms of the 'vector rigid ion' used above, the function $F(\underline{r},\underline{r}')$ must be calculated with sufficient accuracy in metals to reflect the topology of the Fermi surface.

One case in which $F(\underline{r},\underline{r}')$ can be approximated is in the covalent semiconductors diamond, Si and Ge; this is discussed in detail by Claesson et al[26], to which paper the interested reader is referred.

6. Inversion of cohesive energy as function of volume: large displacements

The fundamental theory of force fields for phonon displacements described above is valid only to first order in the displacements $\underline{u}_\ell$. While this is already very valuable, in a number of the examples where we need to utilize force fields we must describe the ground state energy $E(\{\ell\})$ for large displacements from the equilibrium crystal lattice.

Therefore, it is of interest to conclude the present discussion with a summary of the very recent work of Carlsson et al[27]. They make the following observations: (i) for a given crystal structure, the cohesive energy, E_c say, of a monatomic solid, can be expressed as a function of the nearest-neighbour distance a. (ii) that if one assumes, for any distance a, that $E_c(a)$ can be written as a sum over atomic sites of a volume-independent radial pair potential $\phi(r)$, then a systematic inversion procedure can be applied to determine ϕ. In Figure 5(a) and (b), pair potentials extracted in this way for bcc Mo and fcc Cu, using cohesive energies from the self-consistent augmented spherical wave method, plus a local density exchange and correlation approximation, are shown. For Cu, the Bullough-Englert[28] potential is also plotted though the potentials are clearly defined with different applications in mind.

Carlsson et al[27] have used the above potentials to calculate elastic constants, phonon frequencies and the structure dependence of the cohesive energy. Also Esposito et al[29] have used the same potential for Cu metal to calculate its theoretical tensile strength.

Detailed numerical results may be found in the two papers cited above; we merely summarize here the principal conclusions. These are:

(i)   Better results are obtained for elastic constants (long wavelength excitations) than for zone-boundary phonon frequencies (short wavelengths). This, as stressed by Carlsson et al[27], is to be expected since the only input information in constructing their potentials is the cohesive energy change, resulting from uniform dilations.

(ii) The Cauchy relations are satisfied: this is not observed in the experimental elastic constants.

(iii) The potential for bcc Mo predicts the fcc structure as having the minimum energy.

Evidently, such potentials therefore have considerable limitations; but they do embody information about large displacements, which is their main merit. In a metal like Cu, however, the potentials describe energy changes, as the lattice parameter increases, through a metal-insulator transition, which does not seem an entirely satisfactory state of affairs.

Finally, we note that, conventional metal force field theory writes as a first approximation

$$E(\{\underline{\ell}\}) = E_0(\Omega) + \sum_{\underline{\ell}} \phi_\Omega(\underline{\ell}_i - \underline{\ell}_j) + \text{higher-body terms} \qquad (16)$$

and the volume Ω-dependent pair potential, ϕ_Ω, describes therefore structural re-arrangements at constant volume. However, for a given set of $\{\underline{\ell}\}$'s (say fcc), the potential for Carlson et al writes the cohesive energy simply as a sum of

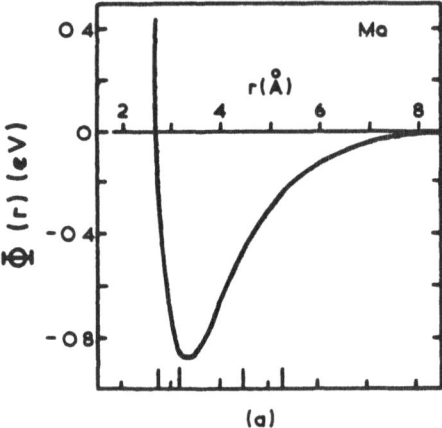

(a)

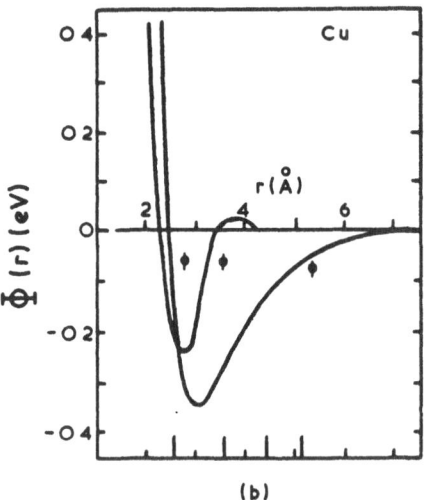

(b)

Figure 5 (a) Pair potential for bcc Mo derived by inversion of cohesive energy[27].
         (b) Pair potential for fcc Cu derived by inversion of cohesive energy[27].
Potential with repulsive hump is due to Bullough and Englert[28].

Ω-independent pair potentials. Within this, admittedly limited, framework, knowledge of the Carlsson et al[20] potential, plus Φ_Ω, would allow the volume dependent, structure independent energy $E_0(\Omega)$ to be constructed.

7. Summary

The following points have been stressed in this brief survey:

(i) The ground-state energy E is very naturally considered as $E[\rho(\underline{r}\{\underline{\ell}\})]$, ρ being the electron density and $\{\underline{\ell}\}$ the totality of nuclear sites.

(ii) Whereas the wave equation gives a delocalized $\rho(\underline{r}\{\underline{\ell}\})$ inevitably (and only determines localized distributions uniquely in a special case like NaCl), physical and chemical intuition can go into choice of localized densities, which will then often allow decomposition into pair, and many body terms.

(iii) Even for small displacements from equilibrium, however, as in for example lattice vibrations, such localized electron distributions must deform as the nuclei move.

(iv) For the classical ionic model, the localized picture corresponds to the Heitler-London crystal wave function of cations and anions, which leads in turn to a superposition property of first-order density matrices. The additivity property of polarizabilities follows as a consequence.

(v) Bonds afford a valuable description of electron density and hence force fields in group IV semiconductors, partially ionic 3-5 compounds, and metals with directional charge distribution (e.g. Be[30-32], Fe, Cr).

(vi) For small displacements, an exact formal theory exists. But this requires the concept of a tensor charge density; this has not yet been made quantitative though.

(vii) Some progress is possible for large displacements, but presently only through limited cohesive energy inversion at a first-principle level. Otherwise, modelling or phenomenology is still essential in this important area. This latter problem will be taken up in several later chapters in the book.

References

1. Castman, B.,Pettersson, G., Vallin, J. and Calais, J.L., Phys. Scripta 3, 35 (1971).

2. March, N.H. Self-consistent fields in atoms. (Oxford: Pergamon, 1973).

3. Hohenberg, P. and Kohn, W., Phys. Rev. 136, B864 (1964).

4. See, for example, March, N.H. Orbital theories of molecules and solids. (Oxford: Clarendon Press), p.107, (1974).

5. Gordon, R.G. and Kim, Y.S., J. Chem. Phys. 60, 4332 (1974).

6. Rahman, A., J. Chem. Phys. 65, 4845 (1976).

7. Ruffa, A.R., Phys. Rev. 130, 1412 (1963).

8. Tessman, J.R., Kahn, A.H. and Shockley, W., Phys. Rev. 92, 890 (1953).

9. Shanker, J. and Verma, M.P., Phys. Rev. B12, 3449 (1975).

10. Adams, W.H., J. Chem. Phys. 34, 89 (1961); ibid 37, 2009 (1962).

11. Gilbert, T.L. Molecular orbitals in chemistry, physics and biology. (Eds. Pullman and Löwdin) (New York: Academic).

12. Anderson, P.W., Phys. Rev. 181, 25 (1969).

13. Bullett, D.W., J. Phys. C8, 3108 (1975).

14. Stenhouse, B., Grout, P.J., March, N.H. and Wenzel, J., Phil. Mag. 36, 129 (1977).

15. Musgrave, M.J.P. and Pople, J.A., Proc. Roy. Soc. A268, 474 (1962).

16. McMurray, H.L., Solbrig, A.W., Boyter, J.K. and Noble, C., J. Phys. Chem. Solids 28, 2359 (1967).

17. Larkins, F.P. and Stoneham, A.M., J. Phys. C4, 143 and 154 (1971).

18. Tubino, R. and Piseri, L., J. Phys. C13, 1197 (1980).

19. Rustagi, K.C. and Weber, W., Solid State Commun. 18, 673 (1976).

20. Matthai, C.C., Grout, P.J. and March, N.H., J. Phys. Chem. Solids 42, 317 (1981).

21. Walter, J.P. and Cohen, M.L., Phys. Rev. B2, 1821 (1970).

22. Altmann, S.L., Coulson, C.A. and Hume-Rothery, W., Proc. Roy. Soc. A240, 145 (1957).

23. Jones, W. and March, N.H., Proc. Roy. Soc. A317, 359 (1970).

24. March, N.H. and Wilkins, S.W., Acta. Crystl. A34, 19 (1978).

25. Flores, F., March, N.H., Ohmura, Y. and Stoneham, A.M., J. Phys. Chem. Solids 40, 531 (1979).

26. Claesson, A., Jones, W., Chell, G.G. and March, N.H., Int. J. Quantum Chem. Symp. No. 7. (Editor Löwdin, P.O.) (Wiley: New York), 629 (1973).

27. Carlsson, A.E., Gelatt, C.D. and Ehrenreich, H., Phil. Mag. A41, 241 (1980).

28. Englert, A., Tompa, H. and Bullough, R. Fundamental aspects of dislocation theory. (Washington NBS: Spec. Publ. 317), p.273 (1970).

29. Esposito, E., Carlsson, A.E., Ling, D.D., Ehrenreich, H. and Gelatt, C.D., Phil. Mag. A41, 251 (1980).

30. Brown, P.J., Phil. Mag. 26, 1377 (1972).

31. Yang, Y.W. and Coppens, P., Acta. Cryst. A34, 61 (1978).

32. Matthai, C.C., Grout, P.J. and March, N.H., J. Phys. F. 10, 1621 (1980).

CHAPTER 9

POTENTIALS IN METALS

by

J.E. Inglesfield
Science Research Council, Daresbury Laboratory,
Daresbury, Warrington WA4 4AD

1. Introduction

A full understanding of metallic cohesion, including the energy and structure of perfect crystals, and of defects such as vacancies, dislocations, grain boundaries and surfaces requires the solution of the Schrödinger equation for approximately 10^{23} electrons, all interacting with one another and with the positive ions. In the case of the perfect crystal, Bloch's theorem means that we need only solve the equation in one unit cell, and we can greatly simplify the electron-electron interaction problem by assuming that each electron moves in the average Hartree field of the other electrons, together with an exchange-correlation potential determined by the local electron density. Self-consistent calculations have been carried out in this way for a whole range of metal crystals, and reproduce very well cohesive energies, atomic radii and bulk moduli[1]. We should note that cohesive energies are the differences between very large terms, so these calculations have to be highly accurate. This fully self-consistent approach is now being extended to alloys[2] and surfaces[3]; it represents the most sophisticated level of presently available theory. But for many purposes we want a more qualitative picture of metallic cohesion, as it is not always useful or even possible to carry out the full solution of the Schrödinger equation. It is the aim of this article to show the uses and limitations of an approach to metallic cohesion which divides the cohesive energy into a large volume-dependent term which provides the major component of the cohesion and a second smaller term representing the interaction between conduction band electrons and ion cores which to second order in perturbation theory can be represented by pair potentials between the ions; as such this latter term is structure sensitive.

This approach, we will see, enjoys considerable success in treatments of the perfect lattice properties of simple metals in which the valence electrons are of s or p type. Its limitations, however, become evident when it is applied to defect properties; and as discussed in the final sections of the article, the method breaks down when applied to transition metals.

2. Cohesion in Metals

Metals fall conveniently into two classes - sp bonded metals like Na, Mg and Al and the 3d, 4d and 5d transition metals like Cr, Mo, W. In the sp-bonded metals the electrons behave as if they were free electrons, outside the ion cores, with wavefunctions of the type, $\psi(\underline{r}) = \exp(i\underline{k}\cdot\underline{r})$.

which give an energy, E(k):

$$E(k) = \frac{\hbar^2 k^2}{2m} \qquad (1)$$

This is shown very clearly by the nearly-free-electron-like band structure $E(\underline{k})$ of Al - see Figure 1. The energy gaps are due to the weak scattering by the atomic potentials, which we can reproduce with a much weaker pseudopotential. The pseudo-potential can then be treated as a perturbation on the free-electron gas, and to first order the cohesive energy, U_0, relative to separated valence electrons and ions is given by[4]:

$$U_0 = U_{KE} + U_{ES} + U_{XC} \qquad (2)$$

where U_{KE} is the kinetic energy of free electrons, U_{ES} is the electrostatic energy of the uniform electron density interacting with the pseudoions, and U_{XC}, the exchange-correlation energy of the uniform electron gas. This volume-dependent, structure-independent expression gives nearly all the cohesive energy of the simple metals; results are summarised in Table 1. On going to second order in perturbation theory, which describes two scatterings of the electrons by the weak pseudo-ion, we introduce the much smaller structure-dependent terms.

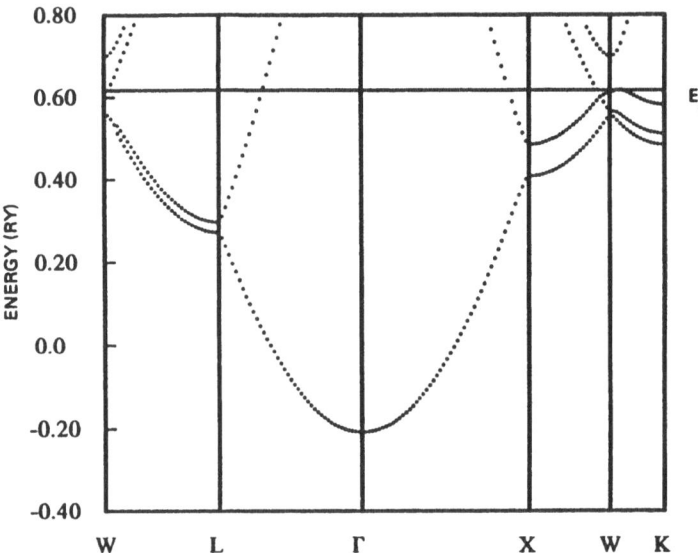

Figure 1   Al band structure showing nearly-free-electron bands with small energy gaps at the Brillouin zone boundaries[1].

Table 1

Observed and calculated values of atomic radius R_a and cohesive energy. Atomic units are used, and the energy is measured (per ion) relative to separated ions and electrons[4]

Element	R_a		-U	
	obs.	calc.	obs.	calc.
Li	3.26	3.76	0.256	0.256
Na	3.93	4.24	0.230	0.228
K	4.86	5.36	0.194	0.184
Be	2.35	3.07	1.130	0.995
Mg	3.34	3.70	0.822	0.852
Zn	2.90	3.09	1.050	0.991
Cd	3.26	3.30	0.933	0.937
Hg	3.35	2.88	1.100	1.050
Ca	4.12	4.48	0.733	0.722
Ba	4.66	5.41	0.617	0.613
Al	2.98	3.26	2.070	1.980
Ga	3.15	3.09	2.205	2.070
In	3.47	3.38	2.040	1.920
Tl	3.58	3.09	2.145	2.070
Si	3.18	2.99	3.920	3.640
Ge	3.31	2.96	3.940	3.660
Sn	3.51	3.26	3.540	3.380
Pb	3.65	3.18	3.620	3.460
As	3.26	2.87	6.375	5.725
Sb	3.65	3.14	5.600	5.300
Bi	3.85	3.18	5.525	5.250
Se	3.51	2.71	8.075	7.100
Te	3.79	3.03	7.820	6.450

Bonding in the transition metals, on the other hand, is dominated by the tightly-bound 3d, 4d or 5d electrons[5]. When the atoms are brought together to make a solid, the atomic d levels broaden out into five bands as shown in Figure 2. Most of the binding energy comes from the change in one-electron energy; maximum binding is predicted where the atom has five d-electrons, in rough agreement with observation. The d-electron wavefunctions and energies can be calculated using tight-binding methods, which provide the zeroth order approximation for the electronic structure of transition metals. By filling up these tight-binding energy bands with the appropriate number of electrons, and adding up the one-electron energies we can then explain the structural trends in the transition metals as illustrated in

Figure 3[6].

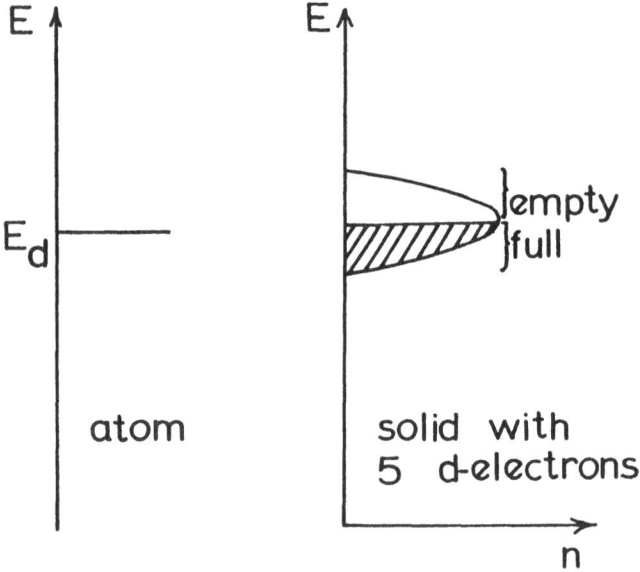

Figure 2 Atomic d levels broaden into bands in the solid. With five or less
d-electrons per atom, all the electrons in the solid have lower energy
than in the isolated atom, but with more than five some are raised
in energy.

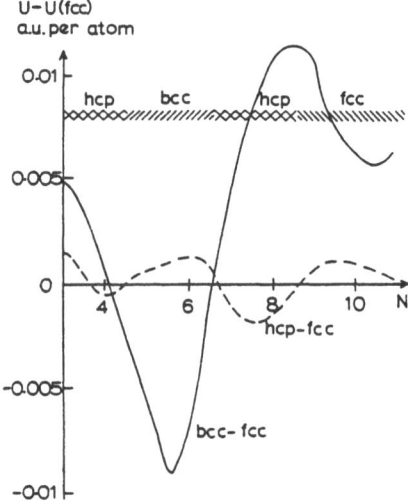

Figure 3 Total energy of bcc and hcp structures relative to fcc in a transition
metal, calculated in a band structure approach, as a function of the
number of d electrons. This reproduces the observed structural
trends[6].

The structure-dependent part of the energy of sp bonded metals, but not the transition metals, can be written in terms of pair potentials between the atoms. Let us start by putting a single perturbing potential in the free-electron gas; this is screened via the wavevector-dependent dielectric function $\varepsilon(q)$, giving a potential in Fourier transform

$$v(q) = v_{ps}(q)/\varepsilon(q) \tag{3}$$

where $v_{ps}(q)$ is the wavevector-dependent potential of the 'pseudo-ion' providing the perturbation. The potential in equation (3) corresponds to the charge density

$$\rho(q) = \frac{q^2}{4\pi} \, v_{ps}(q)/\varepsilon(q) \; . \tag{4}$$

If we now bring up a second pseudo-ion, its energy of interaction with the charge density of the first, screened pseudo-ion is given by:

$$\Phi(r) = \frac{\Omega}{8\pi^3} \int d^3q \, \rho_1(\underline{q}) \, v_2(\underline{q}) \tag{5}$$

where Ω is the atomic volume, ρ_1 is the charge density of the first pseudo-ion, and v_2 is the bare potential of the second pseudo ion. If the second pseudo-ion is at a distance r

$$v_2(\underline{q}) = \exp(i \, \underline{q} \cdot \underline{r}) \, v_{ps}(q) \tag{6}$$

and the interaction energy becomes

$$\Phi(r) = \frac{\Omega}{8\pi^3} \int d^3q \, \frac{q^2}{4\pi} \exp(i \, \underline{q} \cdot \underline{r}) \, v_{ps}(q)^2/\varepsilon(q) \; . \tag{7}$$

This behaves like an effective potential between the two ions, and the structure-dependent part of the energy of a simple metal is given, to second order in perturbation theory, by

$$U_{str} = \tfrac{1}{2} \sum_{i \neq j} \Phi(|\underline{r}_i - \underline{r}_j|) \; . \tag{8}$$

Asymptotically $\Phi(r)$, and the screening charge density $\rho(r)$, show Friedel oscillations, coming from the singularity in ε due to the sharp Fermi surface[4]

$$\rho(r) \sim A \cos(2 \, k_F r)/(2 \, k_F r)^3$$
$$\Phi(r) \sim B \cos(2 \, k_F r)/(2 \, k_F r)^3$$

where k_F is the Fermi wavevector. The Friedel oscillations in the screening charge around point defects like impurities or vacancies can be measured using for example NMR[7]. Similar spin-polarised charge density oscillations surround magnetic atoms in an otherwise non-magnetic substrate, and this gives rise to an oscillatory magnetic interaction between such atoms, for example between Mn atoms in the Heusler alloy Pd_2MnSn. Experimental confirmation of the oscillatory interaction can be obtained by neutron diffraction studies of the spin-wave dispersion. Results are illustrated in Figure 4.

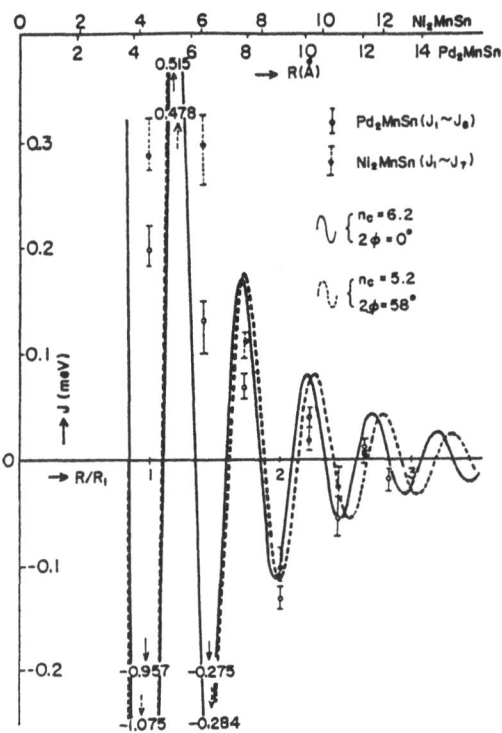

Figure 4 Oscillatory exchange parameter between Mn atoms in Pd_2MnSn and Ni_2MnSn, fitted to spin-wave dispersion results[8].

3. Metallic Structures

In a crystal of a simple metal we have seen that the energy consists of a large volume-dependent, structure-independent term plus a smaller structure-dependent term which can be described, to second order in perturbation theory, in terms of an effective pair potential. The stable crystal structure minimises this pair potential part of the energy at the volume determined by the structure-dependent term. Calculations of the crystal structure of simple metals using this second order perturbation theory approach have been successful for many systems. In practice it is usually easier to calculate the structure-dependent part of the energy in reciprocal space in a perfect crystal, using the Fourier transform of $\Phi(r)$ because of the long-range oscillations in the real-space potential. Some results for the structural energies of the divalent metals are shown in Table 2. These give correctly the hcp structure for Be, Mg, Zn and Cd with the correct trend in c/a ratio[9] (Hg is predicted to have a very distorted hcp structure rather than the observed structure). The pair potentials calculated by Weaire[4] appear to be useful for explaining the trends in c/a ratios: Mg has an ideal hcp structure because the nearest neighbours fall in the minimum of the pair potential at equilibrium volume, whereas in Cd the minimum is at a shorter internuclear separation, and the ideal hcp structure distorts to lower the energy at fixed volume, as shown

in Figure 5[4].   Similarly in the trivalent metals it is found that fcc Al is
stable, corresponding to nearest neighbours falling in the first minimum of the
pair potential at the equilibrium lattice spacing[9,10] - see Figure 6. On the other
hand, the minimum in the pair potential of Ga is at a shorter internuclear spacing
as shown in Figure 7, and as the atoms cannot simply move into the minimum and
maintain the fcc structure at the equilibrium volume, a complicated crystal
structure develops which puts atoms in the pair potential minimum[11].   Pseudopotential perturbation theory has been successfully used to explain the pressure-
induced phase transition in Ga[11,12], and also in Sn[13].

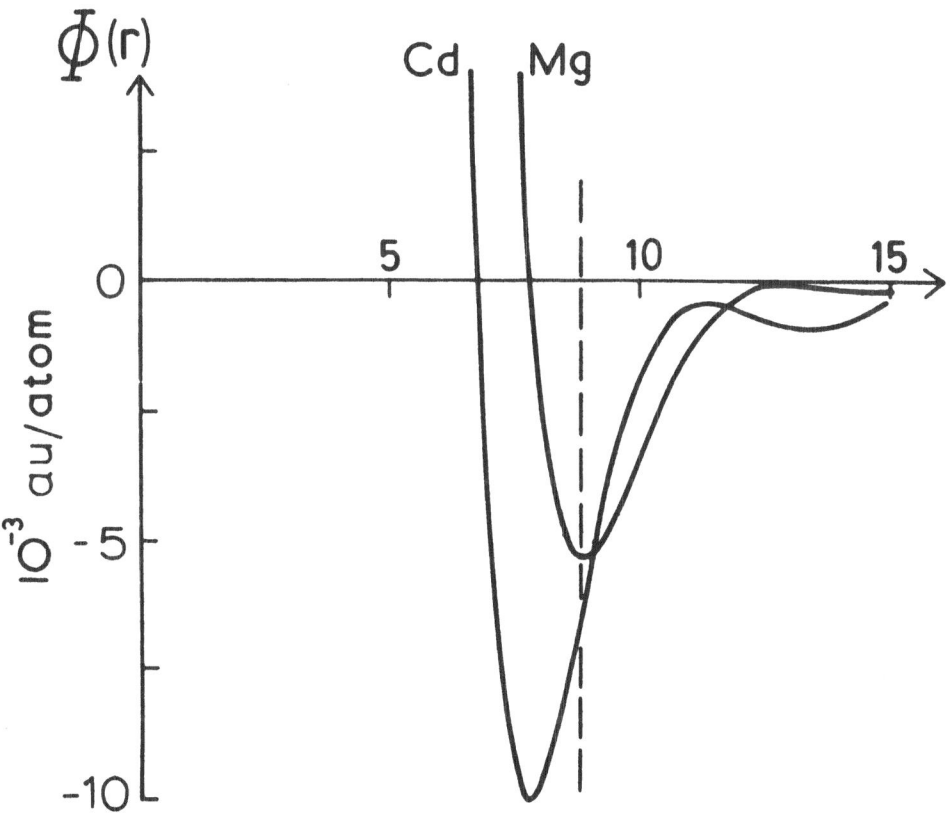

Figure 5 $\Phi(r)$ for Mg and Cd[4].   The dotted line shows the nearest neighbour
distances for hcp with the ideal axial ratio.

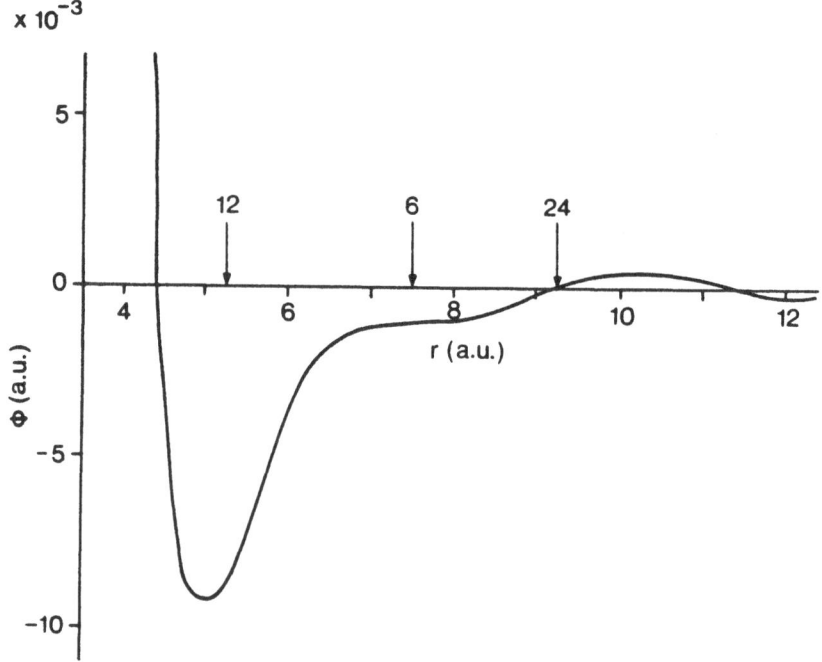

<u>Figure 6</u> $\Phi(r)$ for Al, with the interatomic distances in the observed fcc structure, at the observed atomic volume for Al[10].

Table 2

Observed structure and hcp c/a ratio of the divalent metals, with calculated structural energies and c/a ratio at which hcp energy is minimised[9]. Energies are in units of $Z^2/2R_a$, where Z is the valence and R_a is the atomic radius.

	Observed	Calculated				
		hcp	c/a	fcc	bcc	A10
Be	hcp 1.568	-1.97887	1.584	-1.95657	-1.96850	-1.91649
Mg	hcp 1.6235	-1.80446	1.625	-1.80333	-1.80225	-1.79627
Zn	hcp 1.85	-1.80676	1.684	-1.80663	-1.80451	-1.80406
Cd	hcp 1.886	-1.81606	2.30	-1.81404	-1.81164	-1.81354
Hg	A10	-1.87396	3.2	-1.85668	-1.85002	-1.85954

4.   Alloys

Trends in alloying behaviour of simple metals can be explained using pseudo-potentials and second order perturbation theory. For example the Hume-Rothery rules relating alloy crystal structure to the electron-atom ratio come from the asymptotic oscillations in Φ[4,14,15]. Hafner has shown that the Laves-type structures of some AB_2 alloys like Na_2K, K_2Cs and $CaMg_2$ are stable because, as shown

123

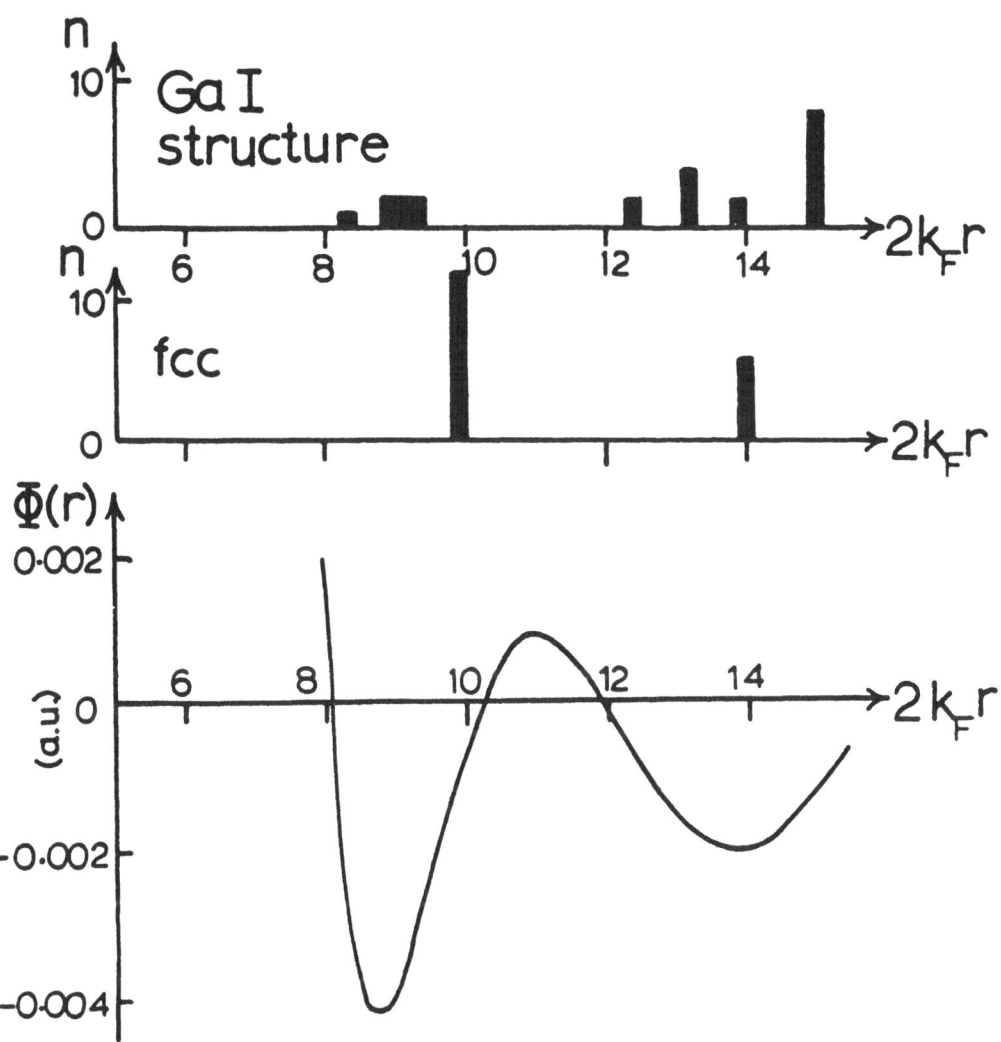

Figure 7   Φ(r) for Ga, with the interatomic distances in
fcc and the Ga I structure, at the observed
atomic volume[11].

in Figure 8, the nearest neighbours fall right in the minima of the pair potentials[16,17]. In fact, using the pseudopotential approach he has been able to reconstruct the whole Ca-Mg phase diagram, explaining the propensity of this system to form metallic glasses. The ordering part of the energy of an alloy can be described in terms of the "alloying potential" Φ_a, which depends on the difference pseudopotential[18]

$$U_{ord} = c^2 \sum_{A,A'} \Phi_a(|\underline{r}_A - \underline{r}'_A|) + (1-c)^2 \sum_{B,B'} \Phi_a(|\underline{r}_B - \underline{r}'_B|)$$

$$- 2c(1-c) \sum_{A,B} \Phi_a(|\underline{r}_A - \underline{r}_B|) . \tag{9}$$

A positive value of Φ_a at the nearest neighbour distance obviously favours an ordered alloy with unlike atoms as neighbours, as in the Hg-Mg system shown in Figure 9[19]. In terms of Φ_a Inglesfield explained the increasing tendency to form intermetallic compounds as the electronegativity difference increases in the alloys Cd-Hg, Cd-Mg and Hg-Mg, and also trends in ordering energy[20].

5. Defects and Distortions

As well as explaining perfect crystal structures, pair potentials can be used to study deviations from the perfect lattice; in this context, however, two points should be noted:

(i) The potentials refer to systems at fixed volume

(ii) Potentials are obtained from second order perturbation theory, and are only valid as long as the electron density is reasonably uniform.

Lattice vibrations are an example of a constant volume distortion of the perfect crystal structure, and the first principles pseudopotentials give good phonon spectra as illustrated in Figure 10[4,10,21]. Phonon spectra are usually calculated with the Fourier transform of Φ; but the real space potential is useful for calculating anharmonic phonons using molecular dynamics simulations[22]. The atomic displacements need not be small, as long as the density is fairly constant, and liquid structure factors found from these pair interactions are also in good agreement with experiment as illustrated in Figure 11[23]; structure factors are obtained either using many-body liquid theory[23], or again using molecular dynamics simulations[22].

Stacking fault energies can be calculated quite successfully[7,14], as these planar defects do not greatly affect the electron density; for example the pair potential for Al derived by Dagens et al[21], which gives a good phonon spectrum, has been successfully used in computer simulation experiments to calculated the stacking fault energy and the structure of grain boundaries in Al[24]. On the other hand the core of an edge dislocation is quite a drastic perturbation on the system, for which the validity of second order perturbation theory is not clearly demonstrated.

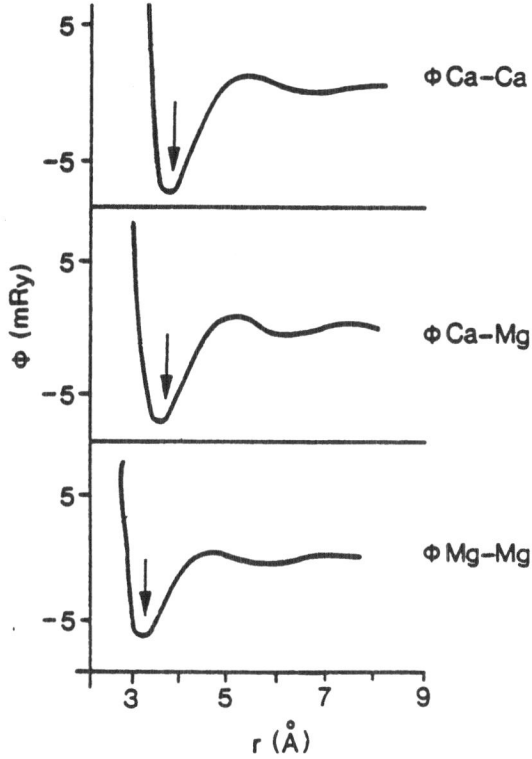

Figure 8 Pair potentials for Ca-Ca, Ca-Mg and Mg-Mg interactions in $CaMg_2$[17]. Arrows show the nearest neighbour distances in the Laves phase at this composition.

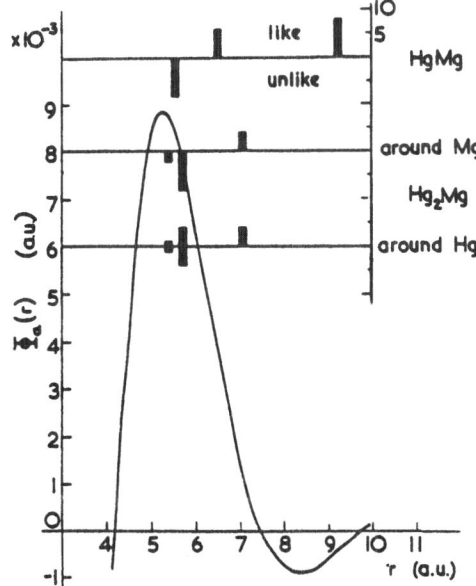

Figure 9 $\Phi_a(r)$ for Hg-Mg with the number and positions of like and unlike neighbours in Hg Mg and Hg_2 Mg[19].

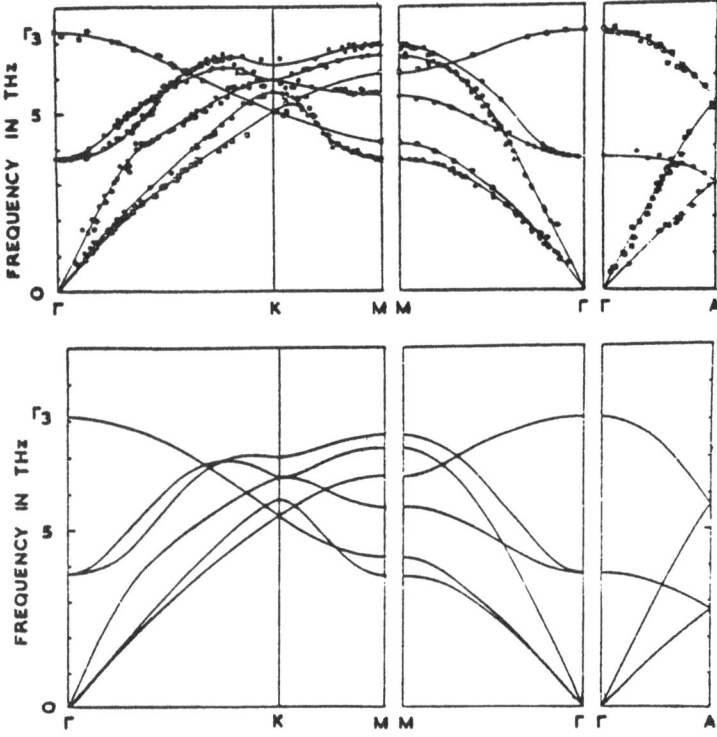

Figure 10   Phonon spectrum for Mg$^{(4)}$.   Top - observed; bottom - calculated.

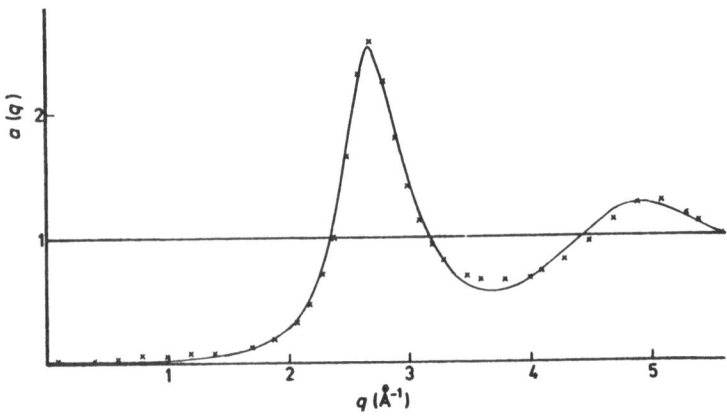

Figure 11   Liquid structure factor for Al$^{(23)}$.   Solid line - experiment;
crosses - calculated.

Vacancies, in which an atom is removed from a lattice site, have been extensively studied using pseudopotentials and pair interactions[25,26,27]. The calculations are best considered in three stages. First, we remove an atom from an (N+1)-atom crystal, and calculate the energy of the unrelaxed N-atom crystal with a vacancy; as the electron density is less than in a perfect N-atom crystal, there is a contribution to the energy change from volume-dependent forces as well as from the pair interactions. Next, we allow the atoms to relax around the vacancy under the influence of the pair forces, and finally we allow the volume to change. Using a simplified pseudopotential fitted to the equilibrium lattice spacing and bulk modulus, Popović et al calculated vacancy formation energies in this way which are in good agreement with experiment[26]. In a more recent molecular dynamics calculation using first-principle pair potentials, good agreement was obtained for the vacancy formation energy of the alkali metals[24] but rather poor agreement was found for Al[22]. Results are presented in Table 3. The reason for the discrepancy in the case of Al is that a vacancy represents a major perturbation on the system with relatively large charge inhomogeneity; this leads to a breakdown in second-order perturbation theory, and a change in the screening of the atomic pseudopotentials. The result is that pair interactions appropriate to bulk Al are no longer applicable. However, possibly the best example of a defect with large charge inhomogeneity is a metal surface, and pair interactions are quite incapable of explaining the surface contraction of about 10% on open metal surfaces. This is due to a change in shape of the screening cloud at the surface, resulting in an inward electrostatic force on the surface atoms[28,29].

Table 3
Vacancy formation energy in eV[22,27]

	Theory	Experiment
Na	0.25	0.36
Al	0.19	0.66

6. Discussion and Summary

The approach to metallic cohesion presented in this article has rested largely on pseudopotential theory, which gives rise, in a second order perturbation treatment, to pair potential models. However, we should emphasize that the pseudo-potential v_{ps} in simple metals and hence the pair potential, is not unique, as there are many different pseudopotentials which scatter the electrons in the same way[30,31]. OPW-type pseudopotentials can be used, in which the rapid oscillations in the valence wavefunctions in the core region of the atom are removed by projecting out the core wavefunctions[32]; model pseudopotentials in which energy- and angular momentum-dependent square wells are fitted to atomic spectra are also widely used[9]. All pseudopotentials which scatter the electrons correctly must give the same answer if calculations are carried out to all orders of perturbation theory, but if we stop at second order - as in pair potential calculations - one potential may well be the best, and the ability to reproduce phonon spectra seems a reasonable criterion. In the one-electron theory to which these pseudopotentials are applicable, the interaction between the conduction electrons is taken into account by screening v_{ps} with the dielectric function ε. (In fact in the full theory each angular momentum component must be screened separately[33]). There are several dielectric functions available which go beyond the Hartree approximation, and it turns out that the pair potential is very sensitive to the choice of ε[10,22,34]; the first minimum in Φ for Al can actually go positive with some choice of ε. The phonon spectrum is more sensitive than the crystal structure to the effects of the dielectric function[10].

A pair potential obtained from a properly screened pseudopotential has a firm theoretical basis in simple metals, and is greatly to be preferred to empirical interatomic potentials. Unfortunately there is no such basis for using pair potentials in transition metals, and their crystal structures, which result from multiple d-electron scattering can certainly not be treated in this way. However, there are some cases involving small perturbations on the perfect crystal in which pair potentials can, perhaps, be used in transition metals. Small atomic displacements at constant volume can be written in terms of phonon amplitudes, by Fourier decomposition, and a pair potential fitted to phonon frequencies should give the energy change correctly. In general this will not hold for arbitrary atomic displacements, as in liquids or defects, though model calculations based on pair potentials may be useful to show the possibilities for behaviour in real systems[35].

References

1. Moruzzi, V.L., Janak, J.F. and Williams, A.R., Calculated Electronic Properties of Metals (New York: Pergamon Press, 1978).

2. Williams, A.R., Kübler, J. and Gelatt, C.D., Phys. Rev. B19, 6094 (1979).

3. Kerber, C.P., Ho, K.M. and Cohen, M.L., Phys. Rev. B18, 5473 (1978).

4. Heine, V. and Weaire, D., Solid State Phys., Eds. F. Seitz and D. Turnbull, 24, 249 (1970).

5. Friedel, J., in The Physics of Metals. Ed. J.M. Ziman (Cambridge: Cambridge University Press, 1970).

6. Pettifor, D.G., J. Phys. C., Solid State Phys. 3, 367 (1970).

7. Blandin, A., Phys. Bull. 31, 93 (1980).

8. Noda, Y. and Ishikawa, Y., J. Phys. Soc. Japan 40, 690 (1976).

9. Williams, A.R. and Appapillai, M., J. Phys. F: Metal Phys. $\underline{3}$, 772 (1973).

10. Hafner, J. and Schmuck, P., Phys. Rev. $\underline{B9}$, 4138 (1974).

11. Inglesfield, J.E., J. Phys. C. Solid State Phys. $\underline{1}$, 1337 (1968).

12. Hafner, J., Anzeiger der Österr. Akad. der Wissensch. $\underline{4}$, 37 (1975).

13. Hafner, J., Phys. Rev. $\underline{B10}$, 4151 (1974).

14. Blandin, A., in Phase Stability in Metals and Allows, Eds. P.S. Rudman, J. Stringer and R.I. Jaffee (New York: McGraw-Hill, 1967).

15. Hafner, J., J. Phys. F: Metal Phys. $\underline{6}$, 1243 (1976).

16. Hafner, J., Phys. Rev. $\underline{B15}$, 617 (1977).

17. Hafner, J., Phys. Rev. $\underline{B21}$, 406 (1980).

18. Inglesfield, J.E., J. Phys. C. Solid State Phys. $\underline{2}$, 1285 (1969).

19. Inglesfield, J.E., J. Phys. C. Solid State Phys. $\underline{2}$, 1293 (1969).

20. Inglesfield, J.E., Acta. Met. $\underline{17}$, 1395 (1969).

21. Dagens, L., Rasolt, M. and Taylor, R., Phys. Rev. $\underline{B11}$, 2726 (1975).

22. Jacucci, G., Taylor, R., Tenenbaum, A. and van Doan, N., J. Phys. F: Metal Phys. $\underline{11}$, 793 (1981).

23. Kumaravadivel, R. and Evans, R., J. Phys. C: Solid State Phys. $\underline{9}$, 3877 (1976).

24. Pond, R.C. and Vitek, V., Proc. Roy. Soc. $\underline{A357}$, 453 (1977).

25. Ho, P.S., in Interatomic Potentials and Simulation of Lattice Defects. Eds. P.C. Gehlen, J.R. Beller and R.I. Jaffee (New York: Plenum Press, 1972).

26. Popović, Z.D., Carbotte, J.P. and Piercy, G.R., J. Phys. F: Metal Phys. $\underline{4}$, 351 (1974).

27. Jacucci, G. and Taylor, R., J. Phys. F: Metal Phys. $\underline{9}$, 1489 (1979).

28. Evans, R. and Finnis, M.W., J. Phys. F: Metal Phys. $\underline{6}$, 483 (1976).

29. Finnis, M.W. and Heine, V., J. Phys. F: Metal Phys. $\underline{4}$, L37 (1974).

30. Heine, V., Solid State Phys., Eds. F. Seitz and D. Turnbull, $\underline{24}$, 1 (1970).

31. Cohen, M.L. and Heine, V., Solid State Phys., Eds. F. Seitz and D. Turnbull, $\underline{24}$, 37 (1970).

32. Hafner, J. and Nowotny, H., Phys. Letts. $\underline{37A}$, 335 (1971).

33. Shaw, R.W. and Harrison, W.A., Phys. Rev. $\underline{163}$, 604 (1967).

34. Jacucci, G. and Taylor, R., J. Phys. F: Metal Phys. $\underline{11}$, 787 (1981).

35. Nemanich, R.J., Tsai, C.C. and Connell, G.A.N., Phys. Rev. Letts. $\underline{44}$, 273 (1980).

CHAPTER 10

INTERIONIC POTENTIALS IN IONIC SOLIDS

by

C.R.A. Catlow
Department of Chemistry, University College, London
20 Gordon Street, London WC1H 0AJ

M. Dixon
Theoretical Physics Department, University of Oxford
attached to Theoretical Physics Division, A.E.R.E. Harwell, Oxon OX11 ORA

and

W.C. Mackrodt
I.C.I., PLC, The Heath, Runcorn, Cheshire WA7 4QE

1. Introduction

Lattice modelling techniques have enjoyed greatest success in studies of ionic
and semi-ionic materials.   This is primarily due to the development for these
systems of potential models which can account for most of the observed properties
of the crystal and which have a sufficiently simple functional form to be
incorporated in computer codes of the type discussed in Chapter (1). For these
materials, therefore, we can make a detailed comparison of the calculated perfect
and defect lattice properties obtained using different types of potential model;
and we can examine particular effects such as the inclusion of many-body terms
and ionic polarisability.   This chapter aims to provide such a comparison after
outlining the detailed features of the different models in current usage. We start,
therefore, by describing the basic features of potentials used in modelling ionic
crystals, concentrating on pair potential models, although we present a brief
account of many body terms.   This is followed by a detailed discussion of the
treatment of ionic polarisability, after which the status of currently available
models for halide and oxide systems is reviewed. We concentrate throughout on the
types of potential model that may be implemented most successfully in the computer
codes of the type discussed in Chapter (1).

2. Basic features of potential models

Ionic solids are most often described by models which assign charges (which
are integral for the 'fully' ionic description) to point entities between which
short range interactions are specified; the latter describe the effects of the
overlap of the electron charge clouds of the interacting ions.   Polarisability
can be included using a variety of methods which are discussed later in
section 6.   To date, the only models that have been successfully incorporated in
computer codes of the type discussed in Chapters (1) and (5) are of the
central-force type, in which the total potential energy, $V(\underline{r}_1...\underline{r}_N)$ of an assembly
of N particles with coordinates $r_1...r_N$, is written as

$$V(\underline{r}_1 \cdots \underline{r}_N) = \sum_{i>j} V_{ij}(|\underline{r}_i - \underline{r}_j|) \ . \tag{1}$$

Thus V is taken as a sum of <u>pair interaction</u> terms each of which is <u>dependent only on the distance between the particles</u>. The omission of angle-dependent forces and many-body effects generally results in a severe restriction on the application of these models to more covalent systems as discussed in Chapters (12) and (15). We return to this problem in section 5.

The short range component of the two-body potential is commonly described by a simple Born-Mayer or Buckingham function

$$V(r) = A \exp(-r/\rho) - Cr^{-6} \tag{2}$$

although functional forms such as the Morse potential have been used. The use of such functions necessitates fixing the parameters such as A, ρ and C in equation (2), and the procedures used in determining their optimum values are discussed in the next section.

3. Derivation of short range potentials

Two basic procedures are generally employed here. The first uses empirical data for the perfect lattice to determine the parameters in equation (2); the second attempts to calculate the interaction between the ions directly using approximate quantum mechanical methods. Both have been extensively used in studies of ionic crystals.

3.1 Empirical parameterisation

Given a specified set of appropriate pair potentials for the various inter-actions in a crystal, it is possible, as discussed in Chapter (1), to calculate straightforwardly its elastic, dielectric and piezo-electric constants as well as the phonon dispersion curves. Certain of the quantities, notably the dielectric constants, require specification of the polarisation parameters, the determination of which we discuss in section 6. The methods used for calculating crystal properties are described in detail in Chapter (1). It is important to note that the elastic and dielectric constants and the lattice vibrational frequencies do not depend on the interatomic potentials <u>directly</u>, but rather on the first and second derivatives of the potentials with respect to the interionic separation. The extraction of information on the potential itself, therefore, requires that the analytical forms used to represent the interaction, reliably describe the variation of the potential with interionic separation. Cohesive energies do, of course, include direct information about the potential, but this is of limited value for the extraction of short range potentials, for the lattice energy is normally dominated by the Coulomb term.

The procedure adopted in the empirical parameterisation of potentials is to 'invert' the calculation of crystal data from specified functional forms - a procedure which is also referred to in section 6 of Chapter (8). Thus the potential parameters are adjusted, generally by means of a least squares fitting procedure, to reproduce as closely as possible the measured crystal data. This procedure has now been automated by linking the PLUTO code, discussed in Chapter (1), to appropriate least squares fitting routines. In practice, elastic and dielectric constants are most commonly used in such fitting procedures, and the resulting potentials then checked by calculating the phonon dispersion curves.

Potentials fitted in this way can give an impressive measure of agreement with experimental data, as demonstrated in Tables 1 and 2, which compare calculated and experimental crystal properties for CaF_2 and for TiO_2 and $\alpha-Al_2O_3$. The last two are complex, low symmetry materials the accurate description of which increases our confidence in the reliability of these procedures. Further information on potentials can be obtained from structural data: for in addition to the properties discussed above, the minimisation of bulk and internal strains acting on the unit cell may be calculated. Such strains will arise when the potential used is not fully compatible with the observed structure. Thus the variation of short range parameters to minimise these strains corresponds essentially to 'fitting' to the crystal structure. This procedure is particularly powerful in the case of low symmetry materials where the complexity of the structure provides a rich source of information since a given pair interaction is effectively sampled at several interionic separations in the perfect lattice. This has been exploited in recent work on mineral systems[1], as discussed in Chapter (15).

Empirical parameterisations have now been applied successfully to a wide range of oxide and halide crystals, including the alkali[2] and alkaline earth halides[3] and transition metal[4], alkaline-earth[5] and actinide oxides[6]; and a compilation of empirical (and other) potentials for ionic crystals is now available[7]. The method has, however, one weakness, for it samples ionic interactions only at their values in the perfect lattice whereas studies of defective lattices require accurate potentials for a greater range of internuclear separations. The reliability of the potentials at distances which are appropriate to the defective lattice, therefore, depends on the validity of particular functional forms for the interionic potentials. The problem is less acute for low symmetry materials where, as commented above, a given interaction is sampled over a range of separations in the perfect lattice. A similar effect can be achieved for high symmetry structures when a whole class of isostructural crystals is considered, enabling the same potential to be sampled over a range of separations in different crystals. A good example of this is provided by the alkali halides, where, for example, the lattice $Cl^-...Cl^-$ separation varies from 3.63Å in LiCl to 4.65Å in RbCl.

Table 1

Calculated and observed crystal properties for CaF_2 (after Catlow and Norgett[22]

who give references to experimental data)

	Expt	Calculated value
r_0	2.722	(2.722)
C_{11}	17.124	(16.9)
C_{12}	4.675	(4.80)
C_{44}	3.624	3.23
ε_0	6.47	(6.42)
ε_∞	2.05	(2.01)
ω_{TO}	270.0	(259.2)
ω_R	330.5	310.7
α_+	0.979	(0.984)
α_-	0.759	(0.765)
C_{111}	-124.6	-107.8
C_{112}	-40.0	-33.8
C_{123}	-25.4	-17.5
C_{144}	-12.4	-9.3
C_{166}	-21.4	-23.2
C_{456}	-7.5	-7.8
H_L	-26.76	-28.06

Key: r_0 - lattice constant (Angstroms)

C_{11}, C_{12}, C_{44} - second order elastic constants (10^{11} dyne cm^{-2})

$\varepsilon_0, \varepsilon_\infty$ - dielectric constants at zero and high frequency

ω_{TO}, ω_R - transverse optic and Raman frequencies (cm^{-1})

α_+, α_- - cation and anion polarizabilities ($Å^3$)

$C_{111}, C_{112}, C_{123}, C_{144}, C_{166}, C_{456}$ - third order elastic constants

(10^{11} dyne cm^{-2})

H_L - lattice formation energy (eV).

(Bracketed values used in fitting)

Table 2

Calculated and observed crystal properties for Al_2O_3 and TiO_2 (after Catlow

et al[44] who give references to experimental data)

Lattice	α-Al_2O_3		TiO_2	
Properties/System	Calculated	Observed	Calculated*	Observed
Lattice energy (eV)	-160.21	-160.4	-109.90	126.0
$C_{11}(10^{11}$ dyn cm$^2)$	42.96	49.69	25.33	27.01
C_{12}	15.48	16.36	17.80	17.66
elastic $\quad$ C_{13}	12.72	11.09	20.90	14.80
constants $\quad$ C_{33}	50.23	49.8	77.92	48.19
C_{14}	-2.99	-2.35	-	-
C_{44}	16.66	14.74	9.22	12.39
C_{66}	13.70	$(C_{11}-C_{12})/2$	22.12	19.30
dielectric $\quad$ ε_{11}^0	9.38	9.34	94.76	86.
constants $\quad$ ε_{33}^0	11.52	11.54	157.32	170.
ε_{11}^∞	2.08	3.1	6.28	6.83
ε_{33}^∞	2.02		7.99	8.43

*Calculated value for fitted potentials

This has been exploited by Catlow et al[2] to derive an empirical $Cl^-...Cl^-$ potential which is valid over this range of separations. The method does, of course, assume transferability of the potential between different crystal environments.

However, the lack of information on potentials over a wide range of inter-nuclear separations remains a basic inadequacy of empirical parameterisation. This has led to the development of non-empirical procedures as discussed in the next section.

3.2 Non-empirical methods

As pointed out in Chapter (8), pair potentials as such in solids, cannot be obtained directly from full quantum mechanical calculation. However, they can be extracted from approximate formulations. The most useful of these for ionic interactions (in the solid state) is that based on the density functional treatment of the uniform electron gas[8-11]. There have been a number of more or less equivalent approaches to this problem, and perhaps the most convenient, particularly from a computational point of view, is that suggested by Gordon and Kim[12]. Recalling equations (2) and (5) in Chapter (8), the energy of a (closed shell) ion (or atom) is written as

$$E[\rho] = C_k \int [\rho(\underline{r})]^{5/3} \, d\underline{r} + C_e \int [\rho(\underline{r})]^{4/3} \, d\underline{r} - Z \int \rho(\underline{r})/r \, d\underline{r}$$

$$+ \tfrac{1}{2} \int \int \rho(\underline{r}) \, \rho(\underline{r}')/|\underline{r}-\underline{r}'| \, d\underline{r} \, d\underline{r}' + \int \epsilon_c \, [\rho(\underline{r})] \, \rho(\underline{r}) \, d\underline{r} \qquad (3)$$

in which Z is the nuclear charge and

$$C_k = (3/10)(3\pi^2)^{2/3}; \quad C_e = -(3/4)(3/\pi)^{1/3} \quad .$$

The first four terms represent the kinetic, exchange and coulomb energies respectively, while the last term is an approximation to the electron pair correlation energy which Gordon and Kim[12] suggest can be determined simply by interpolating between the high and low density limits for a homogeneous electron gas. For a pair of ions, AB, the total electron density, $\rho_{AB}(r)$ is assumed to be the superposable <u>sum</u> of the separated ion densities, $\rho_A(r)$ and $\rho_B(r)$. The total energy of AB, therefore, is given by

$$E_{AB} = E[\rho_{AB}] = E[\rho_A + \rho_B] \qquad (4)$$

so that the <u>interaction energies</u>, V_{AB}, can be written as

$$V_{AB} = E[\rho_A + \rho_B] - E[\rho_A] - E[\rho_B] \quad . \qquad (5)$$

Now V_{AB}, particularly at or near the minimum energy separation, is a small difference ($\sim$ 1 eV) between two considerably larger quantities ($\sim$ 6000-9000 eV). To minimise numerical errors in the evaluation of V_{AB}, therefore, Gordon and Kim[12] proposed a re-arrangement of the integral expressions in (3) to give, the final form for V_{AB} as:

$$V_{AB}(R) = Z_A Z_B/R - Z_B \int \rho_A(\underline{r}_1)/r_{1B} \, d\underline{r}_1 - Z_A \int \rho_B(\underline{r}_2)/r_{2A} \, d\underline{r}_2$$

$$+ \int \int \rho_A(\underline{r}_1) \, \rho_B(\underline{r}_2)/r_{12} \, d\underline{r}_1 \, d\underline{r}_2$$

$$+ \int \{ [\rho_A(\underline{r}) + \rho_B(\underline{r})] \, E_G[\rho_A(\underline{r}) + \rho_B(\underline{r})]$$

$$- \rho_A(\underline{r}) \, E_G[\rho_A(\underline{r})] - \rho_B(\underline{r}) \, E_G[\rho_B(\underline{r})] \} \, d\underline{r} \qquad (6)$$

in which

$$E_G[\rho(\underline{r})] = C_k[\rho(\underline{r})]^{2/3} + C_e[\rho(\underline{r})]^{1/3} + E_c[\rho(\underline{r})] \qquad (7)$$

where r, r_{1B}, r_{2A} and r_{12} are functions of the internuclear separation, R. These are the expressions most commonly used.

Subsequently, there have been a number of refinements to this approach. Rae[13] has pointed out that the original expression for the exchange energy includes a self-energy contribution, which though negligible for an infinite electron gas, is significant for a small, finite number of electrons. However, a simple correction for this can be made by replacing the exchange term in the Gordon and Kim scheme, V_e, by a modified contribution, V_e', given by[13]

$$V_e' = V_e [1 - 8/3 \ \delta + 2\delta^2 + 1/3\delta^4] \tag{8}$$

in which δ is a solution of

$$(4N)^{-1} = \delta^3 (1 - 9/8 \ \delta + 1/4 \ \delta^3) \tag{9}$$

and N is the number of electrons. This modification gives rise to a number of interesting effects, including the removal of spurious long-range minima which appear in the non-coulombic contribution to the interaction potentials for closed shell systems[14]. It also leads to a reduction in the calculated lattice cohesive energy of crystals such as NaCl, CaF_2 and MgO by up to 10% if free-ion densities are used.

Mackrodt and Stewart[15] emphasised that for anions, and in particular for O^{2-} which is <u>unbound</u> as a free ion, electron densities appropriate to the crystal should be used. A simple procedure for obtaining these is to solve the corresponding atomic Hartree-Fock equations (numerically or otherwise) in an external spherical potential the strength of which is equal to the Madelung potential at the anion sub-lattice for the crystal in question. To a certain extent, therefore, potentials derived from such densities are 'solid state' pair potentials which should be more appropriate for the calculation of perfect lattice and defect properties than those derived from free ion densities. To date a large number of potentials have been obtained using this procedure, mainly for oxides: the majority are contained in two recent compilations by Stoneham[7] and Colbourn et al[16]. We discuss the current status of such potentials in section 7.

4.   van der Waals interactions

Dispersive interactions, whose importance has been recognised for nearly fifty years now, arise from the correlated motions of electrons on different atomic (or molecular) centres. This correlation, which is due to the Coulombic inter- action of the electrons, results in an <u>instantaneous</u> dipole on each of the interacting species. The interaction of these dipoles and of the higher-order multipoles which they induce gives rise to what is usually referred to as the van der Waals energy, Φ_v. It can be written within the framework of perturbation theory as an asymptotic expansion of the form

$$\Phi_v(r) = r^{-6} \sum_{n=0} C_n \ r^{-2n} \tag{10}$$

where C_n are the so-called van der Waals coefficients.

From the above description of its origins, it is clear that the van der Waals interaction is essentially a polarisation phenomenon: indeed, the lowest order, or 'London' energy, $\Phi_L(r)$ can be written (exactly) in the form[17]

$$\Phi_L(r) = (3\hbar/\pi r^6) \int_0^\infty \alpha_A(i\omega) \, \alpha_B(i\omega) \, d\omega \qquad (11)$$

in which α_A and α_B are the polarisabilities of the two species at imaginary frequency ($i\omega$). Despite this, however, Φ_V is most often (and most conveniently) added to the short range (or strictly speaking the non-coulombic) part of the total interaction, V, which derives from exchange forces. Moreover, in practice, it is often difficult to separate the van der Waals from other attractive terms in the potential.

Estimates of the coefficients, C_n, have been available for some time, following the initial work of Mayer[18]: in general, however, the values are poorly known. Early work attempted to determine these coefficients from optical adsorption data, but there are two difficulties with such an approach. First, the energies of the optical excitations of anions and cations often overlap; second, it is uncertain as to how much adjustment should be made for 'local field' corrections in an ionic solid. Empirical estimates based on the well-known formula

$$\Phi_L(r) = -3/2 \left[\frac{\delta_A \, \delta_B}{\delta_A + \delta_B} \right] \frac{\alpha_A \alpha_B}{r^6} \qquad (12)$$

can be made, but these depend on the choice made for the average polarisabilities, α_A and α_B and the effective 'excitation' energies δ_A and δ_B. Consequently, there are large variations between the van der Waals coefficients reported by various authors using this approach.

It appears at present that methods based on 'effective' ionic polarisabilities and excitation energies, are unable to yield realistic values of the van der Waals coefficients. However, work is still in progress on this problem. A hopeful development is the application of ab initio quantum mechanical methods which include the calculation of correlation effects by means of Configuration Interaction and other techniques. With the introduction of even more powerful computers such calculations now become feasible, at least for crystals containing the lighter elements. Recent work along these lines has been reported by Andzelm and Piela[20] on LiF and NaF.

In view of the problem associated with the reliable theoretical estimation of van der Waals energies, many recent studies have treated the coefficients, C_n, as variable parameters which are determined by fitting to empirical data[25], as mentioned above, or to a calculated interionic potential[20]. As noted in section 3 the coefficient of the r^{-6} term in the Buckingham potential is generally treated purely as an empirical parameter. In this case it is highly doubtful whether the resulting parameters reflect the contribution of genuine dispersion effects; it is more likely that they include a number of contributions including covalency. Moreover, the coefficients of r^{-6} terms derived from fitting to calculated Hartree-Fock potentials[21] _cannot_ refer to dispersion forces since electron correlation is not included in calculations at this level of approximation. Electron correlation is included in some formulations of the electron gas model (see Chapter (8) and section (3.2) of the present chapter), but it is by no means clear as to how dispersion energies can be extracted from potentials based on this method. In this connection it is worth noting that a common feature of empirical potentials is the occurrence of large coefficients of the r^{-6} term for second neighbour interactions. Examples of this are provided by CaF_2[3] and UO_2[6], for which the attractive term for the $F^-...F^-$ and $0^{2-}...0^{2-}$ interaction respectively are approximately ten times the Mayer values. No clear explanation of this phenomenon is available, and it has been speculated[32] that there is possible contribution of covalency to the anion-anion short range potential.

In summary, then, the study of dispersion energies in ionic solids is far from complete and at present the picture is confused. Often the very magnitudes are in doubt, and in the case of empirical potentials there is no convincing interpretation of the r^{-6} term itself. Reliable and accurate theoretical calculations would clearly be of considerable value.

5.   Many body effects

Chapter (8) has stressed the importance of many body effects in covalent and metallic crystals. Here we wish to draw attention to the ways in which these effects have been treated in ionic crystals, where, although they are of lesser importance, they do have an observable influence on elastic[23] and dynamic properties[24]. Our account is brief, as, to date, the various models used in the study of elastic and dynamic properties have not been successfully incorporated into the most recent lattice simulation codes of the type referred to in Chapter (1). We should emphasise, however, that the extension of these codes and the procedures therein to more covalent systems will require the development of simulation techniques which include many body effects.

The treatment of many body effects is closely related to the models used to describe ionic polarisability, a topic to which we return later in this chapter. Here we draw attention to two important approaches to the many body problem itself.   The first is the so-called charge transfer model, the physical basis of which is the deformation of the electron charge clouds by ionic displacements, which in turn causes a transfer of charge between the ions thereby modifying the interaction of a given ion with all the others.   This approach has enjoyed considerable success when applied to the alkali halides[24,25], and is easily combined with suitable treatments of ionic polarisability.

The second apporach to which we refer is the bond charge model, due originally to Phillips[26], and designed specifically for covalent systems; reference has already been made to this model in Chapter (8). In essence it represents the bonding electron density by a charge midway between the bonded atoms (or ions). The model has been adapted to lattice dynamical calculations by Martin[27,28] who has applied it to a study of MgO. However, it is of potentially greater value in treating semi-ionic systems such as quartz and the silicates[1].

This concludes our account of short range interactions. We continue with a discussion of the methods used to describe the electronic polarisation of ions. An extensive and detailed review of many-body contributions to the lattice dynamical properties of ionic crystals, is given by Singh[29]. As noted the treatment of many-body theory is intimately related to the models used to describe the electronic polarisation of ions.

6. Ionic polarisation

The account of ion polarisation presented here will concentrate on phenomenological models. Such models are justified largely by the extent to which they can be used both to interpret and predict experimental data. However, we begin by considering the physical origins of electronic polarisability and its macroscopic manifestations.

For present purposes polarisability is related to the induction of a dipole moment in a crystal by the application of an electric field. The dipole moment per unit volume is defined as the polarisation, the magnitude of which depends on the frequency of the applied field which can be expressed in terms of the frequency dependent susceptibility $\chi(\omega)$. The dielectric function, ε, in turn is frequency dependent and given by

$$\varepsilon(\omega) = 1 + 4\pi \chi(\omega). \tag{13}$$

Figure 1 gives a schematic representation of the frequency dependence of $\varepsilon(\omega)$. The variation of $\varepsilon(\omega)$ with ω can be represented by the following relationship

$$\varepsilon(\omega) = \varepsilon_\infty + \frac{(\varepsilon_0 - \varepsilon_\infty)\, \omega_{TO}^2}{(\omega_{TO}^2 - \omega^2)} \tag{14}$$

where ε_∞ is the high frequency asymptotic value of ε and ω_{TO} is known as the transverse optic frequency. Equation (14) is only valid when $\hbar\omega$ is less than the excitation energy required for an electronic transition. In our subsequent discussion we will be concerned primarily with ε_∞ which arises solely from electronic displacements, nuclei being unable to follow the high frequency field. At lower frequencies both nuclear and electronic displacements contribute to the dielectric function. We note that there is a singularity in $\varepsilon(\omega)$ at $\omega=\omega_{TO}$. In addition, the dielectric function is zero at a frequency which we identify as ω_{LO}. This frequency is associated with a longitudinal lattice vibration and is known as the longitudinal optic frequency. From equation (14) we deduce that

$$\frac{\omega_{LO}^2}{\omega_{TO}^2} = \frac{\varepsilon_0}{\varepsilon_\infty} \qquad (15)$$

which is the well-known Lyddane-Sachs-Teller (LST) relationship. It is obeyed by ionic crystals. NaI provides a typical example; the ratio $\varepsilon_0/\varepsilon_\infty$ at $0^{\circ}K$ is 2.12, while $(\omega_{LO}/\omega_{TO})^2$ is 2.0.

This brief discussion emphasises the role which dynamic data can play in providing information on ionic polarisability: indeed, we shall see that comparison between calculated and experimental dispersion curves provides one of the most critical tests of the reliability of phenomenological models, although it is often the case that polarisation parameters are obtained merely by fitting to ε_0 and ε_∞. The following section will consider both the nature of these models and their parameterisation. In principle ionic polarisabilities should be amenable to quantum mechanical calculation via second-order perturbation theory[30]; reliable values, however, remain sparse[31]. As a result phenomenological approaches are widely used both in lattice dynamics and in calculations of static properties of crystal lattices (see Chapter 1).

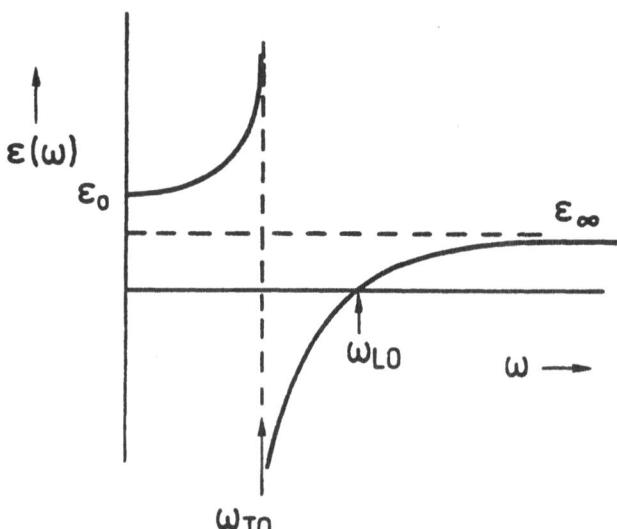

Figure 1   Schematic plot of the frequency dependence of $\varepsilon(\omega)$

6.1 Rigid ion models

The simplest class of ionic model, namely the rigid ion model, ignores polarisation effects completely*, that is to say it assumes $\varepsilon_\infty = 1$. From the LST relationship, therefore, it is clearly impossible for such models to provide an adequate description of the dynamical properties of the lattice. Thus in Figure 2 we show a comparison of the dispersion curves for NaI obtained from inelastic neutron scattering data[25] with the predictions of a rigid ion model. We see that, although fair agreement is obtained for the acoustic branches, for the optic branches the predictions of the rigid ion model are most unsatisfactory.

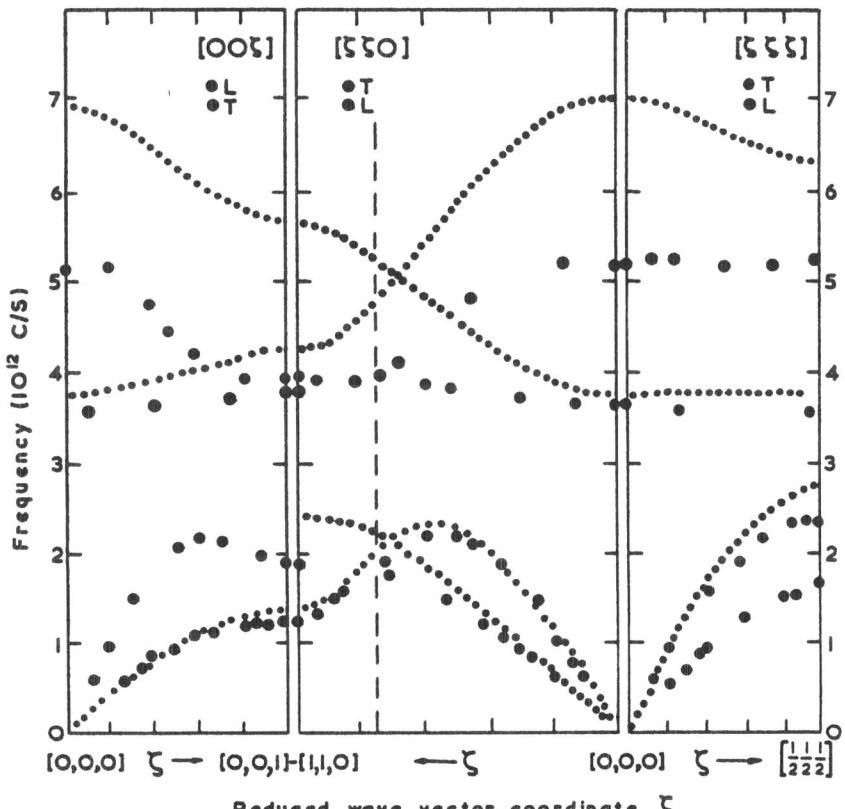

Figure 2 Experimental dispersion curves[25] for NaI compared with the predictions of the rigid ion model (....) at 100°K

---

*We should note that one effect of ionic polarisability, namely the existence of dispersion (or van der Waals) terms, is often included in rigid ion potentials, as described in section 4.

Rigid ion models also fail badly when applied to the calculation of defect energies, giving results which may be in error by up to 50%. For defect studies, however, it does seem possible to adjust the model to represent, at least partially, the effects of electronic polarisation. The simplest procedure is to adjust the short-range (repulsive) parameters so that the <u>static dielectric constant</u> of the crystal is reproduced correctly. In view of the importance of the dielectric response of the lattice to charged defects, a correct description of the static dielectric properties is the most important single requirement of a model in defect studies - a point which has been emphasised in several papers by the present authors[6,22]. Rigid ion models which correctly reproduce the static dielectric constant do so at the expense of other crystal properties. However, in a number of cases, notably the fluorite-structured crystals CaF_2 [32][33] and UO_2[34] they do appear to give adequate defect energies. Rigid ion models of this type, therefore, would seem to be appropriate in areas such as molecular dynamics, where the inclusion of polarisability is very costly in computer time but where potential models which correctly describe the defect properties are needed. Example of this are discussed in Chapters (5) and (18). In general, however, it is clear that some explicit representation of ionic polarisability is necessary and we now discuss the simplest model which includes this effect.

6.2 Point polarisable ion (PPI) models

In the PPI model, ionic polarisability is introduced via the dipole, μ, induced by an electric field, E, thus

$$\mu = \alpha E \qquad (16)$$

where α is the polarisability. Lyddane and Herzfield introduced the model into lattice dynamic studies[24], though the results, in general, were found to be unsatisfactory. Indeed, for certain wave vectors imaginary lattice frequencies were calculated[25] indicating a fundamental instability in the model. This was also found to be the case when PPI models were applied to defect studies[35,26], the form of the instability being a 'polarisation catastrophe' in which the dipole moments of pairs of ions increase without bound due to their polarisation fields. Moreover, even when calculations of this type appeared to converge, it was found that the resulting defect energies were considerably lower than the experimental values[37,28]. The reason for this discrepancy can be traced to a further flaw in the PPI predictions for perfect lattice properties, namely that excessively large static dielectric constants are obtained by these models when used in conjunction with short-range parameters obtained by fitting to <u>elastic</u> constants or acoustic phonon data.

The fundamental inadequacy of the PPI model is that it fails to allow for any interdependence of short-range forces and ionic polarisabilities. In a polarisation field, whether internal or external, there is a displacement of electrons which alters the short-range interactions between the ions due to over-lapping charge clouds.   It is essential that this effect is taken into account, and the simplest model which includes this is the shell model discussed in the next section.

6.3 The shell model

The shell model of solids is a simple mechanical model[39] which couples ionic polarisation to the 'effective' overlap forces.   It is illustrated in Figure 3.   Each ion consists of two components, a core X and a shell Y, such that the total ionic charge is the sum of the charges of X and Y.   The mass of the ion is centred at the core while the short range or overlap forces act through the mass-less shell. X and Y are coupled by a harmonic spring with a force constant, k, so that the polarisability of the ion, α, is given by

$$\alpha = Y^2/K \ . \tag{17}$$

The parameters Y and K are generally obtained by fitting to dielectric data, although both elastic and phonon properties are sometimes included.

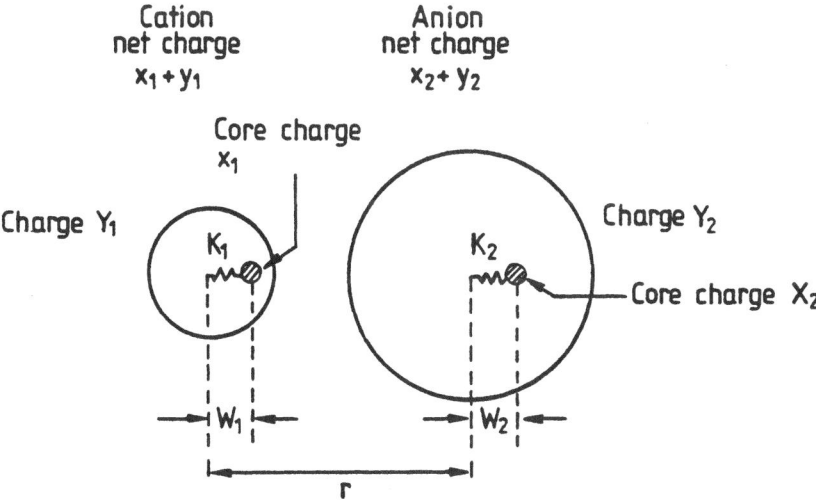

Figure 3 Simple shell model

Figure 4 shows a comparison of experimental dispersion curves for NaI with those calculated by a shell model. As shown, the agreement is good over a wide range of wavevectors; and it is certainly clear that the instabilities and inadequacies of the PPI model are eliminated. The same is true for defect studies. Polarisation catastrophes rarely occur with shell model calculations, which have now achieved a remarkable degree of quantitative success, as is clear from Chapters (12) and (15).

The inadequacies that remain in shell model treatments of ionic polarisation are due largely to the intrinsic short-comings of central force potentials. Thus the agreement with experimental dispersion curves, for example, can never be complete; in rock salt structured crystals the longitudinal optic mode at K = (1/2, 1/2, 1/2) will be incorrectly calculated whatever the model for ionic polarisation. Furthermore, central force potentials cannot account for the Cauchy violation for f.c.c. crystals since they always yield $C_{12} = C_{44}$ at $0°K$. For this reason, the simple shell model has been extended to allow for the distortion of the shells by neighbouring ions. Such refinements can, at least in part, take account of many body effects. Thus Schroder has introduced the 'breathing' shell model[40] which allows for spherically symmetric distortion of the shells. By including this additional degree of freedom it is possible to get good agreement with experiment

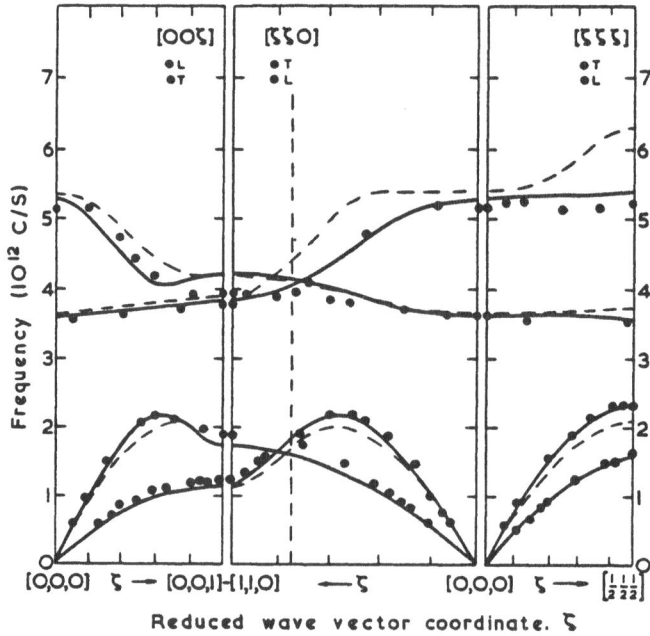

Figure 4   Experimental dispersion curves for NaI[25] compared with predictions of shell model (----- simple shell model; ———— breathing shell model)

over the whole Brillouin zone: a detailed study is reported for NaI[40] (see Figure 4). However, there are still problems with Cauchy violation, for breathing shell models only allow $C_{12} < C_{44}$, whereas violation in the reverse sense is observed in crystals such as silver chloride. For this reason Sangster[41] has extended the model to include ellipsoidal deformations - an appropriate mode for the d^{10} shell of the Ag^+ ion - from which it is possible to account for $C_{12} > C_{44}$. We recall from section 5 that an alternative approach to the explicit inclusion of many body effects is by the charge transfer model of Verma and Singh[23]. Shell model equations have been adapted for this model and the resulting agreement with the experimental phonon dispersion curves found to be very satisfactory[42].

It is clear, then, that shell models can accurately reproduce properties of the lattice such as phonon dispersion curves, and also elastic and dielectric properties as shown in Tables 1 and 2. This is equally true for complex oxides such as α-Al_2O_3[43,44] and $BaTiO_3$[45] as it is for the simpler halides. Furthermore, they can be applied to the calculation of a wide range of defect properties as mentioned elsewhere in this book. Despite this apparent success, however, the model has been criticised recently by Murrel et al[46] who have carried out molecular orbital calculations of clusters for Li^+ and F^- ions at spacings appropriate to the LiF lattice. From these calculations they propose that the most appropriate form of the potential is of the point ion type with two and three body interactions. While quantum mechanical calculations of this type of undoubted value, so far they have not been converted into interionic potentials of the type that can be used to study ionic solids. It is also worthwhile noting that within the formulation of the shell model, the work of Verma and Singh[23] suggests that additional corrections to the dynamical and electronic properties of ionic crystals from charge transfer effects are quite small.

It is important to recall that the shell model is a phenomological model, the use of which is justified by its success. Moreover, improvements are clearly needed; in particular it would be desirable to develop reliable non-empirical methods for deriving shell charges and spring constants, for those derived empirically often appear to be devoid of physical significance. However, it is worth emphasising that the shell model is the only available approach which can reproduce accurately the structural, dynamic and defect properties of ionic and semi-ionic crystals.

This completes our account of the general features of currently available potential models for ionic crystals. We conclude this chapter with a detailed discussion of currently available potentials for halides and oxides.

7. State of current models

The discussion above has demonstrated the general viability of shell model potentials. Previous sections, however, have drawn attention to problems involved in their parameterisation. These arise from the failure of empirical fitting methods to sample other than perfect lattice spacings, and the uncertainty as to the van der Waals coefficients. The discussion presented here will highlight those areas where most uncertainty remains. We consider first halide and then oxide systems.

7.1 Halides

7.1.1 Alkali halide potentials

Potentials for this class of material have been widely studied, and indeed the alkali halides have become a test bed for new potential models. Here we shall concentrate on the recently derived empirical potentials of Catlow et al[2] and Sangster et al[47][60], and the non-empirical model of Mackrodt and Stewart[15]

In essence both Catlow et al[2] and Sangster et al[47][60] extended an earlier empirical approach of Fumi and Tosi[48] by explicitly including polarisation effects in their potentials by means of the shell model and by using a wider range of data in the empirical fitting procedure. Thus both sets of workers determined shell parameters by fitting to ε_0, ε_∞ and ω_{TO}; in the case of Sangster et al[47] the van der Waals parameters were estimated from the resulting polarisabilities. Catlow et al[2], on the other hand, included the van der Waals terms only between second neighbours (i.e. cation-cation and anion-anion interactions) and treated the coefficients as variables in the empirical fitting procedure. The repulsive components of the second neighbour potentials were estimated by the electron-gas method described previously both in this Chapter and in Chapter (8). The parameters for the nearest neighbour potential were determined by an empirical fitting to data which included elastic and dielectric constants in addition to the lattice parameter. Sangster et al[47], in contrast, included only the lattice parameter and the cohesive energy. However, the latter potentials have the advantage that the parameters are specified in terms of the constituent ions and can be used, therefore, for mixtures of alkali halides.

In the work of Mackrodt and Stewart[15] all short range parameters were determined from electron gas calculations, although the shell parameterisation still relied on fitting to dielectric data. To this extent, therefore, the method is best described as semi-empirical. When the resulting potentials are compared with those of Catlow et al[2] and Sangster et al[47] there is found to be reasonable agreement in the case of the nearest neighbour potential, but there is a considerable spread in the parameters for the second neighbour interactions. Table 3 compares the values calculated with the three potentials for a variety of crystal properties with the experimental data. In this respect, the potentials of Catlow et al[2] perform best, as the widest range of data was used in the parameterisation; those of Sangster et al[47][60] perform least well, giving rise to errors typically up to 30% for C_{11}. When applied to the calculation of phonon dispersion curves both the empirical potentials lead to satisfactory results. Detailed calculations based on the electron gas potentials have not been reported, but in general the agreement with experiment is less satisfactory than for empirical potentials[49]. Sangster et al[47] extended their model to include 'breathing' shell effects but found only slight improvement to the calculated dispersion curves.

Table 3
Calculated and experimental crystal properties for NaCl

Crystal Property	Experimental value	Calculated values		
		Catlow et al[2]	Mackrodt and Stewart[15]	Sangster et al[47]
W_L (eV)	-7.91	-7.93	-7.91	-8.01
a_0 (Å)	2.82	2.79	2.89	2.79
C_{11} (10^{11} dyne/cm^{-2})	5.73	5.73	5.45	7.244
C_{12} "	1.12	1.12	1.56	1.901
C_{44} "	1.33	1.33	1.56	1.901
ε_0	5.45	5.45	5.45	5.44
ε_∞	2.35	2.35	2.35	2.36

All three potentials have been used in defect simulation studies. They yield energies for Schottky defect formation and vacancy migration that are in good agreement with those deduced from the analysis of conductivity data[15,49]; further details of these calculations are given in Chapter (12). Moreover, the calculation of saddle point energies for vacancy migration samples interatomic spacings which differ appreciably from those in the perfect lattice. Consequently the results suggest that the empirical potentials are reliable over a wider range of interionic separations than was used in their initial parameterisation. A further source of information on the reliability of potential models can be obtained from scattering studies of the structural and dynamical properties of the molten salts. Problems associated with the early potentials of Fumi and Tosi[48] are emphasised by the calculation of diffusion coefficients which are too low, due to difficulties in maintaining charge screening. The performance of more recent potentials has not been evaluated fully, although it has been shown that the inclusion of ionic polarisability results in a quite definite, if small improvement. Further details are given in the review by Sangster and Dixon[51].

In summary, then, the potentials discussed in this Chapter all appear to perform well when applied to the calculation of both perfect lattice and point defect properties. However, there are significant differences of detail between them, particularly with regard to second neighbour interactions. It is an open question as to which are the most reliable over a range of separations. Eggenhoffer et al[52] have recently drawn attention to the different parameters derived by different workers, and it is unlikely that these differences can be effectively resolved on the basis of perfect lattice properties alone. But the inclusion of defect properties and the behaviour of the molten salt should lead to a more stringent test of different potential models.

7.1.2 Other halides

The only extensively studied materials here are the alkaline earth fluorides CaF_2, SrF_2 and BaF_2 for which potentials have been reported by Catlow and Norgett[22] and more recently by Catlow, Norgett and Ross[3]. In both cases a shell model potential was fitted to elastic and dielectric properties of the crystal. The experimental data for these materials are accurately reproduced by the two models which, in addition, perform well in calculating phonon dispersion curves and a very diverse range of defect properties. The more recent model differs from that of the earlier study largely in the treatment of the second neighbour F^-....F^- interaction, an important feature of which is the existence of a large attractive term, as discussed in section (4); they are far larger than can be accounted for by dispersive interactions. In both models the attractive term is described by an r^{-6} term, but in the more recent study this is _splined_ to the repulsive Born-Mayer term at a given separation. That is to say the repulsive and attractive terms separately operate only over specified ranges of the inter-nuclear separation. This procedure appears to generate more reliable potentials, and the parameters derived by Catlow, Norgett and Ross[53] show definite, if small, improvements when applied to defect studies.

Shell model potentials parameterised from elastic and dielectric data have also been reported for $SrCl_2$[53,54] and MnF_2[55]. The models perform adequately when applied to perfect lattice and defect properties although a critical evaluation of their status is not possible at present.

7.2 Oxides

As indicated in section 2 of Chapter (12) potentials have been obtained for a wide range of oxides from MgO to spinel and δ-Bi_2O_3, based on both empirical and non-empirical procedures. For the present we concentrate on the details of the more recent models; but in so doing we acknowledge the value of earlier work listed in reference 56, which stimulated much of the later development.

In their study of MgO, Catlow, Faux and Norgett[5] laid the foundation of much that has followed in the derivation of empirical potentials for oxides. They assumed the non-Coulombic contributions to be of the form

$$V_{+-}(r) = A_{+-} \exp(-r/\rho_{+-}) \tag{18a}$$

$$V_{--}(r) = A_{--} \exp(-r/\rho_{--}) - C_{--}/r^6 \tag{18b}$$

and

$$V_{++}(r) = 0 \tag{18c}$$

in which the suffices ,+-,-- and ++, have the obvious interpretation. Following earlier work by Sangster[56], ionic charges were taken to be ±2. A_{--} and ρ_{--} were obtained by fitting to the calculated Hartree-Fock potential for O_2^{2-} in a truncated point ion field,[57,6] and the remaining parameters, A_{+-}, ρ_{+-} and C_{--} then fitted to the equilibrium equation (lattice parameter) and the experimental values of C_{44} and $C_{11}-C_{12}$. Owing to its small size, Mg^{2+} was assumed to be non-polarisable and the oxygen shell parameters fitted to the two principal dielectric constants, ε_0 and ε_∞. This basic model was then extended to include 'breathing shell' effects of the type already discussed in section 6.3, and a further extension made to simulate many-body terms by re-fitting parameters to the lattice parameter and C_{11} separately. In Table 4 we summarise some of the calculated defect energies for the three potentials. From this we see that the inclusion of breathing shell effects makes very little difference to the calculated defect energies, though as we have commented earlier, it does influence the Cauchy violation and phonon dispersion curves. The results obtained for Model III, in which many-body terms are included, differ by about 10% from the other two in the vacancy formation energies, though the calculated Schottky energy is much the same. Overall, the results are in good agreement with the known thermodynamic and diffusion data[5].

Similar potentials were used by Catlow et al[4] in their study of the transition metal oxides, MnO, FeO, NiO and CoO. Once again a completely ionic model was used, together with the somewhat more questionable assumption that only the anion is polarisable. V_{--} for all four oxides was taken to be that in MgO as was the anion shell charge. The remaining parameters, A_{+-}, ρ_{+-} and k_-, the anion spring constant, were then fitted to the lattice parameter, cohesive energy and static dielectric constant. These potentials have been used to investigate electronic conduction[4] and vacancy diffusion mechanisms[4] and in both cases there is substantial agreement with experiment. Non-cubic oxides have received far less attention, though as we have mentioned before, the increased complexity of these materials in fact affords more structural information, which can be utilised in empirical potentials, than is available for the simpler cubic materials.

Table 4

Calculated defect energies for MgO based on the potentials of Catlow, Faux and Norgett[5]

Defect	Energy (eV 1)		
	Model I	Model II	Model III
Cation vacancy	23.82	23.74	20.21
Anion vacancy	24.70	24.63	27.01
Schottky	7.7	7.5	7.6
Divacancy	5.3	-	5.2
Activation energy for migration			
(i) Anion vacancy	2.1	1.3	2.0
(ii) Cation vacancy	2.1	1.8	1.9

Model I   Parameter fit to a_0, $C_{11}-C_{12}$, C_{44}
Model II   Breathing shell extension to I
Model III Parameter fit to a_0, C_{11}, C_{12}.

Catlow et al[44] have derived potentials for α-Al$_2$O$_3$ and TiO$_2$ and the latter, although problematic with regard to the treatment of polarisation, have found fruitful application in the study of shear plane formation and stability[58].

The most recent set of empirical potentials is that reported by Sangster and Stoneham[59] for the alkaline-earth and transition metal monoxides. A notable feature of this work is the derivation of consistent transferable cation polarisabilities similar to that previously obtained by Sangster, Schroder and Atwood[60] for the alkali halides. As before cation-cation interactions were assumed to be purely Coulombic, while V_{--} was taken to be the same in all the crystals and identical to that found by Catlow, Faux and Norgett[5] for MgO. Sangster and Stoneham[59] made the further assumption that the oxygen polarisability and shell charge are the same in all the oxides and equal to their values for MgO. The remaining parameters were obtained by fitting to ε_0, ε_∞, ω_{TO} and the condition that the cohesive energy is a minimum at the equilibrium lattice spacing. Perfect lattice and defect properties have been calculated using these potentials, and given their simplicity, there is remarkably good agreement with experiment. In Table 5 we list the cation polarisabilities and calculated cohesive energies and elastic constants for all eight oxides and in Figure 5 give the phonon dispersive curves for NiO and CaO. We return to both these sets of empirical potentials later in this section.

Table 5

Perfect lattice properties based on the potentials of Sangster and Stoneham[59]

| | | Cohesive energy (eV) | | Elastic constants (10^{12} dyn cm^2) | | | | |
				C_{11}	$C_{12}=C_{44}$	C_{11}	C_{12}	C_{44}
	$\alpha_+(\mathring{A}^3)$	Calc.	Expt.*	Calc.		Expt.		
MgO	0	-41.0	-40.4	3.74	1.57	2.89	0.88	1.55
CaO	1.510	-36.0	-36.0	2.22	0.94	2.26	0.62	0.81
SrO	2.555	-33.4	-33.8	1.56	0.71	1.73	0.45	0.56
BaO	3.303	-31.1	-31.9	1.12	0.55	1.26	0.50	0.34
MnO	1.803	-38.7	-39.5	2.81	1.28	2.23	1.2	0.79
FeO	2.010	-40.1	-40.7	3.40	1.44	3.59	1.56	0.56
CoO	1.529	-41.0	-41.4	4.22	1.50	2.56	1.44	0.80
NiO	1.593	-41.9	-42.3	4.59	1.64	2.70	1.25	1.05

*The references for the experimental data are given in full in the paper by Sangster and Stoneham[59].

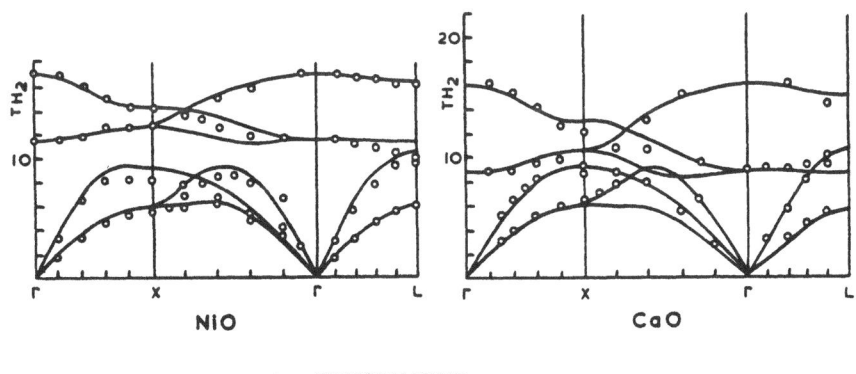

o   EXPERIMENT

—   CALCULATION

Figure 5   The calculated and experimental phonon dispersion curves for NiO and CaO[59].

We turn now to non-empirical potentials, or to be more precise semi-empirical potentials for as we have noted earlier, the use of the 'shell-model' to account for electronic polarisation renders all potentials of this type empirical to some extent. Dienes et al[61] developed a polarisable point-ion shell model for α-Al_2O_3 in which the short-range repulsive interactions were calculated on the basis of the electron-gas approximation for the density functional outlined previously in this chapter and by March in Chapter (8). Free ion wave functions were used for Al^{3+} and those calculated by Watson[62] for an ion in a spherical well of charge (+1) for O^{2-}. Dispersion forces between anions were included in which the interaction constant (C_0 of equation (10)) was evaluated using the Slater-Kirkwood formula[63]. As an alternative to the Dick-Overhauser model for the electronic polarisation, Dienes et al[61] used an approach of the type mentioned in section 6.2; that is to say, the <u>effective repulsive interaction</u> at a separation r_{ij} between charge clouds distorted by internal electric fields was approximated by the interaction between undistorted ions at a <u>effective separation</u>, r_{ij}^{eff}.

Thus

$$\phi^{(distorted)}(r_{ij}) = \phi^{(undistorted)}(r_{ij}^{eff}) \qquad (19)$$

in which r_{ij}^{eff} is given by

$$\underline{r}_{ij}^{eff} = \underline{r}_{ij} + \frac{\underline{u}_i}{Q_i} - \frac{\underline{u}_j}{Q_j} \qquad (20)$$

where $\underline{u}_i$ and $\underline{u}_j$ are the equilibrium dipole moments of the ions i and j, and Q_i and Q_j empirical constants. $\underline{u}_i$ and $\underline{u}_j$ were obtained by minimising the equilibrium lattice energy of the polarised crystal with respect to the individual components of $\underline{u}_i$ and $\underline{u}_j$.

Subsequently Mackrodt and Stewart[15] derived potentials for a wide variety of oxides based on a variation of the Gordon and Kim[12] procedure outlined in section (3.2), which included the exchange correction proposed by Rae[13], and separate O^{2-} wavefunctions calculated for each crystal structure. No specific allowance was made for van der Waals forces as was done by Dienes et al[61]; although we believe that it is a questionable point, as to whether the inclusion of the correlation energy density $\varepsilon_c[\rho(\underline{r})]$ in equation (3) introduces some measure of dispersive interaction in the resulting potentials. Shell parameters of the Dick-Overhauser model were obtained by fitting to the measured high frequency and static dielectric constants. An important feature of these (and other semi-empirical) potentials, particularly with regard to their use in defect calculations, is that they lead to strained unit cells for the measured lattice structures. Consequently, they predict equilibrium (i.e. strain-free) lattice structures which differ from experiment, though in most cases the resulting distortions are small. A summary of calculated lattice energies, lattice parameters and compressibilities based on these potentials is given in Table 6.

Table 6
Calculated lattice energies, lattice parameters and compressibilities for a range of metal oxides*

Oxide	W_L (eV)		a_0 (Å)		Compressibility	
	Calc.	Expt.	Calc.	Expt.	Calc.	Expt.
Li_2O	-31.6	-30.1	2.30	2.31	0.68	-
BeO	-49.4	-46.9,-47.7	2.72	2.70	-	-
MgO	-40.75	-40.8,-40.4	2.18	2.11	0.43	0.64
α-Al_2O_3	-161.9	-160.4(b)	13.2	13.0	0.25	0.40,0.32
CaO	-36.0	-37.0,-36.1	2.46	2.41	0.70	0.88
TiO_2	-122.4	-126.0	4.52	4.59	-	-
MnO	-37.49	-39.54	2.36	2.22	0.57	0.65
α-Fe_2O_3	-148.7	-156.3	14.3	13.7	-	-
Ga_2O_3	-149.7	-157.4,-161.18	14.3	13.4	-	-
SrO	-33.9	-34.3,-33.8	2.62	2.58	0.86	1.14
CdO	-34.7	-40.2	2.58	2.35	-	-
SnO_2	-113.6	-123.0,-117.8	4.87	4.74	0.32	0.49
BaO	-32.0	-32.4,-31.9	2.78	2.76	1.10	1.63
CeO_2	-107.7	-107.2	2.72	2.71	0.33	-
PbO_2	-109.7	-121.8,-119.9	5.05	4.95	-	-
ThO_2	-101.5	-103.0,-104.7	2.87	2.80	0.43	0.52,0.42
UO_2	-102.4	-106.7	2.84	2.73	0.40	0.47
$CaTiO_3$	-159.1	-161.9	3.90	3.84	0.33	-
$BaTiO_3$	-154.5	-158.2	4.04	4.01	-	-
$MgAl_2O_4$	-203.6	-200.7	4.12	4.04	0.33	0.48

*Experimental values taken from reference (65).

Electron gas potentials of the type discussed above inevitably omit covalency, and in an attempt to include this effect, Campbell et al[64] have derived potentials for MgO, MnO and ZnO based on ab initio molecular orbital methods. Details of the method are described elsewhere[64,65], but the principal features to emerge from this work are that for MgO and MnO, perfect lattice and defect properties are very much in line with those from both empirical and electron-gas potentials[15], but for ZnO there are radical differences. Whereas electron-gas potentials fail to give the correct structure, ab initio potentials not only predict this to within 2-3% of the measured crystal geometry, but the resulting defect

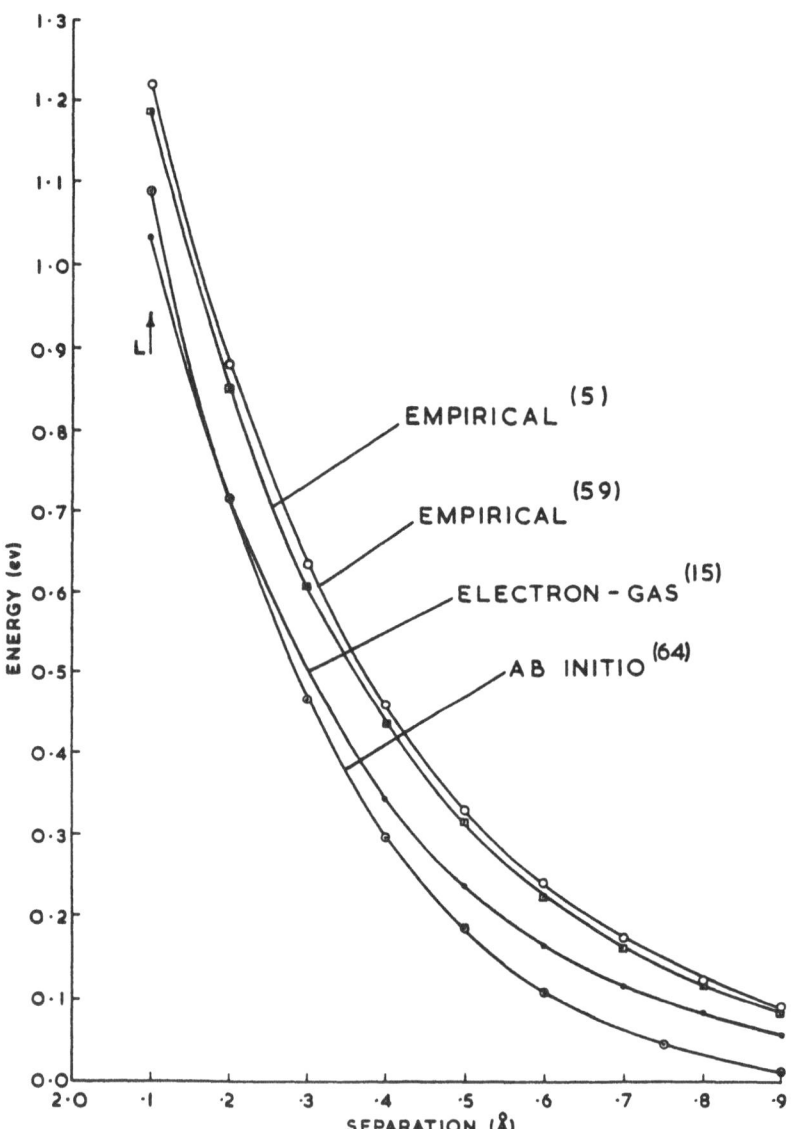

Figure 6 A comparison of non-Coulombic potentials for Mg^{2+} - O^{2-} in MgO

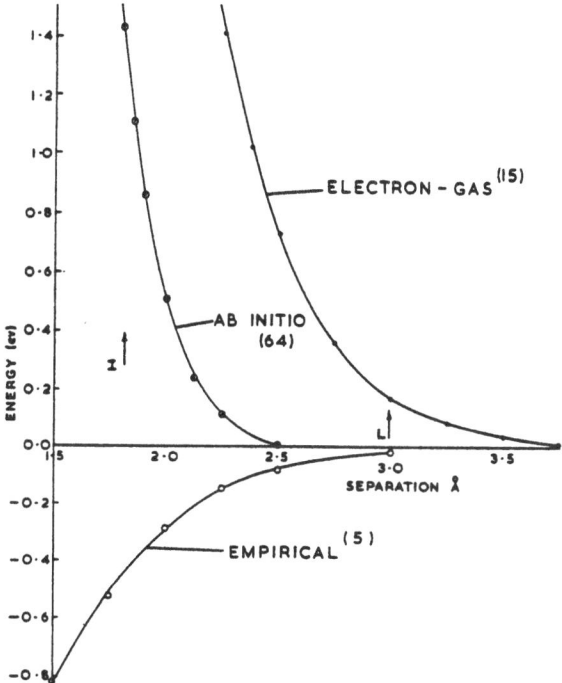

<u>Figure 7</u> A comparison of non-Coulombic potentials for O^{2-}-O^{2-} in MgO.

structure[66] is in good agreement with that deduced by Kroger[67] from optical and electrical measurements.

How then do these various potentials compare, both in form, and in the crystal properties they predict?   In Chapter (12) a brief comparison of defect energies is made for α-Al_2O_3[44]; to end this present chapter we examine the differences in these potentials in greater detail with specific reference to MgO and α-Al_2O_3. Figures 6 and 7 illustrate the non-Coulombic components of the potentials for Mg^{2+}-O^{2-} and O^{2-}-O^{2-} derived from empirical[5,59], electron-gas[15] and an initio[64] procedures. At the normal lattice spacing (indicated by L in Figures 6 and 7), the differences between the potentials are small.   There is a spread of about 0.2 eV in the cation-anion potentials and slightly less than this in the anion-anion interaction. As shown in Table 7 these lead to cohesive energies which are within a volt of each other and experiment.   The calculated elastic constants are similarly close, except for the value of C_{11} obtained from <u>ab-initio</u> potentials.   The calculated vacancy energies are also similar though the ordering of the energies based on electron-gas potentials differs from the other three, a feature that Sangster and Rowell[68] have recently discussed. There are, however, major differences in the potentials at internuclear separations corresponding to the interstitial position, for the empirical potentials for O^{2-}-O^{2-} are attractive here whereas it is everywhere repulsive in the case of

electron-gas and <u>ab-initio</u> potentials.   The electron-gas potential in particular
is extremely repulsive at these separations and this leads to differences of over
5 eV in the anion interstitial energy and corresponding differences in the
Frenkel energy.

The situation is much the same in α-Al_2O_3 as Figures 8 and 9 show.   There
is remarkably close agreement between the three potentials for the Al^{3+}-O^{2-}
interaction throughout the range, but appreciable differences in the anion-anion
potentials similar to those found previously for MgO.   The short-range
contribution to the O^{2-}-O^{2-} found by Dienes et al[61] is broadly similar to the
more recent electron-gas potential[15];   the differences between them serve to
illustrate the sensitivity of the interaction to the O^{2-} wavefunction.   However,
the most important feature of both the empirical and earlier non-empirical
potentials is the larger r^{-6} term, which in the case of the latter contributes
-3.6 eV to the O^{2-} octahedral interstitial energy compared with the electrostatic
energy of 5.9 eV.   If dispersion forces can, indeed, make such an appreciable
contribution to defect energies in oxides there is clearly a need for the development
of more reliable methods for calculating them, a point which we have previously
emphasised.

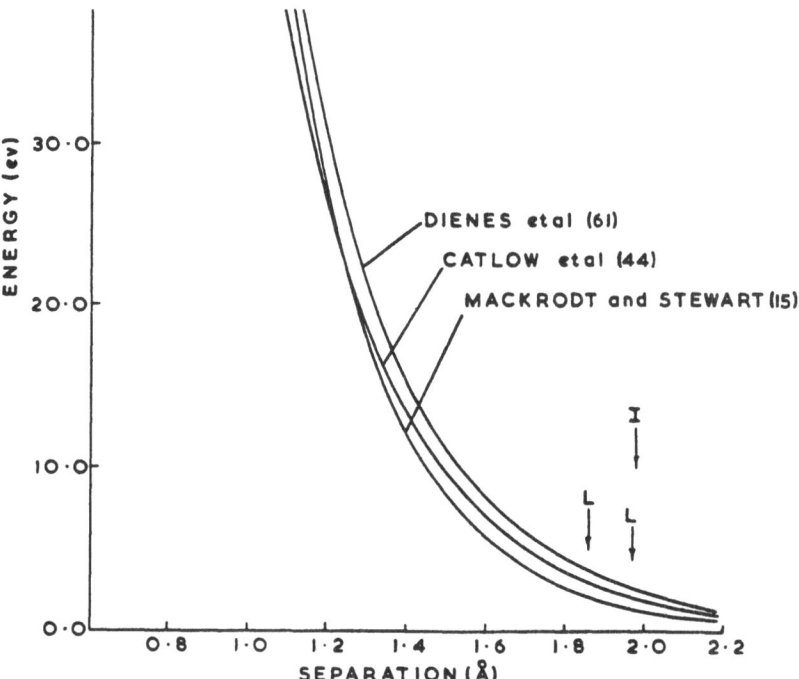

Figure 8   A comparison of non-Coulombic potentials for Al^{3+}-O^{2-} in α-Al_2O_3

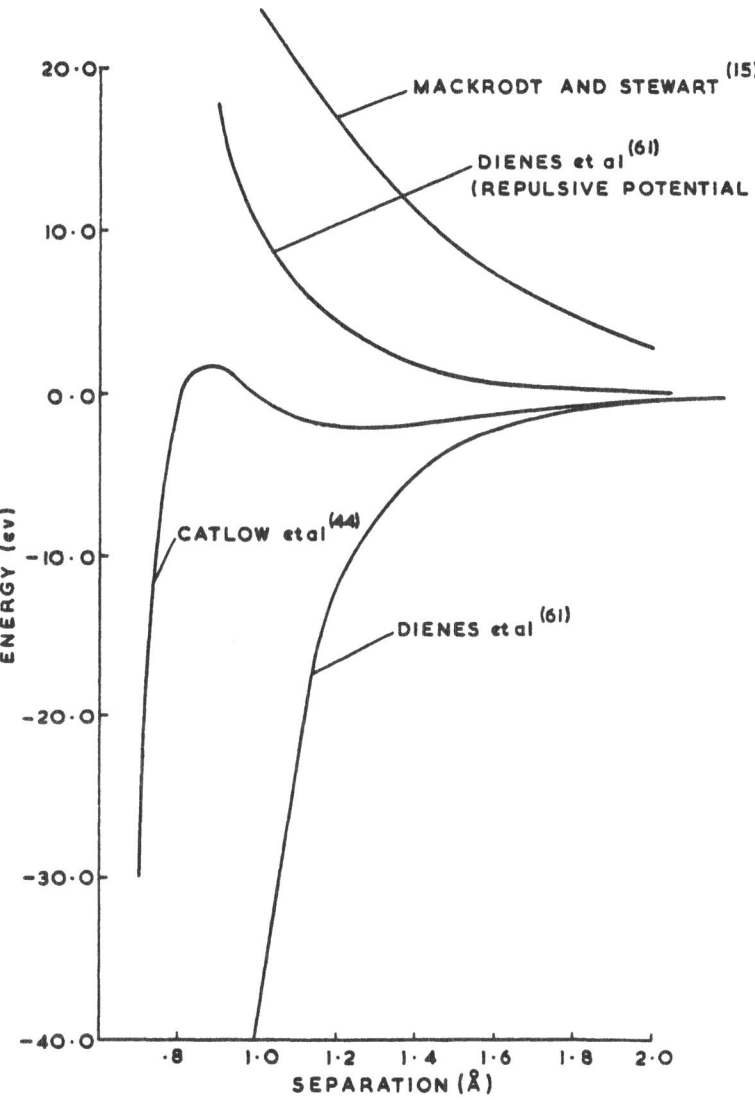

Figure 9 A comparison of non-Coulombic potentials for $O^{2-} - O^{2-}$ in α-Al_2O_3

Table 7

Comparison of perfect lattice properties and defect energies of MgO for
different empirical and non-empirical potentials

Property/Defect Energy (eV)	Potential				
	A	B	C	D	Expt.
Cohesive energy (eV)	-40.85	-40.75	-41.0	-41.61	-40.35 to -40.79[69]
C_{11} (10^{11} dyn cm^{-2})	35.6	31.9	-37.4	50.9	31.0[70]
C_{44} "	15.5	19.1	15.7	15.0	16.0[70]
C_{12} "	15.5	19.1	15.7	15.0	9.6[70]
Cation vacancy energy	23.8	25.4	24.0	24.5	-
Anion vacancy energy	24.7	22.9	24.7	25.9	-
Schottky energy	7.7	7.5	7.7	8.8	-
Cation interstitial energy	-11.4	13.5	-11.6	-12.2	-
Anion interstitial energy	-12.6	-7.7	-12.4	-12.0	-
Cation Frenkel energy	12.4	11.9	12.4	12.3	-
Anion Frenkel energy	12.1	15.2	12.3	13.9	-
Cation vacancy migration energy	2.1	2.16	2.07	2.6	2.2[71]; 2.28±0.2[72]
Anion vacancy migration energy	2.1	2.38	2.11	3.0	2.42[73]; 2.61[74]

A (Empirical)   Catlow, Faux and Norgett[5]
B (Electron gas)   Mackrodt and Stewart[15]
C (Empirical)   Sangster and Stoneham[59]
D (ab initio)   Mackrodt et al[64]

References

1.   Catlow, C.R.A., Thomas, J.M., Jefferson, D.A. and Parker, S.C. Nature 295, 658 (1982).

2.   Catlow, C.R.A., Diller, K.M. and Norgett, M.J., J. Phys. C10, 1395 (1977).

3.   Catlow, C.R.A., Norgett, M.J. and Ross, T.A., J. Phys. C10, 1627 (1977).

4.   Catlow, C.R.A., Mackrodt, W.C., Norgett, M.J. and Stoneham, A.M., Phil. Mag. 35, 177 (1977); 40, 16 (1979).

5.   Catlow, C.R.A., Faux, I.D. and Norgett, M.J., J. Phys. C9, 419 (1976).

6.   Catlow, C.R.A., Proc. Roy. Soc. A353, 533 (1977).

7. Stoneham, A.M., AERE Report, AERE-R.9598 (1981).

8. Jensen, K., Z. Phys. 101, 141 (1936).

9. Gombas, P., Handbuch der Physik (Springer-Verlag, Berlin), 36, 109 (1956).

10. Wedepohl, P.T., Proc. Phys. Soc. 92, 79 (1967).

11. Wedepohl, P.T., J. Phys. B1, 307 (1968).

12. Gordon, R.G. and Kim, Y.S., J. Chem. Phys. 56, 3122 (1972)

13. Rae, A.I.M., Chem. Phys. Lett. 18, 574 (1973).

14. Cohen, J.S. and Pack, R.T., J. Chem. Phys. 61, 2372 (1974)

15. Mackrodt, W.C. and Stewart, R.F., J. Phys. C12, 431 (1979).

16. Colbourn, E.A., Kendrick, J. and Mackrodt, W.C., ICI Corporate Laboratory Report, CL-R/81/1637/A (1981).

17. McLachlan, A.D., Proc. Roy. Soc. A271, 387 (1963).

18. Mayer, J.E., J. Chem. Phys. 1, 270 (1933).

19. Fumi, F.G. and Murthy, C.S.N. Private communication.

20. Andzelm, J. and Piela, C., J. Phys. C10, 2269 (1977); 11, 2695 (1978).

21. Catlow, C.R.A. and Hayns, M.R., J. Phys. C5, L237 (1972).

22. Catlow, C.R.A. and Norgett, M.J., J. Phys. C6, 1325 (1973).

23. Verma, S.M.P. and Singh, R.K., Phys. Stat. Sol. 36, 335 (1969).

24. Lyddane, S.M.P. and Herzfield, . Phys. Rev. 54, 846 (1938).

25. Woods, A.D.B., Cochran, W. and Brockhouse, B.N., Phys. Rev. 119, 980 (1960).

26. Phillips, J.C., in "Covalent Bonding in Crystals, Molecules and Polymers", (Chicago University Press), 1960.

27. Martin, R.M., Adv. Phys. 14, 39 (1965).

28. Martin, R.M., Phys. Rev. 186, 871 (1969).

29. Singh, R., Phys. Report. To be published.

30. Cohen, M.D. and Roothaan, J. Chem. Phys. 43, 534 (1968).

31. Werner, H.J. and Meyer, W., Mol. Phys. 31, 855 (1976).

32. Gillan, M.J. and Dixon, M., J. Phys. C13, 1901 (1980)

33. Dixon, M. and Gillan, M.J., J. de Physique Colloq. C6-24 (1980) Suppl. Vol 41 (7).

34. Walker, J.R. and Catlow, C.R.A., J. Phys. C14, L979, (1981).

35.   Tharmalingam, K., Phil. Mag. 23, 181 (1971).

36.   Faux, I.D., J. Phys. C4, L211 (1971).

37.   Boswarva, I. and Lidiard, A.B., Phil. Mag. 16, 805 (1967).

38.   Norgett, M.J., J. Phys. C4, 298 (1971).

39.   Dick, B.G. and Overhauser, A.W., Phys. Rev. 112, 90 (1958).

40.   Schroder, U., Solid State Commun. 4, 347 (1966).

41.   Sangster, M.J., J. Phys. Chem. Solids 35, 195 (1974).

42.   Verma, M.P. and Singh, R.K., Phys. Stat. Sol. 36, 335 (1969).

43.   James, R., Ph.D. Thesis, University of London (1979).

44.   Catlow, C.R.A., James, R., Mackrodt, W.C. and Stewart, R.F., Phys. Rev.
      B25, 1006 (1982)

45.   Lewis, G. and Catlow, C.R.A. To be published.

46.   Murrel, J.N., Tennyson, J. and Kamel, M.A., Molec. Phys. 42, 747 (1981).

47.   Sangster, M.J., Schroder, U. and Atwood, R.M., J. Phys. C11, 1523 (1978).

48.   Fumi, F.G. and Tosi, M.P., J. Phys. Chem. Solids 25, 31 (1964).

49.   Colbourn, E.A. and Mackrodt, W.C.   Unpublished results.

50.   Catlow, C.R.A., Corish, J., Diller, K.M., Jacobs, P.W.M. and Norgett, M.J.,
J. Phys. C12, 451 (1979).

51.   Sangster, M.J. and Dixon, M., Adv. Phys. 25, 247 (1976).

52.   Eggenhoffer, R.G., Fumi, F.G. and Murthy, C.S.N., J. Phys. Chem. Solids.
In press.

53.   Bendall, P.J., D. Phil. Thesis, University of Oxford (1980).

54.   Bendall, P.J., Catlow, C.R.A. and Fender, B.E.F., J. Phys. C14, 4377 (1981).

55.   Catlow, C.R.A., James, R. and Norgett, M.J., J. de Physique, 37, C7-443 (1976).

56.   Sangster, M.J., J. Phys. Chem. Solids 34, 355 (1973).

57.   Catlow, C.R.A., D. Phil. Thesis, University of Oxford (1974).

58.   Catlow, C.R.A. and James, R., Nature 272, 603 (1978).

59.   Sangster, M.J. and Stoneham, A.M., Phil. Mag. B43, 597 (1981).

60.   Sangster, M.J. and Atwood, R.M., J. Phys. C11, 1541 (1978).

61.   Dienes, G.J., Welch, D.O., Fischer, C.R., Hatcher, R.D., Lazareth, O. and
Samberg, M., Phys. Rev. B11, 3060 (1975).

62.   Watson, R.E., Phys. Rev. 111, 1108 (1958).

63.   Slater, J.C. in "Quantum Theory of Molecules and Solids", Vol.3, Appendix 5, (McGraw-Hill, New York, 1967).

64.   Campbell, J.C., Hillier, I.H. and Mackrodt, W.C.   To be published.

65.   Mackrodt, W.C., in "Mass Transport in Solids" (Eds. Bénière, F. and Catlow, C.R.A.), Plenum Press, New York, 1982.

66.   Mackrodt, W.C. and Stewart, R.F., J. Phys. C$\underline{12}$, 5015 (1979).

67.   Kroger, F.A. "The Chemistry of Imperfect Crystals", p.691 (North Holland: Amsterdam, 1964).

68.   Sangster, M.J. and Rowell, D.J., Phil. Mag. A$\underline{44}$, 613 (1981).

69.   Samsonov, G.V. "The Oxide Handbook" (IFI/Plenum) (1973).

70.   Peckham, G., Proc. Phys. Soc. $\underline{90}$, 657 (1967).

71.   Duclot, M. and Departes, C., J. Solid State Chem. $\underline{31}$, 377 (1980).

72.   Sempolinski, D.R. and Kingery, W.D., J. Amer. Ceram. Soc. $\underline{63}$, 664 (1980).

73.   Shirasaki, S. and Harma, M., Chem. Phys. Lett. $\underline{20}$, 361 (1973).

74.   Shirasaki, S. and Yamamura, H., Jap. J. App. Phys. $\underline{12}$, 1654 (1970).

CHAPTER 11   INTERATOMIC POTENTIALS IN COVALENT

AND SEMI-COVALENT SOLIDS

by

A.M. Stoneham and J.H. Harding
Theoretical Physics Division, A.E.R.E. Harwell, Oxon OX11 ORA

1. Introduction

In ionic solids, rare-gas solids, and even in metals (at least at constant
volume) little confusion results from talking of "interatomic potentials".   The
traditional picture of pairwise interactions is close enough to reality to be an
effective starting point;   see, for example, Chapters (9) and (10). This is not so for
covalent systems, where neither the form nor the content of such potentials is
agreed.   Our aim in this chapter is to assess the advantages and disadvantages of
some of the different formulations which can be used.   In an ideal situation one
would be able to express the total energy of a system (including those containing
defects) both accurately and conveniently.   Different models make different
compromises, and it is these features which we examine.   Applications to the physics
of semiconductors and other covalent systems are discussed in Chapter (14).

2. Systems

Covalency is a much-abused word, often used to hide areas of ignorance.   An
extended review of the various definitions of covalency has been given by
Phillips[1].   For present purposes, covalency implies the formation of bonds, i.e.
enhanced charge densities between atoms, instead of the charge transfer in simple
ionic crystals.   Among the many covalent systems we might include the following:

(i)   Elements and compounds with open structures because of the strong,
       directed bonds.   Examples are produced by diamond, Si, Ge, S, Se, Te
       and compounds like SiC, the III-V's, some II-VI's, As_2Se_3 and SiO_2.
       When one is dealing with simple compounds, the covalent systems lie in
       a well-defined region of the periodic table.   For binary compounds, the
       relevant region is shown in Figure 1.

(ii)   Molecular crystals, like oxygen (where the unit is O_2).

(iii) Crystals with strongly-bonded radicals, e.g. $(WO_3)^{3-}$, $(SO_4)^{2-}$.

(iv)   Isolated molecular ions or radicals in ionic or inert-gas matrices;
       the V_k centre has been especially fully studied.

We shall concentrate almost exclusively on category (i), since it is much easier to
find workable models for the other instances, e.g. the "molecule in a crystal"
model for class (iv).[2-4]

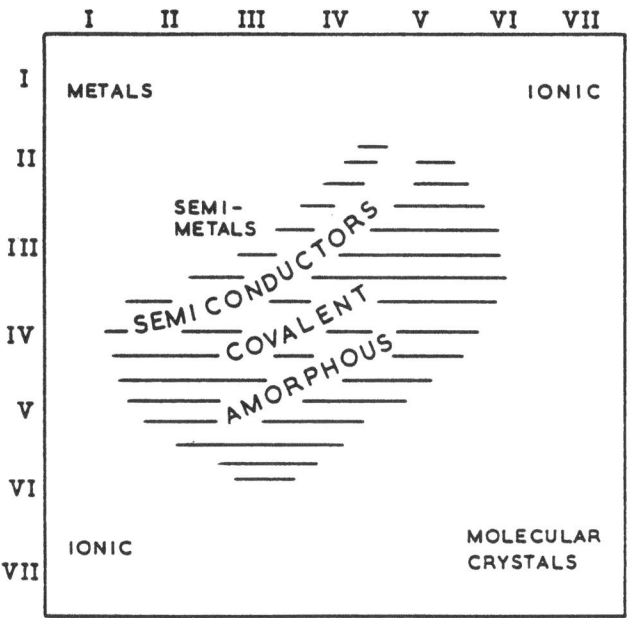

Figure 1

Let us consider one of the crystals of class (i), stabilised by covalency into an open structure. We can see at once why potentials present problems. First of all, directed bonds imply three-body (or more complicated) forces, even at constant volume. A major problem is thus the representation of such few-body forces. Fortunately, this is simplified by the general rule: three-body forces are short ranged. If one excludes straightforward polarisation effects, the interactions between ions A and B depend significantly on only a few other ions C.

Secondly, we may wish to calculate the energies of simple impurity interstitials. If they do not form bonds (e.g. closed shell interstitials) we shall need the short-range repulsive forces. But lattice parameters are determined principally by the covalent bonds and not (unlike simple ionic crystals) by the balance between Coulomb forces and the short-range repulsions from core-core overlaps. Thus for covalent solutes the properties of a crystal near equilibrium (cohesive energy, lattice dynamics, elastic and dielectric constants) do not give any information on short-range repulsions. These repulsions must therefore be calculated by some other method, possibly an electron-gas model of the type discussed in Chapter (10) in which one puts suitable estimates of electron densities.

Thirdly, the most important types of defect involve the addition or removal of atoms covalently bonded to others. When a bond is broken, or interrupted by an impurity, the new forces between atoms are not normally related to the old ones in any simple way. Further, a solution (at least approximate) of the Schrödinger equation will often be needed to estimate the new forces.   The combination of solutions of the Schrödinger equation for very small regions with simpler descriptions at large distances is not trivial.

Finally, covalency does not mean that long-range Coulomb interactions are absent or insignificant.   Moreover, since charge has been transferred away from ions towards the bond centres, there can be problems in defining the charges of the component ions and bonds.   It is worth stressing that most defect calculations are concerned with the dipole moment per unit displacement and not with the charge within some chosen region around an ion. Further, one must include correctly any Coulomb and overlap interactions between bonds as well as between atoms or ions.

3.   Types of approach

Any really satisfactory interatomic potential should satisfy two criteria. First, it should define with acceptable accuracy the total energy as a function of geometry. Secondly, it should be expressed in a "convenient" form, i.e. there should exist some simple algorithm so that the potential can be used in a lattice simulation program.

It need not be necessary to have a universal potential for all possible geometries. For example, most calculations assume constant volume, and the range of interatomic spacings, even near defects, can be quite restricted. Likewise, some special configurations may occur only rarely, and it may suffice to be told whether they arise, rather than attempting a general form to cover them. As an example, there seem to be some geometries for H in Na for which two-body forces are inadequate[5] even at constant volume, despite which, however, two-body forces suffice for most important configurations. Another example is the "polarisation catastrophe" which occurs in calculations on ionic crystals.

These broad criteria mentioned are by no means all. Others include essential constraints (e.g. symmetry) and desirable constraints (e.g. that interactions are transferable for one system to another). We now look at some of the models used, with their advantages and disadvantages. We classify the methods according to whether they attempt to solve the Schrödinger equation, or whether they parameterise a general lattice model.

Class I: Methods which use the Schrödinger equation explicitly

We include here cases in which the Schrödinger equation is systematically approximated, possibly using empirical parameters; it may be used rather than solved. We also stress that it is important to use total energies rather than a sum of one-electron eigenvalues. The difference is especially important for bond-centre interstitials, where the doubly-counting of electron-electron interactions and omission of nuclear-nuclear repulsions may be grossly misleading[6-7].

I.A Dielectric models

These use the fact that, given a pseudopotential for the ion cores, and given a dielectric function $\varepsilon(q)$ of the valence electrons, one can write down the energy for distorted geometries by perturbation theory.   Detailed discussions of this method are given in references (8-11).

I.B Cluster models

These have been widely used for defect calculations of energy as a function of geometry, though they are rarely used to obtain potentials.   Examples are given in references [5,12,14];   they exclude calculations which do not give satisfactory total energies and calculations which are not self-consistent.
D.W. Bullett's local orbitals work[15] is similar in spirit.   We should note that computed potential energy curves can show artificial discontinuities in some circumstances[16].

Cluster methods provide an extremely powerful approach to total energies and their dependence on geometry, and have considerable flexibility.   It should be said, however, that many calculations fail to achieve anything useful.   These failures, usually avoidable, stem from two main sources. One source is computational, e.g. the failure to ensure full convergence.   The other source is physical or chemical: the boundary conditions may be inadequate, e.g. omission of Madelung terms in ionic crystals, failure to saturate surface bonds in covalent crystals, or failure to keep the bulk and cluster Fermi levels the same in metals; lattice distortions and geometries may be poorly chosen; and empirical parameters (whether in molecular integrals or pseudopotentials) may be inappropriately selected.

I.C Electron gas potentials

This method has been extensively discussed in Chapter (10).   Its general limitations apply to covalent materials. In addition however, we note that the method is normally used with pair potential materials.   The method nevertheless offers a convenient way of obtaining potentials for interaction terms which are inaccessible from empirical sources.

Class II: Empirical methods

Section (3.1) of Chapter (10) described in detail the procedure of extracting potential parameters by fitting variables of a generalised potential model to empirical data.   The chapter also stressed the success which shell model potentials have enjoyed in the study of ionic and semi ionic materials. We recall, however, that such models are again generally of the pair potential type; and that the extensions to include many body terms are difficult to incorporate into lattice simulation codes. One approach which, however, can offer the possibility of representing the many-body effects in a manner which can be used in simulation codes is the bond-charge model, details of which are given in the next section.

II.A Bond charge models

These mean different things to different authors, but all presume bond charges between atoms. For present purposes, we can regard the model as an extension of the shell model, with the bond charge represented by a light "bond shell", with charge $Y_b|e|$. This need not be the same as that deduced from x-ray scattering nor the same as that which Phillips[17] relates to the static dielectric constant. Formally, the shell and bond-charge models may be considered special cases of a more general model. This is clear both from Cochran's review[18] and from the dielectric function formalism of Sinha[19] which specifically discusses extended, as opposed to point, dipoles. References using bond-charge models include (20-23).

The advantage of the model is its formal resemblance to basic theories and to the shell model. This is sufficiently close that bond charge models, in the sense discussed above, can be used in lattice simulation codes of the type described in Chapter (1). Moreover, the model includes three body forces in an automatic and convenient manner.

The disadvantages of the model are that it is not easy to parameterise and that the parameters may be neither unique nor easy to interpret. Special problems are encountered in empirical parameterisations. The first is that it is impossible to use the cohesive energy for the fitting, as the major contribution to it is now the covalent bond energy, which is not modelled. Secondly, we cannot assume that we know the charge distribution. It is reasonable to assume that the charges on the ions are of order unity, and an idea of the size of the bond charge can be obtained from the Phillips-Martin criterion or from consideration of bond polarisability. Thirdly, we do not know (except in homopolar systems) where to put the bond charge. Chemical intuition suggest that it relates to the formal ion charges Z_1, according to the simple relation

$$\frac{r(\text{bond-charge-anion})}{r(\text{bond-charge-cation})} = \frac{Z_{1\,\text{cation}}}{Z_{1\,\text{anion}}} \tag{5}$$

an argument that has some backing from pseudopotential calculations.

These three considerations are enough to show why obtaining satisfactory bond-charge parameters is likely to be a lengthy business; yet the partial failure of the shell model (for example in failure to fit the Σ branch of the phonon-spectrum in II-VI semiconductors) shows that some kind of correction is necessary. The use of second-neighbour and longer-range forces has the disadvantage of lacking

physical reality.   The advantage of the bond-charge model is that, conceptually
at least, it puts in the necessary three-body forces in a physically realistic
manner and, as has been remarked already, is simple to use in the defect
simulation codes, although here we should note that there may be problems in the
case of vacancies.

II.C Valence forces models

This model, discussed in Chapter (8), is one of the oldest approaches; it
writes the total energy in terms of changes in bond lengths Δr_{ij} and changes in
bond angles $\Delta \theta_{ijkl}$.   Various forms are possible depending on how many of the
smaller terms are retained.   The Morse Potential description (ignoring bond
angles) can be regarded as a special case.   References which use valence force
models include (24-32).

The advantages of this approach are that it is relatively easy to get
approximate parameters from molecular data, and that three (and more) body forces
are included simply and clearly.   The model also makes good chemical sense, as
is verified by transferability of parameters.

The disadvantages are that it automatically gives an equilibrium potential,
i.e. each bond is separately in equilibrium and does not cause relaxation if it
is broken.   This is not necessarily a serious problem, for the equilibrium
condition for quite general crystal models leads to a zero volume change about
neutral vacancies to first order, the actual change coming from a balance among
small terms. A further problem is that it is not easy to build the model into
the defect simulation codes, since the lattice search routine is written
differently.   However, effective smaller codes have been used[24]. Finally, we
should note that the models imply a specific coordination
(e.g. four-fold in diamond), and it is not trivial to generalise to other situations.

4.   Validity of models

The majority of models for covalent materials are adjusted empirically so as
to reproduce specific properties of the perfect crystal.   In Class I one may
adjust pseudo-potential parameters, or matrix elements, and fit to lattice parameter,
band widths or gaps, cohesive energy, etc.   In Class II the lattice parameter,
elastic, dielectric and phonon properties are used. How can one decide if the
result is adequate?   We suggest that the four following criteria be applied.

1. The model should predict properties of the perfect lattice other than
those used in deriving the potential parameters. For example, we may investigate
the performance of the model with respect to internal strains, higher order
elastic constants or specific details of phonon spectra or band structure.   It
should be emphasised that the data used in testing the model should be empirical

in origin and not the result of other theories.   In addition we should note that some properties, such as the specific heat, will be predicted well by any theory that is flexible enough to interpolate from experimental data and that the reproduction of such properties tells us little about the viability of the potential model.

2.   The model should predict properties of related lattices, e.g. the properties of the different polytypes of ZnS or SiC.

3.   The model should be consistent with chemical trends. Bond-bending forces and covalency should grow in parallel, and one can use this as a constraint if one uses some specific covalency criterion[1] (e.g. Phillips and Van Vechten, or Pauling, or Coulson) with results for forces in similar crystals.

4.   The model should be consistent with molecular data, with assumptions about transferability, either as a test of an electronic structure method or for preliminary predictions of the major empirical parameters.

Almost all models are fitted to perfect crystal properties. For this reason we stress again that this leaves two cases where there are serious problems, the first occurring when bonds are broken or the coordination is otherwise changed, and the second where critical interactions occur at spacings which are far from crystal equilibrium values.

5.   Three-body forces and beyond

We mentioned earlier the role of three-body forces in covalent crystals. If we omit materials where the volume-dependence of forces is the main feature, there are four main situations.   The first occurs when a third atom or ion weakly perturbs the interaction between the other two, e.g. He_3 or Ne_3[33].   The second is found where there are well-defined directional bonding effects.   These include the bond-bending forces in diamond, in organic molecules, and analogous systems.   The third case arises where there are "ligand field" effects, i.e. where the energy associated with atom A depends on a specific symmetrised combination of the displacements of its neighbours. Examples usually involve Jahn-Teller terms from degenerate states of A or crystal field terms.   Both these contributions tend to be small. Apart from diamond, Jahn-Teller energies rarely exceed 0.1-0.2 eV. Crystal field splittings may be of order 1 eV, but change only modestly with geometry.   This is sufficient to affect elastic constants in oxides[34]. In addition charged defects may have important effects on crystal field energies of nearby host ions; this term has been included in calculations on V^- centres in oxides, for instance.   The final case parallels metallic bonding. Probably the only important examples relevant here are for systems close to a metal-insulator transition (e.g. grey tin, or rare-earth hydrides) and ones involving iodine complexes.   Discussions of Cauchy violations in metals and related aspects of three-body forces are given by C.S.G. Cousins and J.W. Martin[35].

It is improbable that there is a simple, general procedure for dealing with all these types of few-body force. However, there are two encouraging features. First, many-body forces are short-ranged. One can envisage search routines for lattice simulation codes which would work efficiently in locating the ions involved. Secondly, the most difficult part (e.g. when there is a change in bonding pattern) has analogies with molecular reactions. Obviously one can attempt full scale electronic structure calculations for a range of reaction paths. However, it is useful to have an interpolation schemes, and several exist for A + CB → AB + C molecular reactions. There are four relevant schemes: the first is London-Eyring-Polanyi-Sato model[36]. This is very convenient, analytic everywhere, and is easily parameterised from its pairwise limits. It writes the interaction energy of a system combining three atoms or ions A,B and C as

$$[I - (J'-J'')^{\frac{1}{2}}]/(1+\sigma)$$

where $I = (Q_A+Q_B+Q_C)$, $J'=(J^2_{AB}+J^2_{BC}+J^2_{CA})$, $J''=(J_{AB}J_{BC}+J_{BC}J_{CA}+J_{CA}J_{AB})$ and $\sigma=(S^2_{AB}+S^2_{BC}+S^2_{CA})$ in which the Q_i are the Coulomb integrals, the J_{ij} are the exchange integrals and the S_{ij} are the overlap integrals.

The other three schemes involve hyperbolic map functions given a generalised Morse potential[37], switching function[37,38] and bond-order conservation[39,42]. The readers should consult the references for details of these methods, which have not to date been applied to any great extent in solid-state systems.

References

1. Phillips, J.C., Rev. Mod. Phys. 42, 317 (1970). There is also a stimulating exchange of views between Phillips and Pauling in "Physics Today", February 1970 and February 1971.

2. Jette, A.N., Gilbert, T.L. and Das, T.P., Phys. Rev. 184, 884 (1969).

3. Tasker, P.W. and Stoneham, A.M., J. Phys. Chem. Sol. 38, 1185 (1977).

4. Stoneham, A.M. "Theory of Defects in Solids" Oxford University Press, 1975.

5. Mainwood, A. and Stoneham, A.M., J. Less. Comm. Met. 49, 271 (1976).

6. Larkins, F.P., J. Phys. C4, 3065, 3077 (1971).

7. Mainwood, A., Stoneham, A.M. and Larkins, F.P., Sol. St. Electron 21, 1431 (1978).

8. Soma, T., J. Phys. C8, 917 (1975).

9. Turner, R.D. and Inkson, J.C., J. Phys. C11, 3961 (1978).

10. van Kemp, P.E., van Doren, V.E. and Devreese, J.T., Phys. Rev. Lett. <u>42</u>, 1224 (1979).

11. Martin, R.M., Phys. Rev. <u>186</u>, 870 (1969).

12. Anderson, A.B. and Hastings, J.B., Phys. Rev. <u>B15</u>, 2422 (1977).

13. Baraff, G.A. and Schlüter, M., Phys. Rev. Lett. <u>41</u>, 892 (1978).

14. Bernholc, J., Lipari, N.O. and Pantelides, S.T., Phys. Rev. Lett. <u>41</u>, 895 (1978).

15. Bullett, D.W., J. Phys. <u>C8</u>, 3108 (1975).

16. See work by Gregory, A.R., Chem. Phys. Lett. <u>11</u>, 271 (1971), and <u>12</u>, 522 (1972).

17. Phillips, J.C., Phys. Rev. <u>166</u>, 832 (1968).

18. Cochran, W., Crit. Rev. Sol. St. Sci. <u>2</u>, 1 (1971).

19. Sinha, K.P., Crit. Rev. Sol. St. Sci. <u>2</u>, 273 (1971).

20. Go, S., Bilz, H. and Cardona, M., Phys. Rev. Lett. <u>34</u>, 580 (1975).

21. Weber, W., Phys. Rev. <u>B15</u>, 4789 (1977).

22. Nielsen, O.H. and Weber, W., Comp. Phys. Comm. <u>18</u>, 101 (1979).

23. Kunc, K., Balkanski, M. and Nusimovici, M.A., Phys. Stat. Sol.(b), <u>71</u>, 341 (1975), (b) <u>72</u>, 229, 249 (1975).

24. Larkins, F.P. and Stoneham, A.M., J. Phys. <u>C4</u>, 143, 154 (1971).

25. Musgrave, M.J.P. and Pople, J.A., Proc. Roy. Soc. <u>A268</u>, 474 (1962).

26. Keating, P.N., Phys. Rev. <u>145</u>, 637 (1966).

27. McMurray, H.L., Solbrig, A.W. and Boyter, J.K., J. Phys. Chem. Sol. <u>28</u>, 2359 (1967).

28. Tubino, R., Piseri, L. and Zerbi, G., J. Chem. Phys. <u>56</u>, 1022 (1972).

29. Bell, M.I., J. Chem. Phys. <u>62</u>, 3357 (1975).

30. Martin, R.M., Sol. St. Comm. <u>8</u>, 799 (1970).

31. Tubino, R. and Piseri, L., J. Phys. <u>C13</u>, 1197 (1980), discuss the relations between valence and tensor forces, including the relation of valence-bond and Born-Von Karman models.

32. Weber, W., Phys. Rev. <u>B15</u>, 4789 (1977); Sol. St. Comm. <u>18</u>, 673 (1976), combine the valence bond and bond charge models. One can only hope this is unnecessary, since it combined the disadvantages of both models.

33. Navaro, O.A. and Beltran-Lopez, V., J. Chem. Phys. <u>56</u>, 815 (1972); Navaro, O.A. and Nieves, F., J. Chem. Phys. <u>65</u>, 1109 (1976); Bader, R.F.W., Navaro, O.A. and Beltran-Lopez, V., Chem. Phys. Lett. <u>8</u>, 568 (1971). Note, however, that Lloyd, J. and Pugh, D., Chem. Phys. Lett. <u>54</u>, 65 (1978) show that the well-known Kim-Gordon method does not predict 3-body forces well.

34. Solt, G. and Erdös, P., Phys. Rev. $\underline{B22}$, 4718 (1980). For a discussion of parallel effects in rare-earth systems, see Lüthi, B., 1976, Proc. Intermag. Conference.

35. Cousins, C.S.G. and Martin, J.W., J. Phys. $\underline{F8}$, 2279 (1978).

36. Laidler, K.J. and Polanyi, J.C., Prog. React. Kin. $\underline{3}$, 1 (1965).

37. Bunker, D.L. and Blais, N.C., J. Chem. Phys. $\underline{41}$, (1964).

38. Bunker, D.L. and Blais, N.C., J. Chem. Phys. $\underline{37}$, 2713 (1962), $\underline{39}$, 315 (1963).

39. Balint-Kurti, G.G., Adv. Chem. Phys. $\underline{30}$ (1975).

40. Simons, J.P. and Tasker, P.W., J. Chem. Soc. Farad. II, $\underline{70}$, 1496 (1974).

41. Johnston, H.S., Gas Phase Reaction Rate Theory (Ronald, N.Y. (1966)).

42. Bamford, C.H. and Tipper, C.F.H. (Eds.). Comprehensive Chemical Kinetics $\underline{2}$ (Elsevier, Amsterdam, 1969).

SECTION C

APPLICATIONS

CHAPTER 12

DEFECT CALCULATIONS FOR IONIC MATERIALS

by

W.C. Mackrodt

I.C.I., PLC, The Heath, Runcorn, Cheshire WA7 4QE

1.   Introduction

Over the past decade or so, defect calculations for ionic materials have improved substantially in the range, accuracy and subtlety of the information they provide.   In many instances theory can now, with confidence, precede experiment rather than the reverse.   Previous chapters in this volume have dealt with theoretical methods and questions of technique involved in calculations of this type; here we concentrate on some quantitative examples that illustrate the essential features of this development. For the most part (defect) energy calculations have predominated, in part because the underlying theory has been established for longer than that associated with entropy and volume calculations (cf. the discussion by Jacobs et al and Gillan and Lidiard in Chapters (3) and (4)); but principally because experimental data for the latter are comparatively sparse, whereas the bulk of transport, calorimetric and spectroscopic measurements can be related to energy calculations of one sort or another either directly or indirectly. Accordingly, this chapter is devoted largely to energy calculations, though we include important developments in the thermodynamic evaluation of defect parameters.   This is by no means a comprehensive appraisal of what has been done, for this would require a much more extensive account than space here permits: rather, it is intended to convey the essence of what is currently possible from a computational point of view, and more especially of what might be attempted in the near future.

2.   Materials

Before proceeding with our discussion it is, perhaps, worthwhile listing some of the materials that have been examined recently.   These include the alkali halides (1,2), the alkaline-earth fluorides (3,4), $CaCl_2$ (5), AgCl (6,7), γ-CuCl (8), the anion excess fluorites (9), Li_2O (10), the alkaline-earth oxides and CdO (11), the transition metal monoxides (12), α-Al_2O_3 and TiO_2 (13,14), β-Al_2O_3 (15), α-Fe_2O_3 and Cr_2O_3 (16), Fe_3O_4 (17), the perovskite oxides (18), $MgAl_2O_4$, Ga_2O_3 and SnO_2 (10), δ-Bi_2O_3 and CeO_2 (19), UO_2 (20), ZnO (21) and Li_3N (22).   This list is not exhaustive but indicates the type of material that can now be treated.

3. Fundamental defects

For a variety of ionic solids the fundamental point defects comprise vacancies (V) and interstitials (I); and these in combination lead to the four basic types of disorder, viz. Schottky [V+ V-], Frenkel [V+ I+] and [V- I-], and interstitial disorder [I+ I-]. As shown in Table 1, there is theoretical evidence for Schottky and Frenkel defects (of both types) in halides and oxides alike, but as yet no reports of purely interstitial disorder. In most cases the agreement with experiment is good, and for the halides, at least, only minor variation with the type of potential used.

In an exhaustive study Catlow et al[2] have shown that the alkali halides, without exception, are Schottky defective with formation energies ranging from 2.91 eV for NaF to 1.54 eV for LiBr. Frenkel energies are 1-2 eV higher. There is impressive agreement with experiment for the entire range, as exemplified in Table 1, in most cases to within the spread of reported values. The alkaline-earth fluorides, on the other hand, are all anion Frenkel dominant with relative formation energies in the order: anion Frenkel<Schottky<cation Frenkel[3]. Recently, Colbourn et al[5] have calculated that orthorhombic $CaCl_2$ is also anion Frenkel defective with a formation energy of 4.7 eV (compared with 3.0 eV for CaF_2), but suggest that $CaBr_2$ will exhibit Frenkel disorder of both types with almost identical formation energies of about 5.5 eV for cation and anion defects. AgCl and γ-CuCl are predicted to be cation Frenkel defective, while it is speculated that CaI_2 might also be (5). Catlow et al[6] found a formation energy of about 1.4 eV for AgCl in good agreement with the reported data, while Laine[8] in a seldom quoted paper obtained a value of 1.05 eV for cubic (ZnS) structured cuprous chloride compared with 1.11 eV determined experimentally[23].

Oxides generally seem to present more of a problem for many reasons, not least of which is the lack of unambiguous experimental data for a number of materials. The alkaline-earth oxides and even the transition-metal oxides seem to be well-behaved in the sense that quite different potentials lead to the same defect structure which in most cases find experimental support. Thus in MgO, for example, there seems little doubt that vacancies are the dominant intrinsic disorder, low though this may be, with an estimated Schottky energy of between 5-7 eV, in good agreement with the calculated values of 7.5-7.7 eV (4,11). Likewise in FeO, there seems to be little evidence for the existence of free interstitials but strong support for free, doubly-charged vacancies[24] which theory predicts (5,12), despite the complications associated with its non-stoichiometry. The same is true for the other transition metal oxides and even for CdO[4]. In a recent study Mackrodt and Stewart[10] have predicted the antifluorite oxide Li_2O to be cation Frenkel defective (the only example reported so far) with a formation energy of 2.28 eV.

Table 1

Fundamental defect energies

	Theory (eV)	Experiment (eV)
LiF (Schottky)	2.37(2)	2.34-2.68(2)
NaCl (Schottky)	2.22(4)	2.20-2.75(2)
	2.32(2)	-
KBr (Schottky)	2.27(2)	2.37-2.53(2)
RbI (Schottky)	2.16(2)	2.1(2)
MgF_2 (Anion Frenkel)	3.12(4)	-
CaF_2 (Anion Frenkel)	2.75(3)	2.7(3)
SrF_2 (Anion Frenkel)	2.38(4)	2.5(4)
BaF_2 (Anion Frenkel)	1.98(4)	1.91(4)
$CaCl_2$ (Anion Frenkel)	4.7 (5)	-
AgCl (Cation Frenkel)	1.4 (6)	1.45-1.47(6)
γ-CuCl (Cation Frenkel)	1.05(8)	1.11(23)
Li_2O (Cation Frenkel)	2.28(10)	-
MgO (Schottky)	7.5(11)	~ 5-7 (40)
BaO (Schottky)	3.4(11)	
α-Al_2O_3 (Schottky)	5.14(14)*	3.7(26)
α-Fe_2O_3 (Schottky)	4.46(16)*	-
MnO (Schottky)	4.6(12)	-
FeO (Schottky)	6.5(5)	-
$MgAl_2O_4$ (Schottky)	4.15(10)	-
ZnO (Anion Frenkel)	2.51(21)	-
UO_2 (Anion Frenkel)	5.47(20)	5.1(20)

*Energy per defect, i.e. energy to form Schottky quintet divided by five.

However, for other oxides such as α-Al_2O_3, α-Fe_3O_3 and ZnO, not all of which are generally thought of as being partly covalent, the calculated defect structure appears to be extremely sensitive to the exact form of the interatomic potential. Thus in α-Al_2O_3, for example, we find radically different defects predicted from different potentials as shown in Table 2. Experimentally the situation is equally unclear. Studies by Pappis and Kingery[25] and Mohapatra and Kroger[26] support vacancy disorder, whereas a recent investigation of precipitation in star sapphire supports oxygen interstitial defects rather than aluminium vacancies[27]. A recent theoretical study of α-Fe_2O_3 and Cr_2O_3[16] suggests that the situation here might be even more complicated, for its predicts Schottky and Frenkel energies (of both types) to be separated by less than 2 eV in these materials.

Table 2

Fundamental defect energies in $\alpha-Al_2O_3$

Defect	Calculated energy per defect (eV)		
	(a)	(b)	(c)
Schottky	4.18	5.14	5.7
Cation Frenkel	5.22	7.09	10.0
Anion Frenkel	3.79	8.27	7.0

(a) Empirical potential (14)
(b) Electron-gas potential (14)
(c) Electron-gas potential (13)

In the discussion so far it has been assumed that the primary disorder involves solely lattice defects, i.e. vacancies and interstitials. Now for materials such as NaCl, MgF_2, CaF_2, $\alpha-Al_2O_3$ and the like with band gaps of the order of 9-10 eV this is undoubtedly so. However, as reference to Table 3 shows, for a range of oxides the band-gap can be comparable to, and in some instances considerably less than the formation energy of the fundamental lattice defects so that an admixture of electronic and lattice disorder would be expected. Table 4 illustrates this point in the case of zinc oxide for which the singly charged defects are predicted to have appreciably lower formation energies than their doubly-charged counterparts.

In an extensive study of the defect and thermodynamic properties of UO_2[20] Catlow has shown that while anion Frenkel defects are undoubtedly the predominant lattice disorder, electronic disorder represented by the disproportionation reaction:

$$2U_U^x \rightleftharpoons U_U' + U_U^{\cdot}$$

i.e. electron-hole formation, is far more extensive. The calculated energy for this process, ~ 2 eV, is close to the measured thermal band gap of 2-3 eV and appreciably lower than the Frenkel formation energy of 5.4 eV. Furthermore, Harding et al[30] have shown that the entropy contribution from electronic disorder of this type is large and could account for the anomalous specific heat of UO_2, whereas that from Frenkel disorder is much smaller. The transition-metal monoxides provide another example of electronic disorder[12]. Here the basic defects are cation vacancies compensated by holes located on the metal sub-lattice, formed by the reaction

$$1/2O_2(g) \rightleftharpoons V_M'' + O_O^x + 2\dot{h}$$

Table 3

Calculated defect energies compared with the measured band-gaps for a number of halides and oxides

Material	Band-gap (eV)[28]	Fundamental defect energy (eV)
NaCl	8.97	2.22 (4)
CaF_2	10.0	2.75 (4)
MgF_2	11.8	6.24 (4)
MgO	7.77	7.7 (11)
$\alpha\text{-}Al_2O_3$	9.5	5.14 (14)
SrO	5.7	5.9 (11)
MnO	3.7	4.5 (4)
ZnO	3.5	5.9 (21)
UO_2	2.3	5.4 (20)
FeO	~ 2.0	6.6 (5)
VO	0.3	?

Table 4

Fundamental defect energies in ZnO

Defect	Defect energy (eV)	
	(a)	(b)
$(V_{Zn}'', V_O^{\cdot\cdot})$ Schottky	5.88	5.67*
$(V_{Zn}', V_O^{\cdot})$ Schottky	4.76	4.04
(V_{Zn}^X, V_O^X) Schottky	5.66	6.29
$(Zn_i^{\cdot\cdot}, V_{Zn}'')$ Frenkel	7.68	-
$(Zn_i^{\cdot}, V_{Zn}')$ Frenkel	4.46	4.42
$(O_i'', V_O^{\cdot\cdot})$ Frenkel	5.77	-
$(O_i', V_O^{\cdot})$ Frenkel	2.51	-
Small polaron	6.54	6.38*
Large polaron	6.35 - 6.10	
Band edge	-4.40	-
Self-trapped exciton	2.98	-

(a) Calculated energies (21)
(b) Experimental value (29)
* Values obtained from reference (29) based on calculated anion vacancy energy

The nature of the free carrier has long been a source of controversy, though Catlow et al[12] have argued in favour of a small polaron in MnO and a delocalised hole in the reminder of the series. The activation energy for hole hopping was calculated to be 0.2 eV and binding to monovalent cations about 1 eV[12], both of which agree with the available data.

4. Defect interactions

Defect interactions are known to play an important role in many solid state processes, yet the exact nature of these interactions is often ill-understood. We have already mentioned the binding of holes to monovalent cations in transition-metal oxides, for this determines the temperature-dependent semiconductivity of the doped materials; we discuss impurity interactions more fully in section 5. Here we consider three examples of the interaction between intrinsic defects.

We begin by considering what is perhaps the simplest defect interaction, viz. vacancy pair formation. In NaCl the calculated interaction energy of a vacancy pair is 0.9 eV. This is less than half the Schottky energy, 2.32 eV, so that simple vacancies are the major defect. However, vacancy pairs make a significant, though minor contribution to diffusion and Catlow et al[2] have obtained an Arrhenius energy for pair diffusion of 2.32 eV, which agrees well with the experimental values (2.35 eV; 1.96 eV) and (2.37 eV; 2.54 eV) for cation and anion transport. However, at the surface the picture is (calculated to be) rather different. Mackrodt and Stewart[31] obtained a pair formation energy of 1.08 eV at the [001] surface, compared with the corresponding Schottky energy of 2.64 eV; and 0.13 eV at the [011] surface as against a Schottky energy of 0.87 eV. The surface defect structure of NaCl, therefore, is predicted to be dominated by vacancy pairs rather than single vacancies and this should have important consequences for the surface properties. If we now consider MgO, the situation is similar, though more interesting in some respects. The bulk Schottky energy is calculated to be 7.7 eV and the vacancy-pair formation energy 5.2 eV[11]. Isolated vacancies, therefore, are predicted to dominate the intrinsic bulk disorder, although this, of course, will be at an extremely low level even up to the melting point because of the energies involved. At the [001] surface, the Schottky energy is predicted to be 9.3 eV as opposed to 3.3 eV for the formation of vacancy pairs. Not only should the latter predominate then, but the level of disorder should be appreciably greater than in the bulk. This could contribute to properties such as sintering, though as Colbourn and Mackrodt[32] have shown recently, impurities are likely to dominate the surface as much as the bulk. If we now consider the [110] surface, we find a Schottky energy of 4.5 eV and a negative vacancy-pair energy[11]. We would predict, therefore, that the [110] surface is unstable by virtue of ion-pair removal, and this seems to agree well with the thermal faceting of the [110] surface reported by Henrich[33].

The interaction of defects with dislocations and grain-boundaries, particularly in relation to ion diffusion and mass transport, is often invoked as a means of 'explaining' anomalous behaviour of one sort or another. It is seldom, if ever, however that the magnitude, or even the sign of these interactions can be determined to justify such claims. Now it is in situations such as this that calculations can be most useful. Woo et al[34] have calculated the interaction energies of vacancies with an a/2 [110] edge dislocation at various positions shown in Figure 1. Their results are given in Table 5. Not only do they find interesting changes with location about the dislocation core ($\perp$), the magnitude of which might not have been expected, but the results show a remarkable similarity to [110] surface calculations by Mackrodt and Stewart[31] also shown in Table 5. The positions C and D lie on what could be described as an 'internal' [110] surface and the position A one layer below; and as Table 5 shows, the interaction energies for the free and 'internal' surfaces are almost identical. It is often conjectured that dislocations and grain boundaries have properties similar to free surfaces and here we have a persuasive demonstration that this might be so.

We take as our final example in this section recent theoretical studies of defect aggregation in $Fe_{1-x}O$. Both X-ray and neutron diffraction investigations have suggested the presence of vacancy-interstitial clusters in $Fe_{1-x}O$, although the exact nature of these clusters and their mode of formation have not been completely resolved. In an elegant demonstration of the use of lattice calculations Catlow and Fender[35] showed that the basic cluster is a complex of

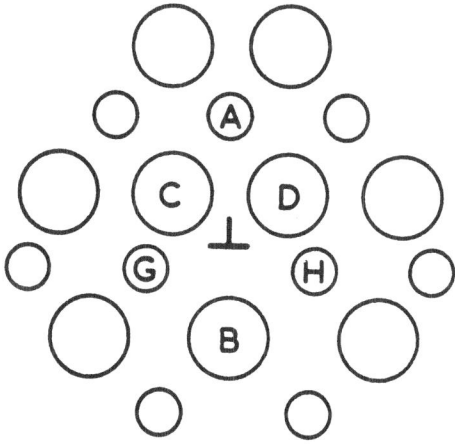

Figure 1   Core configuration of an a/2 [110] edge dislocation in MgO

Table 5

Interaction energies (eV) of vacancies with an a/2 [110] edge dislocation and a free [110] surface in MgO

Defect	[110] surface (31)		a/2 [110] edge dislocation (34)			
	Surface	1 layer below	A	B	C/D	G/H
Cation	-1.6	1.1	1.2	-0.1	-1.0	-0.1
Anion	-1.7	1.2	1.2	-0.1	-1.1	-0.2

four vacancies and an interstitial Fe^{3+} ion, originally proposed by Cheetham et al[36] to account for their neutron results; and furthermore that these clusters aggregate to give larger 'edge-shared' clusters of the type shown in Figure 2 which were shown to be more stable than the structure proposed by Koch and Cohen[37]. More recently, Colbourn et al[5] have confirmed this view and sought to establish the detailed mechanism of cluster formation. Relative to isolated vacancies and $Fe_{Fe}^{\cdot}$ ions, the 4:1 cluster is stable by about 1.4 eV per interstitial and the 6:2 cluster by 1.1 eV, whereas the 13:4 Koch-Cohen complex is unstable by nearly 5 eV per $Fe_{Fe}^{\cdot}$ ion and the smaller 2:1 Koch complex by 2.4 eV. However, the essential problem is concerned with the 'primary' formation of the interstitial, $Fe_i^{\cdot\cdot\cdot}$, from $Fe_{Fe}^{\cdot}$; for the simple process

$$Fe_{Fe}^{\cdot} \rightleftharpoons Fe_i^{\cdot\cdot\cdot} + V_{Fe}''$$

is highly endothermic, requiring about 15 eV. This, of course, is reduced by defect interactions, and in particular Colbourn et al[5] have shown that the formation of the [111] complex involving two vacancies and an $Fe_{Fe}^{\cdot}$ ion by the reaction

$$V_{Fe}'' + \{V_{Fe}'': Fe_{Fe}^{\cdot}\}_{110} \rightleftharpoons \{2V_{Fe}'': Fe_{Fe}^{\cdot}\}_{111}$$

is the most energetic process in the formation of the 4:1 cluster and thereafter the 6:2 complex, requiring about 0.5 eV. The complex $\{2V_{Fe}'': Fe_{Fe}^{\cdot}\}_{111}$ then rearranges to give the 3:1 complex.

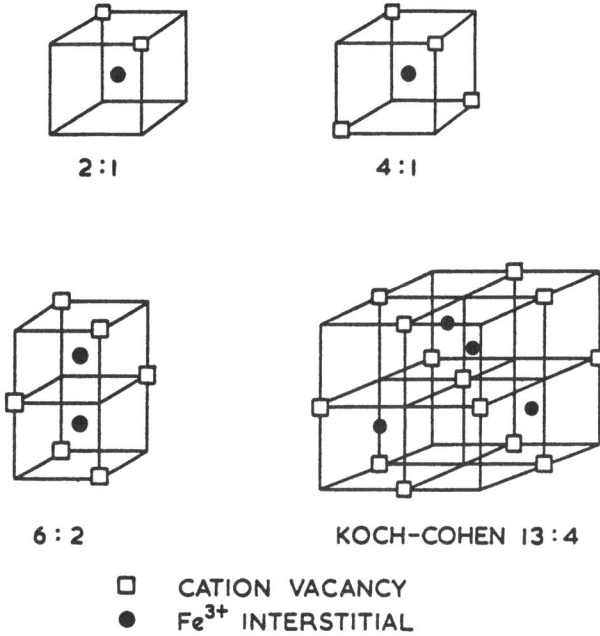

2:1   **4:1**

6:2   **KOCH-COHEN 13:4**

☐ CATION VACANCY
● Fe^{3+} INTERSTITIAL

Figure 2   Vacancy-interstitial complexes in $Fe_{1-x}O$

5. Ion diffusion

Lattice defect calculations have been closely associated with diffusion in ionic materials from the beginning, and in particular with the interpretation and evaluation of the Arrhenius energy, W, which is related to the diffusion coefficient, D(T), by

$$D(T) = D_0 \exp(-W/kT).$$

In general we can identify two contributions to W, viz. the formation energy, U_F, of the diffusing defect and the activation energy of migration, E_A. For diffusion which is intrinsically controlled, U_F is simply the Schottky or Frenkel energy per defect of the dominant disorder.   On the other hand, if disorder is extrinsically controlled as in the generation of cation vacancies by oxygen incorporation, for example

$$1/2 \; O_2(g) \rightleftharpoons V_M'' + O_O^x + 2h^{\cdot}$$

U_F corresponds to the appropriate enthalpy of reaction. Alternatively, if disorder is impurity controlled, U_F is effectively zero and $W = E_A$. Since we are normally concerned with diffusion processes at elevated temperatures, E_A is calculated most simply on the basis of a straightforward classical model. Thus for vacancy migration in a material such as NaCl, for example, E_A is the difference in energy between the transition state, T, and the initial state, I, shown in Figure 3. As we shall see, however, ion migration is often complicated by additional factors, and it is precisely in situations such as these that defect calculations can be most useful.

We begin by considering the simplest of processes, viz. vacancy migration in the alkali halides, which Catlow et al[2] have examined in detail. Taking NaCl as a typical example we find calculated activation energies for migration of 0.67 eV and 0.72 eV respectively for cation and anion vacancies in good agreement with the experimental values of 0.65 eV - 0.80 eV and 0.77 eV - 1.62 eV[2]. With the other alkali halides the trend is much the same. An interesting feature of these calculations first mentioned by Guerin and Laforgue[38] and investigated more fully by Catlow et al[2] is that the transition state saddle points can have large displacements up to a quarter of a lattice spacing out of the plane of the adjacent lattice sites, though the effect on the activation barrier seems to be small.

However, ion migration is seldom as simple as this as we see for MgO, which provides a good illustration of some of the uncertainties associated with even the simplest of materials. Impurity controlled vacancies are undoubtedly the predominant disorder and on this basis Mackrodt and Stewart[11] calculated

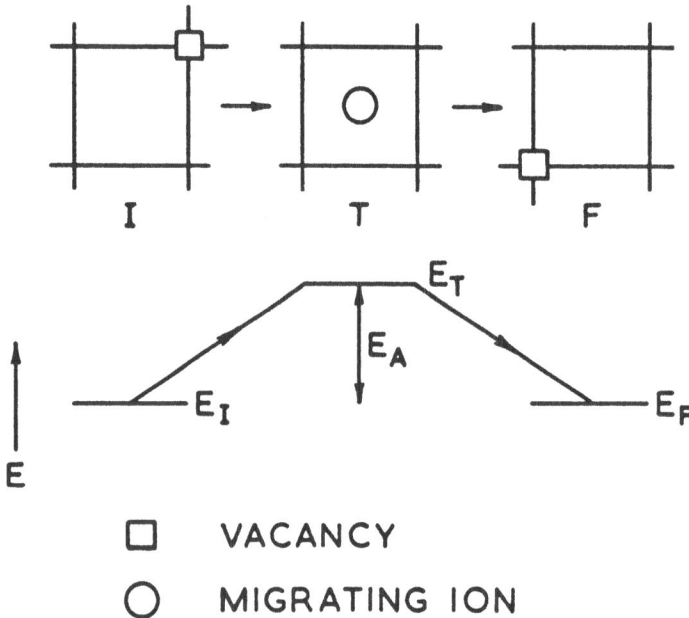

Figure 3 Schematic diagram for vacancy migration in fcc solids

activation energies of 2.18 eV and 2.38 eV respectively for free cation and anion vacancy migration respectively. While the anion value seemed to be in good agreement with the most recent data of Shirasaki et al[39] viz. 2.42 eV and 2.61 eV, the cation value was significantly lower than that of 2.76±0.08 eV reported by Wuensch et al[40]. Sempolinski and Kingery[41] have re-investigated the diffusion of magnesium vacancies and report a value of 2.28 ± 0.2 eV in much better agreement with the theoretical value. Subsequently, Colbourn and Mackrodt[19] have calculated the activation energy for migration in the presence of Ca^{2+}, Al^{3+}, Sc^{3+} and Fe^{3+} (common impurities in MgO) and find an average value of 2.65 eV which neatly fits the range previously found by Wuensch et al[40]. Thus from a combination of experiment and theory we have a more complete picture of cation vacancy migration in MgO, viz. free vacancies require an energy of 2.2 - 2.3 eV for migration but this is increased by 0.5 eV - 0.7 eV in the presence of associated impurities. Surface calculations[31] also indicate the sort of differences that might be expected between single crystal and polycrystalline MgO, for as we have mentioned earlier the properties of internal and external surfaces seem to be similar in some respects. Thus at the [001] and [110] surfaces Mackrodt and Stewart[31] obtained values of 0.7 eV and 1.2 eV respectively for cation vacancy migration so that on this basis alone migration would be expected to be more facile in the polycrystalline material.

CaF_2 is another good example of difficulties associated with an apparently simple material. Anion Frenkel disorder predominates here and for anion vacancy migration there is excellent agreement between calculation and experiment, the corresponding values of W being 1.89 eV and 1.92 eV respectively[4]. Cation migration, however, presents more of a problem for there are three mechanisms by which this can take place, viz. single vacancy, divacancy and trivacancy, with calculated Arrhenius energies of 5.05 eV, 4.72 eV and 5.75 eV respectively[4]. Experimental values of 3.8 eV, 4.15 eV, 4.37 eV and 5.16 eV have been reported, which might narrow the choice to a single or divacancy mechanism, but beyond this we cannot go.

In the previous examples we have seen that to a large extent calculation and experiment can be matched fairly closely provided that complicating factors such as defect interactions can be included adequately in the calculation of E_A. To put the subject in perspective we end this section with an example for which this has not been possible so far. James[42] has recently examined diffusion mechanisms in α-Al_2O_3 in some detail and his work serves as a good illustration of the difficulties that await the analysis of more complex systems. Given Schottky disorder, which seems to be justified on experimental grounds[25,26], anion migration would be expected to occur by a vacancy mechanism. For the corundum structure of α-Al_2O_3, unlike the simpler cubic materials referred to earlier, there are <u>five</u> possible routes that can contribute to conduction (migration) with activation energies varying from 1.77 eV to 9.52 eV. Furthermore James has shown that the mechanism corresponding to the lowest of these should lead to _isotropic_ conductivity. Now a value of 1.77 eV is somewhat lower than, though comparable to the value of 2.5 eV observed by Oishi and Kingsley[43] in their ^{18}O tracer studies, bearing in mind possible impurity interactions, and by Kitazawa and Coble[44] from conductivity measurements. However, Oezkam and Moulson[45] found the conductivity to be highly anisotropic, $\sigma_{||}/\sigma_{\perp} \sim 10$, with an activation energy of ~ 4 eV. More recently Mohapatra and Kroger[26] reported a substantially reduced anisotropy but increased activation energies of 4.84 eV and 4.39 eV for conduction parallel and perpendicular to the c-axis. Now a calculated Schottky energy of 5.14 eV leads to an Arrhenius activation energy for intrinsic diffusion of 6.91 eV which seems to be in fair agreement with that of 6.59 ± 1.08 eV found by Oishi and Kingery[43] and the more recent value 6.37 ± 0.43 eV by Reddy and Cooper[46]. However, Reed and Wuensch[47] have recently reported an Arrhenius energy of 8.15 ± 0.30 eV for anion diffusion which they propose is extrinsically controlled and equal to one third of the Schottky energy plus the migration energy. The corresponding calculated value for this quantity is over 11 eV which in view of the agreement for intrinsic diffusion raises possible doubts as to the mechanism suggested by Reed and Wuensch[47]. James has concluded that a number of these anomalies are due to defect clustering resulting from the heavy doping of the experimental materials, but a fuller analysis along these lines has not been attempted so far.

6. Impurities

The treatment of lattice impurities is perhaps the single most important application of defect calculations, for in very many circumstances, both the defect structure and population are extrinsically controlled, most often by impurities. In the majority of cases cation rather than anion impurities have been the more studied, experimentally as well as theoretically, and it is useful to distinguish isovalent from aliovalent doping. Our examples here relate solely to oxides for the doping of the alkali halides and alkaline-earth halides has been discussed previously. As before they are chosen to illustrate the novelty and scope of the information that calculations can provide rather than give a comprehensive review of what has been done.

We begin by considering isovalent doping which is characterised largely by size effects as illustrated by the doping of MgO by Be^{2+}, Ca^{2+}, Sr^{2+} and Ba^{2+}. There is a near parabolic relationship between the calculated heats of solution and ion size, while the binding energy of the impurity to what may be described as a 'strain-compensated' vacancy is also of the same form. As shown in Table 6, what is surprising here is the magnitude of this interaction, which in the case of Be^{2+} is calculated to be nearly -1.5 eV[11]. Sangster and Stoneham[48] have recently examined divalent transition-metal ions and Mg^{2+} substituted in CaO, SrO and BaO and

Table 6

The doping of MgO by isovalent impurities

Impurity	E_s (eV)	E_b (eV)	E_m (eV)
Be^{2+}	2.69	-1.48	1.48 (1.60, 1.68)*
Ca^{2+}	1.07	-0.09	2.29 (2.13, 2.76)
Sr^{2+}	2.84	-0.20	1.78 (2.91, 3.09)
Ba^{2+}	4.94	-0.31	1.17 (1.6 - 1.8)

E_s - solution energy

E_b - vacancy binding energy

E_m - migration energy

*Experimental values in brackets taken from reference (11).

their results reveal a number of interesting features. All the impurity ions investigated, viz. Mg^{2+}, Mn^{2+}, Fe^{2+}, Co^{2+} and Ni^{2+}, are calculated to remain at normal lattice sites in CaO, but move 'off-centre' in SrO and BaO. The effect is most pronounced in BaO where, for example, they predict Ni^{2+} to move off-centre in the [111] direction by approximately 0.4Å with a gain in energy of 1.42 eV from the normal lattice position. While the experimental data, mainly e.s.r. are limited here, there is evidence for off-centre behaviour for Fe^{2+} and Co^{2+} in SrO and Mn^{2+} in BaO all three of which Sangster and Stoneham correctly predict.

We turn now to aliovalent doping, where the situation is generally more interesting than the isovalent case because of the diversity of effects that can arise. The necessity for charge compensation leads to the possibility of defect-impurity interaction (referred to in section 4), the associated temperature dependence of which is an important factor in determining the properties of many materials. Gourdin and Kingery[49] have reported calculations of Al^{3+} and Fe^{3+} in MgO with particular emphasis on defect-impurity interactions. For both impurities they found the [100] vacancy dimer and trimer more stable than the corresponding [110] complexes with interaction energies per impurity ion varying from about 0.5 eV to 1.1 eV. Their results are in broad agreement, though not always in detail, with similar calculations by Colbourn and Mackrodt[32] who find similar interaction energies but different symmetries for the lower energy complexes in some cases. Gourdin and Kingery[49] have also considered more extensive 'spinel-like' vacancy clusters in relation to the precipitation of magnesium aluminate (spinel) and magnesium ferrite (inverse spinel) from magnesia solid solutions. The calculations here are similar to those by Catlow and Fender[35] for $Fe_{1-x}O$, that is to say they are based on tetrahedrally coordinated interstitial ions (Mg^{2+} in the case of the normal spinel and Fe^{3+} for the inverse) surrounded by four cation vacancies. While the calculations are uncertain in some respects what they demonstrate quite clearly is the enhanced stability of the spinel-like and even more so the inverse spinel-like clusters in relation to isolated defects and vacancy dimers and trimers.

An interesting point to emerge from recent calculations is that two quite different situations seem to arise with regard to the type of charge-compensating defects that impurities can induce. On the one hand we have cases such as the doping of NaCl by Ca^{2+} or MgO by Sc^{3+}, for example, in which the nature of the compensating defects is _independent_ of any primary clustering that might occur. That is to say, in both these examples the extrinsic ions are compensated by vacancies rather than interstitials, and this is predicted to remain so _even allowing for dopant interactions_ either at low temperature or high impurity concentrations. This is in marked contrast with the doping of MgO by Li^+ or α-Al_2O_3 by Mg^{2+} under reducing conditions. Here calculations suggest that the mode of compensation is affected by interactions with the dopant as shown in Table 7, in which H_s is the calculated

Table 7

Impurity-defect interactions in MgO and α-Al_2O_3

	Defect	H_s (eV)	
		High T	Low T
$\underline{Li^+:MgO}$	Anion vacancy	2.78	1.41
	Cation Interstitial	4.90	-
	Self-compensation	3.92	0.72

$$2 Li'_{Mg} + V^{\cdot\cdot}_O + MgO \rightleftharpoons Li^{\cdot\cdot}_i\text{-}V''_{Mg}\text{-}Li^{\cdot\cdot}_i + Mg^x_{Mg}$$

$$\Delta E = -2.06 \text{ eV}$$

	Defect	High T	Low T
$\underline{Mg^{2+}:\alpha\text{-}Al_2O_3}$			
	Anion vacancy	3.15	1.87
	Cation Interstitial	3.59	-
	Self-compensation	3.03	2.22

$$4Mg'_{Al} + 2 Mg^{\cdot\cdot}_i \rightleftharpoons 3\{Mg'_{Al}\text{-}V^{\cdot\cdot}_O\text{-}Mg'_{Al}\} + \alpha\text{-}Al_2O_3$$

$$\Delta E = -1.16 \text{ eV}$$

enthalpy of solution. In what is referred to as the high temperature limit (High T) Li^+ in MgO is thought to give rise to anion vacancies, while Mg^{2+} in α-Al_2O_3 is predicted to lead to both anion vacancies and Mg^{2+} interstitials (the self-compensating mode). In the low temperature limit, on the other hand, that is to say, allowing for primary association, the self-compensation mode is predicted to the most favourable for Li^+ in MgO and anion vacancy compensation for

Mg^{2+} in α-Al_2O_3[14]. The corresponding enthalpies of association per dopant are calculated to be -2.06 eV and -1.16 eV respectively.

We end this section on impurities by returning to surface effects, for as we have noted previously, this is an area where quantitative experimental information is sparse and the potential input that calculations might provide is considerable. It is clearly of importance in interfacial phenomena such as catalysis and corrosion, for example, and is likely to be an active area of interest in the near future. Table 8 lists defect energies, E_s, for a number of dopant ions in the bulk

Table 8
Surface defect energies of impurities in MgO

Defect	E_s (eV)	
	Bulk	[001] Surface
$Li_{Mg}^{\prime}$	16.27	16.40
$Li_{Mg}^{\prime}$-$V_O^{\cdot\cdot}$-$Li_{Mg}^{\prime}$)	52.71	52.52
$Na_{Mg}^{\prime}$	18.64	17.95
($Na_{Mg}^{\prime}$-$V_O^{\cdot\cdot}$-$Na_{Mg}^{\prime}$)	57.56	54.79
Be_{Mg}^{x}	-3.57	-3.80
Ca_{Mg}^{x}	5.82	4.42
$Al_{Mg}^{\cdot}$	-30.29	-29.77
($Al_{Mg}^{\cdot}$-$V_{Mg}^{\prime\prime}$-$Al_{Mg}^{\cdot}$)	-36.12	-35.29
$Sc_{Mg}^{\cdot}$	-22.29	-21.78
($Sc_{Mg}^{\cdot}$-$V_{Mg}^{\prime\prime}$-$Sc_{Mg}^{\cdot}$)	-20.71	-21.59
($Sc_{Mg}^{\cdot}$-$V_{Mg}^{\prime\prime}$)	2.33 ,	1.66 ($\parallel$), 1.77 ($\perp$)*

All values in the table are taken from reference (32)

* $\parallel$ and $\perp$ refer to the defect configurations which
are parallel and perpendicular to the surface respectively.

and at the [001] surface in MgO, recently calculated by Colbourn and Mackrodt[32]. Here we have both size and charge effects, which in combination lead to a complicated situation from which no simple general rules seem to emerge. Thus Li^+ as an isolated defect appears to be slightly more stable in the bulk than at the [001] surface, whereas the neutral defect ($Li_{Mg}^{\prime}$-$V_O^{\cdot\cdot}$-$Li_{Mg}^{\prime}$) has marginally lower energy at the surface. For Na^+, on the other hand, the surface defect energy is appreciably lower than the bulk, irrespective of the charge state of the defect. The differences between bulk and surface are predicted to be small, whereas those for Ca^{2+} are large, with a marked preference in favour of the surface at thermal equilibrium. Al^{3+} in both charged and neutral defects is predicted to have a lower energy in the bulk than at the surface, while Sc^{3+} shows variations depending on the type of defect. Free Sc^{3+} is more stable in the bulk, due to the enhanced

Madelung potential, though the singly-charged dimer and neutral trimer have lower energies at the surface.

7. Oxidation and reduction of oxides

The discussion so far has concentrated almost exclusively on the fundamental properties of defects, though we have made reference to the calculations by Catlow and Fender[35] and Colbourn et al[5] on clustering in $Fe_{1-x}O$, and Gourdin and Kingery[49] on precipitation in magnesia solutions, to indicate the connection with solid state processes. Here we pursue this theme still further by outlining if only in brief the oxidation and reduction of oxides. We consider the following reactions

$$1/2\ O_2(g) \rightleftharpoons V_M'' + O_O^X + 2h^\cdot$$

and

$$O_O^X \rightleftharpoons 1/2\ O_2(g) + V_O^{\cdot\cdot} + 2e'$$

and the calculated energies E_0 and E_R respectively for five oxides as shown in Table 9. We have assumed throughout that conduction band electrons were delocalised and the holes were large polarons. No allowance has been made for the association of electrons and holes with vacancies.   The binding energy of holes to form V centres is small so that

Table 9
Energies of oxidation and reduction

Materials	E_0 (eV)	E_R (eV)	E_0' (eV)
MgO(5)	6.2 (a)	10.0	-1.60
CaO(5)	3.6	9.1	-1.70
α-Al_2O_3(14)	13.0	9.9	4.43 (c)
FeO(5)	-2.7 (b)	5.6	-9.2
ZnO(21)	4.9	2.2	1.7

(a) Experimental value 6.32 ± 0.8 eV (50)

(b) Experimental value -2.5 to -2.9 eV (51)

(c) Experimental value 4.05 eV (52).

E_0 should not be altered much;   however, E_R could be reduced substantially, for the trapping of electrons at anion vacancies is much larger, of the order of 2-4 eV.   Not only do the results confirm the general oxidation/reduction properties of these materials, viz., that α-Al_2O_3 has a very narrow range of stoichiometry, that FeO is unstable with respect to oxygen incorporation, that ZnO is easily reduced and that the alkaline earth oxides do so with difficulty: but where available we get quantitative agreement with experiment. Furthermore, if we consider the doping of the divalent oxides with say Li_2O or Na_2O and

α-Al$_2$O$_3$ with MgO to create oxygen vacancies we can calculate the energy, $E_0^!$, for the reaction

$$1/2 \ O_2(g) + V_O^{\cdot\cdot} \rightleftharpoons 2h^\cdot + O_O^X \quad .$$

The results are also given in Table 9. Thus we would predict an increase in the (p-type) conductivity of the first four oxides, as indeed is the case, whereas in ZnO which is normally n-type, we would predict a decrease in the semi-conductivity which is also found experimentally to be the case. Thus for the simplest oxidation/reduction reactions of admittedly simple materials defect calculations seem to be able to estimate energy changes which are in good agreement with experiment.

8. Thermodynamic relations

In previous sections of this chapter we have made numerous comparisons between calculated and experimental defect energies on the assumption that it is generally valid to do so. In this final section we draw attention to an important piece of work that serves to emphasise both the relationship (and distinction) between theoretically and experimentally derived quantities, and also the increasing accuracy of both that makes such a distinction possible. Recently, Catlow et al[53] have noted that whereas experimental quantities invariably are determined at constant pressure, calculations are usually carried out assuming constant volume, or more correctly at constant lattice parameter(s). With reference to the high temperature ionic conductivity of AgCl they have calculated the temperature dependence of the formation energy of Frenkel defects at constant volume (u_v) and at constant pressure (h_p). u_v and h_p are related by

$$h_p = u_v + (T\beta_p/\kappa_T) \ V_p$$

in which V_p is the volume change at constant pressure, T the temperature, β_p the volume thermal expansion coefficient and κ_T the isothermal compressibility. Thus h_p has an explicit temperature dependence whereas that of u_v derives from the change in the lattice parameter with temperature. The results of this calculation are given in Table 10. Two main points emerge. They are first, that u_v, h_p and V_p all vary appreciably with temperature. The second, and possibly more important point is that u_v and h_p vary with temperature in opposite ways. While it is true that AgCl is perhaps an extreme example of temperature dependent quantities, future analyses for other systems might well have to consider the possible inclusion of these effects.

Table 10

Calculated formation energies of Frenkel defects in AgCl at constant volume (u_v) and

constant pressure (h_p) (reference 53)

T/K	u_v (eV)	h_p (eV)
300	1.371	1.486
350	1.351	1.489
400	1.331	1.500
450	1.309	1.520
500	1.284	1.550
550	1.255	1.592
600	1.222	1.644
650	1.185	1.701
700	1.143	1.756

9.   Conclusions

In conclusion then, this brief survey has sought to bring together just some of the diverse problems that are now tractable from a computational point of view. In the forseeable future two areas stand out as ripe for further progress.   They are:

(i) The development of improved solid state potentials, including many body effects and refinements to the shell model

(ii) The application of defect lattice methods to extended defects such as grain boundaries, stacking faults and shear planes.

The second, in particular, could find wide application in many important areas of solid state science and technology.

References

1. Catlow, C.R.A., Diller, K.M. and Norgett, M.J., J. Phys. C10, 1395 (1977).

2. Catlow, C.R.A., Corish, J., Diller, K.M., Jacobs, P.W.M. and Norgett, M.J. A.E.R.E. Report TP 713 (1977).

3. Catlow, C.R.A., Norgett, M.J. and Ross, T.A., J. Phys. C10, 1627 (1977).

4. Mackrodt, W.C. and Stewart, R.F., J. Phys. C12, 431 (1979).

5. Colbourn, E.A., Kendrick, J. and Mackrodt, W.C.   Unpublished results.

6. Catlow, C.R.A., Corish, J. and Jacobs, P.W.M., J. Phys. C12, 2433 (1979).

7. Jacobs, P.W.M., Corish, J. and Catlow, C.R.A., J. Phys. C13, 1977 (1980).

8. Laine, J., J. Phys. C6, 637 (1973).

9. Bendall, P.J., Catlow, C.R.A. and Fender, B.E.   In 'Computer Simulation in the Physics and Chemistry of Solids'. (1980).   Daresbury Study Weekend 9-11 May 1980. (Daresbury report DL/SC1/R15).

10. Mackrodt, W.C. and Stewart, R.F.   Unpublished results.

11. Mackrodt, W.C. and Stewart, R.F., J. Phys. C12, 5015 (1979).

12. Catlow, C.R.A., Mackrodt, W.C., Norgett, M.J. and Stoneham, A.M., Phil. Mag. 35, 177 (1977); Phil. Mag. 40, 161 (1979).

13. Dienes, G.J., Welch, D.O., Fischer, C.R., Hatcher, R.D., Lazareth, O. and Samberg, M., Phys. Rev. B11, 3060 (1975).

14. Catlow, C.R.A., James, R., Mackrodt, W.C. and Stewart, R.F., Phys. Rev. (1981). To be published.

15. Walker, J.R. In 'Computer Simulation in the Physics and Chemistry of Solids'. (1980).   Daresbury Study Weekend 9-11 May 1980. (Daresbury Report DL/SC1/R15).

16. Catlow, C.R.A., Colbourn, E.A., Mackrodt, W.C. and Thirlby, P. Unpublished results.

17. Thirlby, P.   Unpublished results.

18. Kilner, J.A., Barrow, P., Brook, R.J. and Norgett, M.J., J. Power Sources 3, 67 (1978).

19. Colbourn, E.A. and Mackrodt, W.C.   Unpublished results.

20. Catlow, C.R.A., Proc. Roy. Soc. A353, 533 (1977).

21. Mackrodt, W.C., Stewart, R.F., Campbell, J.C. and Hillier, I.H., J. de Physique, C6, 64 (1980).

22. Catlow, C.R.A. and Walker, J.H. In 'Computer Simulation in the Physics and Chemistry of Solids'. (1980). Daresbury Study Weekend 9-11 May 1980. (Daresbury Report DL/SC1/R15).

23. Matsui, T. and Wagner, J.B., J. Electrochem. Soc. 124, 610 (1977).

24. Chen, W.K. and Peterson, N.L., J. Phys. Chem. Solids, 41, 335 (1980).

25. Pappis, J. and Kingery, W.D., J. Am. Ceram. Soc. 44, 459 (1961).

26. Mohapatra, S.K. and Kroger, F.A., J. Am. Ceram. Soc. 61, 106 (1978).

27. Phillips, D.S., Mitchell, T.E. and Heuer, A.H., Phil. Mag. 42, 417 (1980).

28. Strehlow, W.H. and Cook, E.L., J. Phys. Ref. Data 2, 163 (1973).

29. Kroger, F.A. 'The Chemistry of Imperfect Crystals' (1964). (Amsterdam: North-Holland).

30. Harding, J.H., Masri, P. and Stoneham, A.M., J. Nuclear Mat. 92, 73 (1980).

31. Mackrodt, W.C. and Stewart, R.F., J. Phys. C10, 1431 (1977).

32. Colbourn, E.A. and Mackrodt, W.C.   To be submitted (1982).

33. Henrich, V.E., Surf. Sci. 57, 385 (1976).

34. Woo, C.H., Puls, M.P. and Norgett, M.J., J. de Physique, C7, 557 (1976).

35. Catlow, C.R.A. and Fender, B.E., J. Phys. C8, 3267 (1975).

36. Cheetham, A.K., Fender, B.E. and Taylor, R.I., J. Phys. C4, 2160 (1971).

37. Koch, F. and Cohen, J.B., Acta. Crystall. B25, 275 (1969).

38. Guerin, P. and Laforgue, A., J. de Physique C7, 379 (1976).

39. Shirasaki, S. and Yammamura, H., Jap. J. App. Phys. 12, 1654 (1973).

40. Wuensch, B.J., Steele, W.C. and Vasilos, T., J. Chem. Phys. 58, 5258 (1973).

41. Sempolinski, D.R. and Kingery, W.D., J. Am. Ceram. Soc. 63, 664 (1980).

42. James, R., Ph.D. Thesis, University of London (1979).

43. Oishi, Y. and Kingery, W.D., J. Chem. Phys. 33, 480 (1960).

44. Kitazawa, K. and Coble, R.L., J. Am. Ceram. Soc. 57, 245 (1974).

45. Oezkam, O.T. and Moulson, A.J., J. Phys. D3, 983 (1970).

46. Reddy, K.P.R. and Cooper, A.R., Am. Ceram. Soc. Bull. 57, 306 (1978).

47. Reed, D.J. and Wuensch, B.J., J. Am. Ceram. Soc. 63, 83 (1980).

48. Sangster, M.J.L. and Stoneham, A.M., Phil. Mag. B, 43, 597 (1981).

49. Gourdin, W.H. and Kingery, W.D., J. Mat. Sci. 14, 2053 (1979).

50. Sempolinski, D.R., Kingery, W.D. and Tuller, H.L., J. Am. Ceram. Soc. 63, 669 (1980).

51. Kofstad, Per., 'Nonstoichiometry, Diffusion and Electrical Conductivity in Binary Metal Oxides, p.225 (1972) (New York: Wiley-Interscience.

52. Mohapatra, S.K., Tiku, S.K. and Kroger, F.A., J. Am. Ceram. Soc. 62, 50 (1979).

53. Catlow, C.R.A., Corish, J., Jacobs, P.W.M. and Lidiard, A.B., A.E.R.E. Report TP 873 (1980).

CHAPTER 13 POINT DEFECT CALCULATIONS IN METALS

by

R. Taylor
Physics Division, National Research Council of Canada,
Ottawa K1A OR6

1. Introduction

In metals, the term point defect is usually assumed to refer to vacancies, divacancies and interstitials, although impurities also constitute a major class of
point defects. This chapter will concentrate principally on the problem of the
calculation of properties of vacancies in simple metals. This problem has received
considerable attention in the literature over the last decade; and is,
surprisingly, still not completely understood. Following the discussion of the vacancy
studies, some comment will be made on the extent to which lessons learned from
this problem can be extended to other point defect calculations in metals.

At first sight it would appear that given a suitable pair potential, the
calculation of the formation energy of a vacancy in a metal, E_f^V, is a trivial task.
In removing a metallic ion to the surface, all the neighbour bonds at the vacancy
site would be broken and then half of these would be rejoined at the surface. Hence,
for the unrelaxed vacancy, E_f^V would be given by half the sum over all neighbours of
the pair potential due to the ion at the vacancy site. However the above account
is oversimplified for the vacancy in a metal, since we must also consider the effect
of the conduction electrons. A pair potential between ions in a metal is really
an effective potential generated by the electrons screening the bare Coulomb
potential acting between the ions. The screening potential is a function of the
density of the conduction electrons. When a vacancy is formed, the crystal expands
by an amount equal to the vacancy formation volume, Ω_f^V, thus lowering the electron
density by $\sim N^{-1}$, where N is the number of ions in the system. This in turn alters
the screening potential between all pairs of ions, thus changing the total energy
by an amount of comparable magnitude to E_f^V. Another complication is the fact that
the cohesive energy of a metallic system is derived principally from the structure-
independent cohesive energy of the conduction electron system. This term gives rise
to a pressure on the system, often referred to as the Cauchy pressure, since it is
clearly evident in the failure of the Cauchy relation, $C_{12} \neq C_{44}$, in cubic metals.
When the system expands, work must be done against the Cauchy pressure, giving rise
to yet another contribution to E_f^V. Yet further complications arise from the fact
that when an ion of charge Z is removed from a lattice site, Z electrons go with it.
There is then the possibility of electronic charge spilling over from neighbouring
sites into the newly-formed vacancy, giving rise to charge relaxation effects and
also alterations in the screening potential between vacancy neighbours. Additional
questions that have to be considered are the validity of the pair potential

description of a metal and also the effect of thermal expansion on vacancy formation as a function of temperature. Hence the vacancy problem and, consequently, all point defect problems in metals are much more complicated than a naive approach would suggest.

In simple metals, i.e. nearly-free-electron metals with small ion cores, the effects described above appear to be reasonably well understood. In the following three sections, the problem of the calculation of vacancy properties in simple metals will be reviewed in some detail. Then in the final section other point defect properties and other materials will be discussed more briefly.

2. Vacancy formation in simple metals

For the purposes of this chapter a simple metal is defined as one in which the ion cores do not overlap significantly with the neighbours, so that the conduction electrons can be regarded as nearly-free-electron-like in behaviour. Typical simple metals are Al or K, as discussed in Chapter (9). In such metals the bare electron-ion potential can be replaced by a weak-scattering pseudopotential. Chapter (9) described in detail how by treating the pseudopotential as a perturbation and by writing the total energy of the non-vibrating lattice E_0 to second order in the perturbation, one obtains E_0 in the form[1]

$$E_0 = E_1(\Omega) + \tfrac{1}{2} \sum_{\ell \neq \ell'} V(|\underline{r}_\ell - \underline{r}_{\ell'}||\Omega) \tag{1}$$

where $E_1(\Omega)$ is a function only of the volume Ω of the system and $V(|\underline{r}_\ell - \underline{r}_{\ell'}||\Omega)$ is an effective pair potential describing the interaction between two ions at positions $\underline{r}_\ell$ and $\underline{r}_{\ell'}$. These positions need not be lattice sites. If, for simplicity of discussion, we restrict ourselves to a local pseudopotential formalism, we can write

$$V(r,\Omega) = \frac{(Ze)^2}{r} - \frac{2(Ze)^2}{\pi} \int_0^\infty F(q) \frac{\sin qr}{qr} \, dq \tag{2}$$

where Z is the valence, and the energy wavenumber characteristic is given by

$$F(q) = \left[1 - \frac{1}{\varepsilon(q)}\right] |<\underline{k+q}|v_{ps}|\underline{k}>|^2 . \tag{3}$$

v_{ps} is the pseudopotential and the dielectric function $\varepsilon(q)$ takes the form

$$\varepsilon(q) = 1 + \frac{4\pi e^2}{q^2} \, \pi(q) \tag{4}$$

where $\pi(q)$ is the electron gas polarizability and is a function of the density of the system. It is this quantity that gives rise to the density (and therefore volume) dependence of $V(r,\Omega)$.

For the more general case of a non-local pseudopotential, F(q) is more complicated[2], but reasonably straightforward to evaluate. At this juncture it should be pointed out that, for a complete description of ground state properties (e.g. lattice dynamics, pair potentials, electron transport, etc.), a local pseudopotential is valid for only one material, Na[3]. However it turns out that, for many materials, a reasonable pair potential can be generated using a suitably chosen local pseudopotential.

One can in principle determine E_f^v using equation (1) simply by evaluating it for the perfect crystal and for the defect crystal and taking the difference between these two results. Unfortunately this procedure is not completely straightforward for several reasons. First, upon formation of the vacancy the volume of the system changes and one must include in the calculation the change in $E_1(\Omega)$ which is usually not known with sufficient precision for quantitative calculations. Also in this connection the density-dependence of $V(r,\Omega)$ must be taken into account. A second complication arises from the possibility that, in the region of the vacancy, the pair potential may undergo a change due to the depletion of screening charge between neighbours to the vacancy. This can come about from the fact, noted above, that, when an ion of valence Z is removed to the surface, Z conduction electrons go with it. A third complication comes from the fact that all simple metal potentials exhibit the well-known long-range oscillatory behaviour. In most cases $V(r,\Omega) \rightarrow A \cos 2k_F r/(2k_F r)^3$ as r becomes large, where k_F is the Fermi wavenumber. Since the number of neighbours at distance r is roughly proportional to r^2, summing such a potential over all neighbours leads to very poor convergence. These complications make it clear that one must approach the simple metal vacancy problem with some care.

The long-range oscillation problem is principally a mathematical one. Techniques have been developed either for summing them over all neighbours[4] or damping them to improve convergence[5]. It is important to recognize that truncation of an oscillatory potential at some arbitrary distance can give rise to almost any desired answer. Therefore, a quantitative calculation must ensure that the long-range parts of the potential are summed correctly to convergence.

The problem of the volume-dependence of $E_1(\Omega)$ and $V(r,\Omega)$ can be circumvented by taking advantage of the fact that, at zero temperature, the vacancy formation energy at constant pressure is, to order 1/N, (N is the number of ions in the system), equal to the vacancy formation energy at constant volume[6]. By doing a constant volume calculation, $E_1(\Omega)$ subtracts out of the problem. In fact, by using the equilibrium condition on E_0, it is not difficult to show that the static formation energy is given by[7]

$$E_f^v = - \tfrac{1}{2} \sum_\ell {}' V(r_\ell,\Omega) - \frac{1}{6} \sum_\ell {}' r_\ell \frac{\partial V(r_\ell,\Omega)}{\partial r} \quad . \tag{5}$$

The first term in this expression is just the usual bond-breaking term and the second results from the fact that, after formation of the vacancy, the ions have to be squeezed together a little in order to keep the volume constant. Unfortunately, one has to pay a price when using equation (5) and that comes from the fact that the asymptotic behaviour of the second term is $\sim \sin 2k_F r/(2k_F r)^2$. The convergence of the neighbour sum is very poor for this sum and special techniques must be used[4,5].

Up until now the discussion has been restricted to zero temperature. In practice, measurements are made over a range of finite temperatures and what is actually measured is the vacancy formation enthalpy at constant pressure, which we shall denote as H_f^V. Jacucci and Taylor[8] have pointed out that this quantity takes the form

$$H_f^V = E_f^V + T(\Delta S_p - \Delta S_V) \tag{6}$$

where ΔS_p and ΔS_V are the defect formation entropies at constant pressure and volume respectively. This can easily be rewritten in the form[8]

$$H_f^V = E_f^V + T \alpha_p \, \Omega \, \Delta P_V \tag{7}$$

where α_p is the lattice expansivity and ΔP_V is the pressure increase upon formation of the defect at constant volume. A similar equation was derived independently in the context of rare gas crystals[9] and further discussion is provided by Catlow et al[10].

On very general grounds one expects the formation energy to increase with increasing temperature[11]. However quasi-harmonic calculations of E_f^V as a function of temperature invariably display the opposite behaviour[12]. The entropy term in equation (7) provides the correction which cancels out the decreasing behaviour of E_f^V and gives an increasing H_V^f as a function of temperature[8]. This term is therefore very fundamental to a theory of vacancy formation.

As mentioned earlier, a further complication centres on the question of the distribution of electronic charge in and near the vacancy site. This is closely connected to the question of the validity of a second order perturbation (or linear screening) approach to pair potentials for the study of vacancy formation - a question that was discussed in some detail by Evans and Finnis[13]. These authors concluded that linear screening might very well be valid for monovalent materials like the alkalis but there seemed to be serious problems with that approach for polyvalent materials like Aℓ. The root of the problem seemed to be that the removal of a polyvalent ion to the surface was too large a perturbation to be described by linear screening. Some attempts at non-linear screening failed to improve the situation. Evans and Finnis finally concluded that the good agreement with experiment for E_f^V obtained by previous pair potentials was probably fortuitous.

This question was examined in a different way by Jacucci et al[14] who suggested that a rather subtle re-screening effect was taking place near the vacancy site. Due to removal of electronic charge in that region, each ion would be

sitting in a slightly stronger repulsive field due to its neighbours.   The
contribution of this effect to vacancy formation in Aℓ was crudely estimated to be
about 0.4 eV, a very significant fraction of the experimental value of 0.66 eV[15].
This effect is restricted to materials with a high electron density and is absent
in metals such as the alkalis[14].   Hence the problem of vacancy formation in metals
like Aℓ cannot be handled within a simple pair potential approach and requires a
more elaborate treatment - a conclusion also reached by Inglesfield in Chapter (9).

To summarise the foregoing paragraphs, the application of pseudopotential
theory to simple metals leads, via second order perturbation theory, to a pair
potential description of these materials.   However the long-range oscillatory nature
of the potentials, combined with the fact that they are density-dependent, means
that they must be treated carefully when applied to the vacancy problem.
Additionally, a pair potential description is by itself not sufficient to describe
vacancy formation in materials with a high density of electrons.

Before ending this section we should examine very briefly the questions of a
pair potential approach to (i) vacancy diffusion and (ii) divacancy binding energies.
Let us consider diffusion first.   It is important to note here that whether we use
a static approach and calculate saddle point energies or use a Molecular Dynamics
simulation, we are considering energy differences between systems which already
contain a vacancy.   It follows therefore than we can perform the calculation at
constant volume provided we take the same care with the long-range oscillations as
was necessary for the vacancy formation process.   More importantly, we do not have
to worry about the non-linear re-screening effects that seem to play such an important
role in vacancy formation in materials like Aℓ.   It is clear that, at any stage
in the diffusion process, the effect of the vacancy on near-neighbour interactions
will be almost precisely the same.   Hence any error made in neglecting this effect
will subtract out of the problem.   Paradoxically, then, it appears that a pair
potential description of vacancy diffusion is well justified in all simple metals and
quite possibly in all metals.

Turning to divacancies, we can obtain a good estimate of the binding energy of
these defects by referring to equation (5).   This gives the static energy for a
single vacancy in the system.   If we now introduce a second vacancy it now becomes
unnecessary to break one of the bonds which would be broken in the case of a pair of
isolated vacancies.   It follows that for an unrelaxed pair of vacancies the binding
energy is given by

$$E_{2v}^b = -V(r,\Omega) - \frac{r}{3} \frac{\partial V(r,\Omega)}{\partial r} \qquad (8)$$

where r is the relative separation of the two vacancies.   If this is negative the
two vacancies will bind to form a divacancy.   Relaxation effects may change this
a little,but even in a high electron-density material equation (8) should give a
good estimate of the binding energy.

As a final comment, a pair potential description of interstitial formation and migration should be valid for all simple metals. Due to the very strong repulsion between ions at short distances interstitials are kept well separated from their near neighbours. Consequently there should be no large changes in the electronic charge distribution resulting from an interstitial, even in Aℓ. However strain fields are very much longer ranged than for a vacancy and boundary conditions become very important in this problem.

3. Vacancy calculations in the alkali metals

Using equations (2-4) to generate a pair potential we see that we need to select a dielectric function and a pseudopotential from the bewildering array of both than can be found in the literature. For the dielectric function it is safe to say that no good treatment of exchange and correlation effects was provided prior to 1970. That comment extends even to the Geldart-Vosko[16] form of the Hubbard dielectric function. The choice then narrows down to about three[17,18,19]. Of these, the Toigo-Woodruff dielectric function[18] has been shown[20] to be an amputated form of that due to Geldart and Taylor (GT)[17]. Hence it seems that the GT and Vashishta-Singwi (VS)[19] forms of $\varepsilon(q)$ are the best choices. They give essentially the same results for the alkali metals. If a simple analytic form is desired, the best choice seems to be that due to Taylor[21] which, despite its simplicity, appears to give very similar results to the GT function for all metals, not just the alkalis.

A complete discussion of pseudopotentials requires a separate paper. Ideally a non-local form should be used, particularly if quantitative calculations are desired. However, for all alkalis except Li, a suitable local pseudopotential can usually be found for the pair potential problem. Li is a special case, due to the absence of p-electrons in the core, which means that the p-component of the conduction electrons is scattered much more strongly than the s-component. Hence a local pseudopotential is simply inappropriate for Li.

Probably the most painstaking calculation of vacancy formation enthalpies in the alkalis, performed to date, is that of Jacucci and Taylor[8]. These authors used a first-principles non-local pseudopotential[22], together with the GT dielectric function for their potentials. There were no adjustable parameters to fit to experimental data in their treatment. When their results were combined with the migration energy calculations of Da Fano and Jacucci[23], they obtained remarkably good agreement with the measured self-diffusion energies, Q. These are displayed in Table 1 along with a comparison with experimental vacancy formation enthalpies. As can be seen, the agreement with the experimental values of H_f^v is not nearly so good as for the Q-values. This is very surprising and has led Jacucci and Taylor to suggest that the experimental results leading H_f^v may need to be re-interpreted. In the case of K, the experimental result should be regarded as only

Table 1

Comparison of calculated vacancy formation enthalpies H_f^V and self-diffusion energies Q with experimental values at low temperatures (0) and at the melting temperatures (T_m). The units are electron-volts.

	Li		Na		K	
T	0	T_m	0	T_m	0	T_m
H_f^V(calc)	0.44	0.48	0.25	0.29	0.29	0.31
H_f^V(exp)		0.34[24]		0.36[25]		0.39[26]
Q(calc)	0.55		0.37	0.47	0.38	0.43
Q(exp)	0.55[27]	0.55[27]	0.37[28]	0.45[28]	0.38[28]	0.42[28]

a qualitative estimate^3[29] and, in view of the expected correlation of H_v^f with the melting temperature (T_m), the Li and Na data appear to be inconsistent. Note that the calculated values of H_f^V do correlate with T_m.

An interesting feature of the calculation of Da Fano and Jacucci[23] on the vacancy migration energy in Na was the discovery of the vacancy double jump mechanism in their Molecular Dynamics simulation. At high temperatures they found that the ion motions were sufficiently correlated that, when one ion jumped into a vacancy site, a second one frequently followed it, causing the vacancy to move two lattice sites in one jump. This provided a nice explanation for the observed anomalous isotope effect for self-diffusion in Na[30]. Also this mechanism was crucial to obtaining good agreement with the experimental Q-values at T_m.

The calculations of Jacucci and co-workers[18,23] seem to give a highly consistent picture of vacancy-dominated diffusion in the alkalis. However Brunger et al[31] have recently suggested that observed curvatures in the Arrhenius plot of self-diffusion data may not be due to the vacancy double-jump mechanism but rather to motion of divacancies. They argue that computed correlation factors for vacancy double jumps are inconsistent with diffusion parameters measured by NMR techniques. Instead their results favour divacancy diffusion.

Using equation (8) it is a trivial matter to estimate the binding energy between two vacancies. Using the same pair potentials as for the vacancy calculations described above, E_{2V}^b has been evaluated at a number of neighbour distances for all the alkali metals plus Aℓ, using a lattice parameter appropriate to the melting temperature. The results are displayed in Table 2. A negative value should be interpreted as being repulsive and a positive of one as being attractive.

Table 2

The vacancy binding energy, estimated using equation (8), for vacancy separations corresponding to the first four neighbours distances of the alkali metals and Aℓ. The units are electron-volts.

Neighbour	Li	Na	K	Rb	Cs	Aℓ
1	0.090	0.040	0.044	0.051	0.065	-0.018
2	0.009	-0.002	0.003	0.009	0.020	0.047
3	-0.006	0.002	0.000	-0.003	-0.007	-0.024
4	0.010	0.002	0.001	0.002	0.002	0.016

As is evident from these results, the binding energy of two vacancies at the nearest neighbour distance in the alkalis is an order of magnitude smaller than H_f^v and weakly attractive. At larger distances it is negligibly small. All other alkali metal pair potentials in the literature are qualitatively similar to the ones used here. Hence one is forced to conclude that there is no theoretical support for the concept of a divacancy in an alkali metal crystal. The vacancies should be largely unaware of the presence of other vacancies in the system. Only if they come in contact at the nearest neighbour distances will they attract each other, and that attraction will persist only until one of them makes a jump.

At the time of writing the questions of curvature in Arrhenius plots of self-diffusion coefficients and of diffusion mechanisms in the alkalis still do not seem to be fully resolved. It has been suggested[32] that oxygen contamination of samples may be responsible for the observed curvature in Arrhenius plots. It would be most interesting if further experiments could be performed to shed more light on these questions.

4. Vacancy calculations in aluminium

Aluminium is a metal with a very high density of conduction electrons ($r_s \simeq 2$, where r_s is the radius of a sphere, in atomic units, enclosing one electron). As has already been pointed out, this can lead to severe complications when calculating vacancy properties. Also, in generating a pair potential, the choice of dielectric function is much more critical[33] than in the alkalis for which $3.2 < r_s < 6$. For small r_s, the GT dielectric function[17] is more or less exact and it is therefore the best available choice for Aℓ, although the very simple form suggested by Taylor[21] is a good second choice. The functional form of pair potential generated by the GT dielectric function is quite different from what one normally expects, in that it is repulsive at the first neighbour distance. Nevertheless this potential has been shown to give a very good description of lattice, defect and disorder properties of Aℓ[14]. The origin of the nearest-neighbour repulsive component of the Aℓ potential is a term of the form $\exp(-\eta R)/R$ which is present in some dielectric functions but not others[33]. Hence the choice of $\varepsilon(q)$ can have a drastic effect on the calculated pair potential and it is most important to pay careful attention to this point when performing calculations in materials such as Aℓ.

In view of the comments in section 2 concerning the validity of a pair potential approach to vacancy formation in Aℓ we should not expect to get good results for H_f^v with a well-constructed pair potential. Jacucci et al[14] obtained $H_f^v = 0.19$ eV, as compared with the experimental value of 0.66 eV[15]. A major source of the discrepancy is the neglect of the re-screening effect described earlier and which accounts for about 0.4 eV. On the other hand, these same authors obtained a value of 0.6 eV for the migration energy which compared very well with the experimental value of 0.62 eV[34]. As pointed out, this is not un-expected and is consistent with the argument for re-screening in the vacancy neighbourhood. It follows then that a pair potential that is fitted to the vacancy formation enthalpy of Aℓ is not likely to give a reliable description of diffusion or, for that matter, any other physical property unless it is constrained to do so.

Another approach to vacancy formation is to use a cluster approach within a band structure scheme to calculate the total energy of a system with and without a vacancy present. In principle this procedure should pick up the terms omitted by a pair potential description, but difficulties occur here too. One problem is the large cancellation that can take place between terms, but another more serious one is the difficulty of dealing with interactions between neighbouring cluster cells[35]. This is probably a manifestation of the long-range oscillations in the pair potential and further work remains to be done on the cluster calculation approach to a vacancy in Aℓ.

A third approach to vacancy formation in Aℓ is the LCAO method to determine the total energy. This has been tried by Singhal and Callaway[36] using localized Gaussian orbitals instead of Wannier functions. However, in this case, they have also met with limited success and an accurate evaluation of H_f^v, using a realistic model for Aℓ has still not been reported.

Turning to other point defects, the results of Table 2 lead one to believe that divacancies are not likely to be present in Aℓ. It is true that there is an attraction between two vacancies at the second neighbour distance but once one vacancy of the pair has made a jump, it no longer attracts the second one, and may well move in another direction. In any case the energies are $\lesssim kT_m$ and the divacancy is likely to be lost in the thermal motion of the system.

Finally, as already point out, there should be no problem with a pair potential approach to interstitials, due to the fact that no large local electronic charge perturbations will be associated with this defect. Satisfactory results have recently been reported by Lam et al[37] for Aℓ and Aℓ-Zn alloys.

5. Other systems and other defects

Turning to metals other than the simple ones discussed above, a pair potential becomes much more difficult to justify on a formal basis and only in metals like Cu and Zn is it possible to consider a first principles approach without adjustable parameters. From the point of view of a computer simulation, a pair potential approach to defects is still the only practical way to proceed. Consequently, in noble and transition metals, a great deal of effort has been put into the development of empirical potentials with parameters fitted to several experimental quantities. A commonly used procedure is to include the vacancy formation enthalpy as one of the experimental data in the fitting procedure. In view of the foregoing discussion of Aℓ, this is clearly a questionable procedure when a high density of electrons contributes to the screening of the ion-ion interaction. In a transition metal, it is not at all clear to what extent the loosely bound d-electrons can be regarded as screening electrons in the nearly-free-electron sense. However it is quite likely that the effective r_s for many of these materials will be ≤ 2, putting them in the high electron density limit and thus rendering a pair potential description of vacancy formation inappropriate. Nevertheless, the procedure due to Johnson[38] of generating short-range empirical pair potentials appears to give good results in systems like Fe, Co and Ni[39]. Why this is so is a question that has not yet been resolved and a great deal of work remains to be done on potentials and defect calculations in the transition metals.

Finally, a few comments are necessary concerning impurities in metals which form another class of point defects. In the simple metals, impurity pair potentials can easily be generated via pseudopotential theory (e.g. see reference 22). For homovalent impurities, a great deal of success has been achieved in describing, for example, diffusion of noble metals in alkali hosts[40]. However, for finite (not dilute) concentrations of non-homovalent impurities, problems can develop, due to charge transfer effects, which are not described within the usual pseudo-potential theory. For example, Beauchamp et al[41] examined the Li-Mg system and found that their potentials predicted that this alloy would exhibit long-range order. In fact no long-range order is observed experimentally. The problem lay in the Li-Mg potential which appeared to be too attractive in the near-neighbour region, due to the fact that no account was taken of the extra electronic charge that one would expect to find in the neighbourhood of the Mg^{++} ion. An alternative formulation of the binary alloy problem by Hafner[42] gives rise to concentration dependent pair potentials. Hafner's formulation is based on Orthogonalized Plane Waves and the concentration-dependence enters through the normalization factors. Despite this feature, there does not seem to be any account taken of local charge transfer effects in Hafner's theory and the potentials generated by it for Li-Mg[43] show a close qualitative resemblance to those of Beauchamp et al[41].

It is clearly evident that these potentials would also predict long-range ordering for Li-Mg. The problem of finite concentrations of non-homovalent impurities still awaits a satisfactory treatment, even in the simple metals.

References

1. Harrison, W.A. Pseudopotentials in the Theory of Metals. (New York: Benjamin, 1966).

2. E.g. see Taylor, R. and MacDonald, A.H., J. Phys. F 10, 2387 (1980).

3. Finnis, M.W., J. Phys. F. 4, 1645 (1974).

4. Duesbery, M.S. and Taylor, R., J. Phys. F 7, 47 (1977).

5. Duesbery, M.S., Jacucci, G. and Taylor, R., J. Phys. F 9, 413 (1979).

6. This was first pointed out by Chang, R. and Falicov, L.M., J. Phys. Chem. Solids 32, 465 (1971).

7. Minchin, P., Meyer, A. and Young, W.H., J. Phys. F 4, 2117 (1974).

8. Jacucci, G. and Taylor, R., J. Phys. F 9, 1489 (1979).

9. Chadwick, A.V. and Glyde, H.R. In Rare Gas Solids (Editors Klein, M.L. and Venables, J.A.). (Academic Press: New York), 1151 (1977).

10. Catlow, C.R.A., Corish, J., Jacobs, P.W.M. and Lidiard, A.B., J. Phys. C. 14, L121 (1981).

11. Seeger, A. and Mehrer, H. Vacancies and Interstitials in Metals. (North Holland: Amsterdam, 1969).

12. E.g. see Popovic, Z.D., Carbotte, J.P. and Piercy, G.R., J. Phys. F 4, 351 (1974).

13. Evans, R. and Finnis, M.W., J. Phys. F. 6, 483 (1976).

14. Jacucci, G., Taylor, R., Tenenbaum, A. and van Doan, N., J. Phys. F 11, 793 (1981).

15. Triftshauser, W., Phys. Rev. B 12, 4634 (1975).

16. Geldart, D.J.W. and Vosko, S.H., Can. J. Phys. 44, 2137 (1966).

17. Geldart, D.J.W. and Taylor, R., Can. J. Phys. 48, 167 (1970).

18. Toigo, F. and Woodruff, T.O., Phys. Rev. B 2, 3958 (1970).

19. Vashishta, P. and Singwi, K S., Phys. Rev. B 6, 875 (1972).

20. Dharma-wardana, M.W.C., J. Phys. C 9, 1919 (1976).

21. Taylor, R., J. Phys. F 8, 1699 (1978).

22. Dagens, L., Rasolt, M. and Taylor, R., Phys. Rev. B 11, 2726 (1975).

23. Da Fano, A. and Jacucci, G., Phys. Rev. Lett. 39, 950 (1977).

24. Feder, R., Phys. Rev. B 4, 828 (1970).

25. Adlhart, W., Fritsch, G. and Lüscher, E., J. Phys. Chem. Solids 36, 1405 (1975).

26. MacDonald, D.K.C., J. Chem. Phys. 21, 177 (1953).

27. Thernquist, P. and Lodding, A., Z. Naturf. 23 (a), 627 (1968).

28. Mundy, J.N., Miller, T.E. and Porte, R.J., Phys. Rev. B 3, 2445 (1971).

29. Cook, J.G., Taylor, R. and Laubitz, M.J., J. Phys. F 9, 1503 (1979).

30. Barr, L.W. and Mundy, J.N. Diffusion in Body-Centred Cubic Metals, 171 (American Society of Metals: Ohio 1965).

31. Brünger, G., Kanert, O. and Wolf, D., Phys. Rev. B 22, 4256 (1980).

32. McKee, R.J. Private communication.

33. Jacucci, G. and Taylor, R., J. Phys. F 11, 787 (1981).

34. Seeger, A., Wolf, D. and Mehrer, H., Phys. Stat. Sol.(b) 48, 481 (1971).

35. Chakraborty, B. and Siegel, R.W., Bull. Am. Phys. Soc. 26, 473 (1981).

36. Singhal, S.P. and Callaway, J., Phys. Rev. B 19, 5049 (1979).

37. Lam, N.Q., Doan, N.V. and Adda, Y., J. Phys. F 10, 2359 (1980).

38. Johnson, R.A., Phys. Rev. 134, A1329 (1964).

39. Beeler, J.R. Jnr. and Beeler, M.F., in Interatomic Potentials and Crystalline Defects. Ed. J.K. Lee (TMS-AIME 1981) to be published.

40. Schober, H., Taylor, R., Norgett, M.J. and Stoneham, A.M., J. Phys. F 5, 637 (1975).

41. Beauchamp, P., Taylor, R. and Vitek, V., J. Phys. F 5, 2017 (1975).

42. Hafner, J., J. Phys. F 6, 1243 (1976).

43. Hafner, J., Inst. Phys. Conf. Ser. No. 30, 102 (1977).

CHAPTER 14   DEFECT CALCULATIONS IN SEMICONDUCTORS

by

A.M. Stoneham
Theoretical Physics Division, A.E.R.E. Harwell, Oxon OX11 ORA

1.   Introduction

In Chapter (11) we surveyed the various approaches to interatomic
potentials and their generalisations in semiconductors[1].   In turning to
applications, we shall concentrate on the links between those methods and the defect
properties which are important in semiconductor science. We shall not therefore give
a review of all theoretical work on defects in semiconductors, for comprehensive
surveys already exist[2,3,4]. Rather, we shall try to identify the needs, the
capabilities, and the range of applications in the context of the other work in this
volume.

2.   The needs for defect calculations

The needs can be defined by three questions: Which systems?   Which defects?
Which properties?

2.1 Which systems?

We shall take a liberal view of "semiconductors", corresponding roughly to the
subject matter of our earlier chapter: the systems covered will be any which have
open structures stabilised by bonds with some directional component. Thus we should
include the four-fold coordinated II-VI's, III-V's and Group IV elements; silicates,
both crystalline and glassy; the two-fold coordinated chalcogenides like Te and Se;
molecular crystals, and other systems of varying complexity.   It will become
clear that there are large gaps in work on these systems.

2.2 Which defects?

Technological problems select themselves. They are not chosen by scientists for
their own convenience. Chemical impurities may come from any column of the periodic
table and their consequences may be felt at low levels. Thus in disposal of nuclear
wastes one may have glasses with tens of elements in significant concentrations.
In geological dating one may exploit exceedingly rare species.   Moreover, it is a
general rule that no crystal, whatever its specification, contains less than 1 ppm of
some impurity.   Even in the best silicon one might find C or O at these levels, in
much higher concentrations than electrically-active impurities.   In the end one
may have to handle such extremes as non-bonding closed-shell impurities (e.g.
rare gases, or ionised alkalis) and network-forming species which bond strongly.

Chemical variety is not the only difficulty.   One may have to deal with high
defect concentrations, so that the defects cannot be considered in isolation.
Heavily-doped semiconductors are an example. This situation can be complicated

further, since the distribution in space may not be random, and since one may need properties with excited (non-equilibrium) carrier populations.

Setting aside these various complications, there remain three broad classes of defect. There are point defects, including the coordination defects in amorphous systems. There are line defects, like dislocations, and there are planar defects: surfaces, interfaces with metals, loops and so on. Most work has been done on point defects and, unfortunately, most work on line defects uses methods which do not perform well for point defects.

2.3 Which properties?

The properties fall into four main groups. First we have transition energies. Almost every paper on defect electronic structure will quote one-electron levels for a particular geometry and charge state. For shallow defects, this may suffice to relate theory and experiment; for deeper defects, the connection is less simple.

The second group of properties relates to total energies, namely the thermodynamic properties. A simple sum over one-electron levels $\Sigma_i \epsilon_i$ differs from the total energy by double-counting electron-electron interactions and by omitting nuclear-nuclear interactions, differences which have particularly disturbing effects in bond-centre and similar defect geometries. The thermodynamic energies are frequently the ones measured, and give Gibbs free energies at constant temperature and pressure. Since most theories give internal energies, i.e. constant volume and zero temperature values, the connection involves both the volume dependence of properties and vibrational entropies (the configurational entropy terms are usually straightforward, though possibly complicated). Thermodynamic energies are needed to determine the relative stability of different defects; the energies give for example the temperature dependence of capture cross-sections and segregation coefficients between phases.

The third property is closely related, and is that of energy surfaces, i.e. total energies as a function of geometry. These play a vital role in non-radiative transitions, in determining whether or not luminescence is expected, in deciding which mechanism of recombination-enhanced motion can occur, and in more subtle manifestations like isotope effects.

Finally, there are the properties which depend on defect wavefunctions. Frequently one must go beyond pseudo-wavefunctions and orthogonalise to any core states. Likewise charge-densities alone are usually not enough. And in many cases wave-functions are needed for both ground and excited cases. Examples incluue spin-resonance parameters (a very important check of defect calculations), oscillator strengths, absolute rates for non-radiative transitions, and response to applied fields, including magneto-optic phenomena.

3.   Defect calculations in semiconductors

3.1   Shallow defects

The archetypal shallow defect has a carrier of effectve mass m^* moving in a diffuse orbit, with hydrogen-like energy levels and an effective Rydberg that is small relative to the band gap and changed by a factor m^*/ε^2 (where ε is the dielectric constant) from that of the hydrogen atom. The standard theory is, of course, effective mass theory.   This theory[5,2] continues to be of enormous value, as a theory with well-defined validity and a theory with quite exceptional accuracy in special cases. It is moreover a good zero-order approximation even when its basic assumptions fail.   These basic assumptions are (I) that the impurity potential has no Fourier components outside the first Brillouin zone, and (II) that the envelope-function varies very slowly in space.

Nature complicates the simple scaling of hydrogenic results for solids, though effective methods have been developed to cover more realistic cases.   Conduction bands are multivalleyed[6], valence bands are degenerate[7]; there may be anisotopy[8], and there may be polaron effects[9].

Less secure is the question of central-cell corrections, the short-range corrections to the main Coulomb potential. Here the several approaches fall into three main groups: the few attempts at accurate calculations for specific systems surveyed in references (2-4); the search for systematic trends with chemical features[10]; and the use of the quantum defect method to relate changes in various physical properties to the binding energy[11].

A further aspect of shallow defects lies in quantum chemistry - the notion that energies of complex defects (donor-acceptor pairs; bound excitons, etc.) can be obtained from values for suitable atomic and molecular analogues by judicious scaling with the dielectric constant and effective mass[12].   Moreover, conventional molecular codes can be modified to put this scaling in systematically for the cases where no analogues exist.   This work has produced predictions of very high accuracy[13], though not without empirical components representing the centre-cell terms.

3.2   Non-bonding defects and analogous species

Here closed-shell impurities fit into interstices (or, more rarely, substitutional sites) in the covalent structure. Examples include rare gas and ionised alkali impurities.   The rare gas behaviour is primarily of interest for geological[14] and related problems, including the evolution of the solar system[15]. The alkali metals, however, are more pervasive and intrusive.   Thus Li may be the source of "self-compensation" in ZnSe.

The alkali interstitials tend to be shallow donors, whose electronic structure is readily handled by effective mass theory.   Much harder are calculations relating to where the impurity goes, its diffusion energy, and whether it binds to other dopants. Recently it has become clear that the methods developed for ionic crystals may be extended to the four-fold coordinated II-VI compounds[16,17]. In particular Harding's calculations[17] on Li, Na and K in ZnSe, show the increasing short-range repulsion terms which lead to a change in favoured site along the sequence. Both Neumark's[16] and Harding's[17] work is consistent with the role suggested for Li in self-compensation.

There are many types of analogous systems. One important one involves transition metals, where the 3d or 4f shells are not strongly bonded (though they can be strongly affected by what one may loosely call "crystal field terms"). Substitutional species therefore show both the characteristics of vacancies, from the states associated with unbonded neighbours, and of the free ions, from the 3d or 4f shells[18,19].

3.3 Intrinsic defects and related deep levels

The main recent development in semiconductor defect studies has been the use of self-consistent theories to study the energy surfaces of vacancies and interstitials in diamond and silicon. These, and parallel studies in other systems, have implications for stability of particular charge states, for structure in optical absorption, for non-radiative versus radiative transition probabilities, and for mechanisms of recombination-enhanced motion. Both experimental work and theoretical analyses confirm that the distortion of the surrounding host lattice can have a big effect on defect properties. Some systems for which good experimental data are available are given in Table 1.

Table 1

Host	Defect	Experiment	Jahn-Teller energy
Diamond	Neutral vacancy Substitutional N	GR1,GR2-8 optical abs. EPR	~ 0.6 eV (E) > 0.7 eV (T)
Silicon	Vacancies V^+ V^0 V^-	EPR/stress	0.4 eV (E) 1.5 eV (E) 1.9 eV (E+T)

The E distortions are tetragonal, the T distortions trigonal; E+T indicates a mixed distortion. The studies of silicon vacancy energies[20] make some working approximations about local force constants; the diamond vacancy energy[21] is largely free from such assumptions. There are some obvious conclusions. The Jahn-Teller energies are large, and calculations for the perfect-lattice geometry may be irrelevant. Coupled electronic-structure and lattice-distortion calculations are needed. Moreover, distant neighbouring atoms may move, not just the first neighbours. These factors determine the type of theory needed. Available methods may be classified as to whether self-consistency is or is not included.

A. <u>Non-self-consistent theories</u> When theories up to the early 1970's were reviewed[2], only one calculation attempted self-consistency. Now self-consistency is accepted as an important ingredient. Certainly one should talk of transition energies for well-defined processes, rather than to try to construct one-electron energy-level schemes, which are of only limited value and which should not be used for direct comparison with experiment. The strength of the electron-electron interaction and the importance of correlation will depend on factors such as whether the local one-electron states are genuinely local or are merely resonances, for example. Comparison of a range of different calculations shows reasonable agreement, except for extended Hückel theory. Lannoo[22] has analysed the non-self-consistent theories of vacancies in terms of a Hubbard model, and his work also confirms reasonable consistency between different approaches: the apparent large differences between calculations reflect relatively modest differences in detail, rather than gross divergences. Lannoo's analysis shows there is a critical parameter, the ratio of matrix elements $<a|h|b>/[<aa|aa>-<aa|bb>]$ among the atomic orbitals $|a>,|b>$, which corresponds to the characteristic parameter of the Hubbard model. So far calculations for dislocations have only used non-self-consistent methods, usually (but not exclusively) for unreconstructed geometries obtained from predictions from elasticity theory.

B. <u>Self-consistent theories</u> More recent theories have been self-consistent. As a result, they have tended to be <u>real space</u> theories, describing a solid as built from atoms; for deep defects in solids the <u>reciprocal space</u> (band structure) methods are less helpful. Several types of self-consistent theory are used. First we have semi-empirical molecular orbital methods (CNDO)[23,24]. This approach has been used for a wide range of problems in ionic, metallic and semiconducting systems, notably at Harwell in the U.K. and by Evarestov[25] in the U.S.S.R. The "empirical" component involves three types of parameter: electro-negativity, orbital exponent (atomic size) and propensity to bond. The parameters are chosen to fit molecular data or (using periodic boundary conditions, rather than a cluster) to fit valence band and crystal equilibrium data. Charge redistribution is treated self-consistently, and a total energy (not the sum of one-electron eigenvalues) given. CNDO is a particularly convenient and flexible cluster method, and has been used so widely that its limitations are largely known and understood. It has been used for the vacancy V^0 and for the interstitials I^0, I^+ and I^- in diamond.

In addition to CNDO there are other cluster methods including the valence bond approach to V^0 in diamond and silicon[26] and an extension of a non-self-consistent theory to allow matrix elements to depend on the charge density[27]. The spirit of band theory is followed in two approaches, using self-consistent pseudopotentials (where self-consistency does not have quite the same meaning as in the other papers referred to) for V^0 in Si[26] and V^+ in Si[28]. Finally, we have approaches based on a pseudopotential formulation of the Slater-Koster method. Two variants (broadly equivalent in content and predictions) have been developed by IBM workers[29] and at Bell Laboratories[30]. These consider a small region of crystal in detail, but embed this in an infinite crystal. Cluster methods (as in (a),(b) above) involve a different type of boundary condition on the cluster instead; the main requirement is that there be no surface states with energies close to the Fermi energy.

Excited states prove hard to calculate self-consistently. There is no problem in principle; it is merely the problem of forcing convergence onto the state desired. Only reference (26) quotes excitation energies.

3.3.1 <u>Choice of method: prediction and obsevables</u> The choice depends mainly on what it is one wants to calculate. The CNDO[23] and the Slater-Koster method[29,30] have both been developed for energy surfaces; CNDO has been tested more extensively (see below), but both should be comparably good. Since the merits and demerits of cluster calculations are often disputed, it is useful to have a set of guidelines so that their undoubted advantages can be exploited. A suggested list is given below.

(1) Can the problem be solved by a cluster method? Clearly one should not expect the details of very extended states to be predicted well, for example.

(2) Can the problem be solved by the particular cluster method? If total energies are needed and the method only gives one-electron energies, it is not the <u>cluster</u> method which is failing. And, if there is significant charge transfer, self-consistency must be included.

(3) If there are empirical parameters (pseudopotentials, molecular integrals, etc.) has the right choice been made?

(4) Have the right boundary conditions and geometry been used? Are problems caused by surface states near the Fermi level?

For energy surfaces one needs to know where to move the more distant neighbours.   There are two models:

(A) "Standard model", which moves the nearest neighbours only, or possible selected independent motions beyond.   The basic assumption is that <u>displacements</u> are short-ranged.

(B) Move close neighbours to minimise the energy whilst moving distant neighbours as if the interatomic forces were those of the perfect crystal.   The assumption is that the forces causing displacements are short-ranged.

The second model, first used by Larkins and Stoneham[31], seems to be preferable, and has been adopted in references (23,24,30).

None of the self-consistent methods prove very convenient for estimating hyperfine constants. The CNDO approach is probably the simplest, at least for rough estimates. However, once one knows the equilibrium geometry, there is always the possibility of a full calculation (e.g. beyond Hartree Fock) for that geometry to give the hyperfine parameters and other observables.

3.3.2 <u>Results for Diamond</u>   The comparison given in Table 2 of CNDO predictions with the very extensive experimental data is encouraging. It verifies that this rather simple self-consistent theory can work well when the geometry is handled reasonably and the empirical parameters sensibly chosen.

Table 2

Centre	Property	Theory	Experiment
V^0 (Ref.23) Neutral vacancy	Jahn-Teller energy	0.46 eV	0.63 eV (optical stress response)
			0.3-0.4 eV (side band)
	Static barrier to reorientation	0.22 eV	0.03 eV stress response
	E mode energy Jahn-Teller	0.13 eV	0.13 eV (ref. 21)
	Distortion	0.22Å	0.16Å
Substitutional N (Ref.32)	Jahn-Teller energy	1.64 eV	1.7 eV (ionisation threshold)
	Reorientation energy	<0.99 eV	0.7 eV
Substitutional B (Ref.32)	Ionisation	0.36 eV	0.37 eV
	Vibrational motion	0.16 eV	0.16 eV
Self-interstitial (Ref.24)	Stable form $(I^0, I^+, ?I^-)$	(100) Split	Probably (100) split, if only by analogy
	Activation energy	0.8-1.5 eV depending on mechanism	1.3 eV?
	Translational vibration	0.13 eV	Not known

3.3.3 <u>Results for Silicon</u>   The data are less complete for Si, since one has only spin resonance data and, in the case of the S=0 V^0 centre, only indirect information. Nevertheless, one striking result has emerged[30]: the V^+ centre may be unstable against decay into V^{2+} and V^0: $2V^+ \to V^0 + V^{2+}$. The point is that the Jahn-Teller energy for V^0 is significantly larger than for V^+ (by a factor 4 in the simplest analysis), and this more than compensates for the extra intradefect energy U. There is some experimental evidence for this[33] instability. Watkins' data for V^+ still stand, of course, but the centre must be regarded as metastable only. The theory can be analysed and compared with experiment in this way[30,34]. Results are summarised in Table 3.

Table 3

	Theory	Experiment
V^0 Jahn-Teller energy	0.68 eV	$\lesssim$ 0.68 eV
Intra atomic energy U	0.25 eV	> 0.31 eV (V^0) < 0.31 eV (V^{2+})
Energy for thermal decay $2V^+ \rightarrow V^0 + V^{2+}$	0.16 eV	0.06 eV

Again, the results accord with what is seen, and it is both interesting and important that theory has led to a re-interpretation of the original work. The conclusions are not without complications, however. In reference (30) it was assumed that both the tetragonal Jahn-Teller forces per electron and the elastic restoring forces were independent of charge state for V^+ and V^{2+}. However, Watkins' data (see §27.2c, of reference 2) show that at least one of these assumptions breaks down for the trigonal terms in comparing V^+ and V^-.

4. Theory of processes

The previous sections have concentrated on the properties of specific defects. But these are only of secondary importance in technology. Instead one wants to understand processes, e.g. fracture, growth, radiation damage, and, by understanding the microscopic phenomena, to guide the more effective use of semi-conductors. The role theory plays is twofold[35]. First, some phenomena are not accessible experimentally, perhaps because they take place too rapidly, or because they require conditions not available in the laboratory. Here one can hope to understand what is going on by extrapolation from a theory which works well in a regime where it can be tested. Secondly, it may be hard to distinguish experimentally between two or more microscopic mechanisms. But a theoretical analysis of critical parts of the mechanisms is often able to discriminate between the various possibilities.

In general, there are two types of theory of this sort. There are phenomenological theories, which provide a framework within which results can be understood. These include kinetic theories, where one tries to understand evolution under a variety of coupled processes. There are also atomistic theories, in which critical energies are calculated for various atomic models. Here one hopes that relatively primitive calculations are good enough to be useful.

4.1 Non-radiative transitions and the form of potential energy surfaces

Non-radiative transitions govern many physical phenomena[36]. These include, amongst others, the suppression of luminescence, the reduction of carrier lifetimes and the creation of intrinsic defects. Some of these phenomena can be deduced from the forms of the potential energy surfaces, without the need to calculate transition matrix elements. Examples are given below.

Will luminescence occur or not? When there is strong electron-lattice coupling, luminescence will be seen after optical absorption if the initial excitation (A→B in Figure 1) is below the crossover X[37,38,39]. If b lies above X, there will be little luminescence, and the efficiency depends on whether the decay down the ladder of vibrational states passes to C after leaving X (and gives luminescence) or goes non-radiatively to A. The branching ratio can be shown, rather generally[38,39], to be proportional to the ratio of the energies E_{XC}/E_{XA}. The results are supported by data for F centres in many ionic crystals.

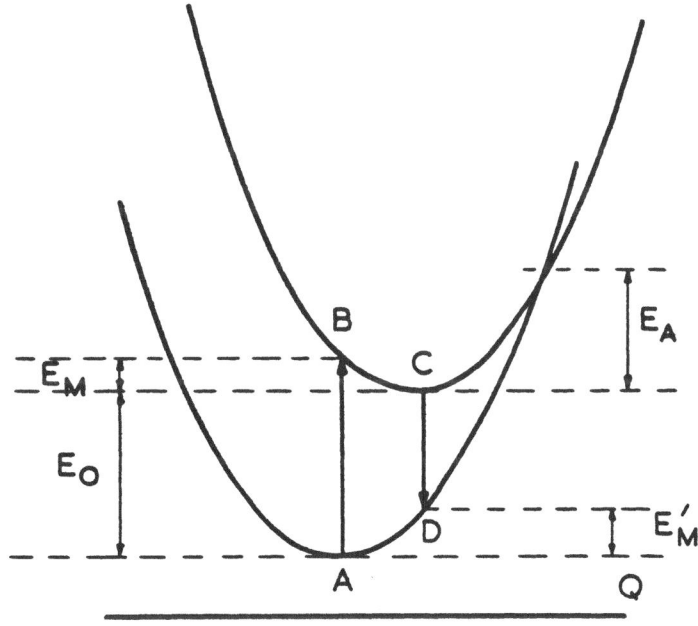

Figure 1   Notation for a simple two-level system. Q is the accepting mode. E_0 is the "zero-phonon" energy. E_M and E_M' are relaxation energies. The absorption energy is E_0+E_M and the emission energy E_0-E_M'. The parameter Λ is $E_M/(E_0+E_M)$.

Persistent photoconductivity[40]   This low-temperature phenomenon can be explained by having the initial absorption from ground state G to an ionised state $X(\equiv D^+ + e^-$, with only a small Stokes shift), with the option of forming a bound state $\tilde{X}$ of slightly lower energy but substantially different distortion (and hence a large Stokes shift)[41].   The various energy barriers and Stokes shifts can be fitted with sensible and consistent values.   Rather similar energy surfaces have been proposed for other metastable excited staes, e.g. for spin resonance in nitrogen-associated centres in diamond[42].

Defect production during radiation damage   The most detailed survey[43] concentrates on the mechanisms of defect production in diamond.   These include the initial process - the charge states and geometries of primary defects and displacement energies - together with some of the subsequent mechanisms giving motion of the defects produced.   The results are largely based on extended Hückel theory estimates of potential energy surfaces.   As mentioned previously, this theory is very approximate.   However, it should be adequate for several important qualitative results needed in understanding the primary defects.

The first important result, bearing in mind earlier reservations, is that the likely interstitial geometry is the split interstitial, in which two carbon atoms are symmetrically placed about one substitutional site. Split interstitials are the common form in alkali halides and metals.   The second result is that the primary process produces charged defects (I^+ and V^-, rather than I^0 and V^0).   This, in turn, modifies the displacement energy, and the calculations suggest that the displacement energy is significantly smaller and more isotropic than previous estimates have suggested.

Whilst there is now a plausible picture of the early stages of radiation damage, there still remain open questions, and there has been little progress in the study of the subsequent evolution to produce larger agglomerates of point defects.   It is, of course, this later stage, where theory would be a particularly important complement to ion-beam studies.

Mechanisms of ionisation and recombination-enhanced processes   Many defect phenomena show enhanced rates when there are high densities of electrons and holes[45,46,35].   The mechanisms can be divided into three main types:

(a) Local excitation, in which energy is given to electronic excitation which makes more rapid transitions in its excited state because the form of the potential energy surface is altered.   An example is the decay of the self-trapped exciton in alkali halides to form F and H centres (the neutral vacancy and interstitial respectively);

(b) Local heating, in which the energy goes to vibrational degrees of freedom. An example is seen in the SiC:H centre[45];   other cases may occur in Si:Al[46] and other semiconductor systems (see a penetrating analysis and the references given in (46)). The mechanism is also known as the "Energy Release Model".

(c) Bourgoin-Corbett mechanism[47] in which a defect occupies different sites in different charge states, and may diffuse by alternate capture of electrons and holes.

For present purposes the energy surfaces from self-consistent calculations can eliminate several possibilities (see reference 24 for a quantitative example). Clearly there must be different sites for different charge states in mechanism (c), or a suitable excited-state energy surface for mechanism (a). Mechanism (b) is more subtle. If there is to be effective local heating, energy must be given to the reaction coordinate efficiently, and this motion must persist long enough for barrier crossing. These features must be satisfied if the reaction coordinate is a genuine local mode (as in SiC:H) or a very local resonance. If the frequency associated with the reaction coordinate was in the middle of a peak in the phonon density of states, neither of these possibilities is likely, and mechanism (b) is not expected to operate.

4.2 Fracture

Most work on fracture has followed the ideas of Griffith[48], or analogous proposals of later workers. These ideas are based on a continuum model, and the central assertion is this: when the stress-intensity factor (which characterizes the crack-opening stress field around a crack tip) exceeds a certain critical value then the crack will extend without limit. In the Griffith criterion, for example, the critical stress-intensity factor is related to the surface energy and to the elastic constants.

Diamond is a brittle material. Since the complications of plasticity can usually be ignored, it is together with silicon a suitable system for study with atomistic calculations[49,50]. In these approaches one sets up a computer model with, for example, a long plane crack involving bond rupture across a (111) face. Within a large central region the atoms interact via the valence force potentials of section III A 1; outside this region the displacements are matched to a continuum solution. The response of this model to forces tending to lengthen the crack can then be monitored. This is, of course, still a formidable job, even when periodic boundary conditions, short-range forces and modern computer methods are fully exploited.

The results introduce a number of new features into the continuum models. Some are related to the geometric differences. For example, cusp-like models of the crack tip are sometimes chosen in continuum treatments, where an atomistic view indicates a better picture is a narrow split held at the end by a line of bonds. A related point concerns the dependence on inter-atomic potential. In continuum models, the elastic constants are the only important material parameters, yet there are infinitely many atomic models which give the same elastic properties. These different models predict different fracture behaviour, and this can be readily investigated. Assessment of these models is also helped by the detailed information produced in the calculations, including bond tensions and atomic positions. There are two obvious examples. Rebonding across the crack, for example, can easily be important, and depends critically on the long-range part of the potential. Chemical effects are another important case, and a natural development of these models would be to predict impurity effects. Other new features are simply natural generalizations of continuum models, for example in modified calculations of critical stresses. But one effect, lattice trapping, is unique to the discrete treatments. In a continuum model, the crack remains immobile until a critical stress is reached, when it rips apart. In a discrete model there is an intermediate stress range, in which the crack merely lengthens by a finite amount as the stress is raised. This is possible because there is a series of stable positions for the crack tip, separated by the lattice parameters.

In addition to studies of the statics of brittle fracture, other related phenomena can be treated by atomistic methods of this sort. Such work includes the thermal nucleation and motion of kinks, and their relation to crack creep.

5. Amorphous systems

So many new features are introduced on going to amorphous systems[51] that only a few broad comments are possible. First, the reference state is altered. In crystalline systems, the perfect lattice provides a reference for defining vacancies, dislocations, etc. In amorphous systems, a topological description, at best, is available[52], defining particular arrangements of saturated bonds. One can still define "danging bond" states, and even processes like slip, but the definitions and vocabulary differ. Secondly, amorphous systems inevitably have inequivalent sites. There is no unique geometry for which to attempt very detailed calculations. The connectivity of sites is not unique, and some may, for example, contribute to solubilities of rare gases but not to permeabilities. Charge transfer instabilities become easier, e.g. $A + B \rightarrow A^+ + B^-$ for localised states of types A and B. Broadly, the tetrahedrally coordinated systems tend to have positive U (endothermic reaction), whereas the chalcogenide glasses tend to have negative U. Thirdly, since amorphous systems tend to come from specific regions of the periodic table (see Figure 1 of reference 1), chemical trends must be anticipated. In particular, the most important defects in chalcogenide glasses are[53] valence-alternation pairs[52]: the positively-charged, 3-fold coordinated chalcogen, and the negatively-charged, singly-coordinated chalcogen. And in amorphous silicon, saturation of danging bonds by substantial levels of hydrogen or fluorine has its own special features. Fourthly, different experiments are standard. There are, of course, parallels which should not be overlooked. However, within the present context, the most important changes in emphasis are (1) a concentration on the density of states, often ignoring wavefunctions, matrix elements, and any but the broadest structural features, and (2) a concentration of topology, rather than geometry. One consequence has been the widespread use of the so-called cluster-Bethe-lattice formulation[54].

References

1.   Stoneham, A.M. and Harding, J.H. Chapter (11) of this book.

2.   Stoneham, A.M. "Theory of Defects in Solids". (Oxford University Press) (1975).

3.   Pantelides, S.T., Rev. Mod. Phys. $\underline{50}$, 797 (1978).

4.   Jaros, M., Adv. Phys. $\underline{29}$, 409 (1980).

5.   Kohn, W., Sol. St. Phys. $\underline{5}$, 257 (1957).

6.   Faulkner, R.A., Phys. Rev. $\underline{184}$, 713 (1969).

7.   Lipari, N.O. and Baldareschi, A., Sol. St. Comm. $\underline{25}$, 665 (1978).

8.   Henry, C.H. and Nassau, K., Phys. Rev. B$\underline{2}$, 997 (1970).

9.   Larsen, D.M., Phys. Rev. $\underline{187}$, 1147 (1969).

10. Hjalmarson, H.J., Vogl, P., Wolford, D. and Dow, J., Phys. Rev. Lett. $\underline{44}$, 810 (1980).

11. Bebb, H.B., Phys. Rev. $\underline{185}$, 1116 (1969).

12. Hopfield, J.J. Proc. Paris Semiconductor Conference, p.725 (1964).

13. Stoneham, A.M. and Harker, A.H., J. Phys. C$\underline{8}$, 1102, 1109 (1975).

14. Jambem, A. and Shelby, J.E. Earth and Planetary Science Lett. $\underline{51}$, 206 (1980).

15. Manuel, O.K. and Sabu, D.D. Earth and Planetary Science Lett. $\underline{51}$, 233 (1980).

16. Neumark, G.F., J. Appl. Phys. $\underline{51}$, 3383 (1980).

17. Harding, J.H., UKAEA Report AERE-TP 881 (1980); J. Phys. C (in press 1981).

18. Hemstreet, L.A., Phys. Rev. B22, 4590 (1980).
     de Leo, G., Watkins, G.D. and Fowler, W.B., Phys. Rev. In Press, 1981.

19. For amphoteric centres like Au and Pt in Si, see Brotherton, S.D. and Lowther, J.E., Phys. Rev. Lett. $\underline{44}$, 606 (1980).

20. Watkins, G.D. Proc. Dubrovnik Conference "Radiation Effects in Semiconductors". (Institute of Physics 1975), and private communication.

21. Stoneham, A.M., Sol. St. Comm. $\underline{21}$, 339 (1977); see also Lannoo, M. and Stoneham, A.M., J. Phys. Chem. Sol. $\underline{29}$, 1987 (1968).

22.   Lannoo, M. Proc. Nice Conference "Radiation Effects in Semiconductors". (Institute of Physics Conf. series 46, 1979, edited J.H. Albany).

23.   Mainwood, A., J. Phys. C11, 2703 (1978).

24.   Mainwood, A., Larkins, F.P. and Stoneham, A.M. Sol. St. Elects. 21, 1431 (1978).

25.   Evarestov, R.A., Phys. Stat. Sol.(b) 72, 569 (1975).

26.   Surratt, G. and Goddard, W.A., Sol. St. Comm. 22, 413 (1977).

27.   Astier, M., Pottier, N. and Bourgoin, J.C., Phys. Rev. B19, 5265 (1979), Louie, S.G., Schlüter, M., Chelikowsky, J.R. and Cohen, M.L., Phys. Rev. B13, 1654 (1976).

28.   Jaros, M., Rodriquez, C.O. and Brand, S., Phys. Rev. B19, 3137 (1979).

29.   Bernholc, J. and Pantelides, S.T., Phys. Rev. B18, 1780 (1978).

30.   Baraff, G.A. and Schlüter, M., Phys. Rev. B19, 4965 (1979) also Baraff, G.A., Kane, E.O. and Schlüter, M., Phys. Rev. B21, 5562 (1980).

31.   Larkins, F.P. and Stoneham, A.M., J. Phys. C4, 143 and 154 (1971).

32.   Mainwood, A., J. Phys. C12, 2543 (1979).

33.   Watkins, G.D. and Troxell, J.R., Phys. Rev. Lett. 44, 593 (1980).

34.   Baraff, G.A., Kane, E.O. and Schlüter, M., Phys. Rev. B21, 3563 (1980).

35.   Stoneham, A.M., Adv. Phys. 28, 457 (1979).

36.   Stoneham, A.M., Phil. Mag. 36, 983 (1977).

37.   Dexter, D.L., Klick, C.C. and Russell, G.A., Phys. Rev. 100, 603 (1955).

38.   Bartram, R.H. and Stoneham, A.M., Sol. St. Comm. 17, 1593 (1975).

39.   Stoneham, A.M. and Bartram, R.H., Sol. St. Elects. 21, 1325 (1978).

40.   Lang, D.V. and Logan, R.A., Phys. Rev. Lett. 39, 635 (1977).

41.   Queisser, H.J., Czech. J. Phys. B30, 365 (1980).

42.   Loubser, J.H.N. and van Wyk, J.A., Rep. Prog. Phys. 41, 1201 (1978).

43.   Corbett, J.W., Bourgoin, J.C. and Weigel, C., 1972. "Radiation Damage and Defects in Semiconductors". (Institute of Physics Conf. Ser. 16, edited J.E. Whitehouse

44.   Kimerling, L.C., Sol. St. Electr. 21, 1391 (1978).

45.   Dean, P.J. and Choyke, W.T., Adv. in Phys. 26, 1 (1977).

46.   Troxell, J.R., Chatterjee, A.P., Watkins, G.D. and Kimerling, L.C., Phys. Rev. B19, 5336 (1979).

47.   Bourgoin, J. and Corbett, J.W., Phys. Lett. 38A, 135 (1972).

48.   Griffith, A.A., Phil. Trans. Roy. Soc. A221, 163 (1920).

49. Sinclair, J.E., J. Phys. C5, L271-274 and Phil. Mag. 31, 647-659 (1972).

50. Sinclair, J.E. and Lawn, B.R., Proc. Roc. Soc. Lond. A329, 83-103 (1972).

51. Mott, N.F. and Davis, E.A. "Electronic Processes in Non-Crystalline Materials". (Oxford University Press, 1971,1979); Mott, N.F., J. Phys. C13, 5433 (1980).

52. Thorpe, M.F. and Weaire, D., Phys. Rev. B4, 2508, 3518 (1971).

53. Kastner, M., Adler, D. and Fritzche, H., Phys. Rev. Lett. 37, 1504 (1976).

54. Pollard, W.B. and Joannapoulos, J.D., Phys. Rev. B19, 4217 (1979). Vanderbilt, D. and Joannapoulos, J.D., Phys. Rev. B22, 2977 (1980).

CHAPTER 15

COMPUTER MODELLING OF COMPLEX AND MASSIVELY DISORDERED CRYSTALLINE SOLIDS

by

C.R.A. Catlow and S.C. Parker
Department of Chemistry, University College London,
20 Gordon Street, London WC1H 0AJ

1. Introduction

This chapter will summarise those recent developments in our understanding of heavily defective materials and complex crystal structures, in which the use of computational techniques has played a significant role. We shall be concerned mainly with the use of the HADES, PLUTO and CASCADE simulation programs, as discussed in Chapter (1), which perform explicit atomistic simulations of both perfect and defective lattices. However, we shall attempt to show where results of these calculations interface with those obtained from other computational techniques and other theoretical methods.

We consider first the properties of heavily defective solids. Large levels of disorder may be introduced into a solid in three ways: first, chemically, i.e. by deviation from the stoichiometric composition or introduction of impurities; second, thermally, by raising the solid to a high temperature; and finally, by irradiation damage. Disorder created in all three ways has been actively studied by computational methods in the last few years. Progress has been made in four main areas:

(i) Heavily doped solids The most important development here has been the demonstration of the inadequacy of models based on small defect-dopant clusters. The most intensively studied materials have continued to be the fluorite structured compounds - both the anion excess fluorites (e.g. CaF_2 doped with Y^{3+}) which have been recently investigated by Catlow et al[1] and Bendall et al[2,3], and the anion deficient compounds (e.g. ZrO_2 doped with Ca^{2+}) which we consider in some detail below. Recent work[1,2] on the anion excess fluorites has shown how the rationalisation of neutron scattering data requires cluster models of increasing complexity. Our account here will describe a more fundamental development, namely the attempt to move away entirely from the concept of discrete cluster models.

(ii) Extended defects in non-stoichiometric oxides have provided a major area for successful application of computational techniques. These have played a vital role in advancing our understanding of the factors controlling the stability of shear planes[4] which form in oxides such as TiO_{2-x}[5] and WO_{3-x}. The nature of the equilibrium between point and extended defcts is also now more clearly understood[6]. These topics are reviewed extensively elsewhere[7,8] and also discussed in Chapter (20). A further intriguing problem, i.e. the ordering of shear planes observed in reduced rutile and WO_{3-x}, has also recently been studied by the computational methods. Results are again presented by Cormack in Chapter (20).

(iii) <u>Superionic conductors</u> have also provided a variety of problems that can be successfully tackled by computational methods. These solids, which show exceptionally high ionic conductivities are discussed in Chapters (5) and (18). Here we wish to draw attention to the intriguing and occasionally transient cluster structures observed in these materials. Examples will be discussed below for the high temperature fluorite structured compounds and for the layer structured material β-Al_2O_3. In addition we should note that owing to the exceptionally high transport coefficients of these solids, dynamic simulation techniques are often more appropriate than the static simulations performed by the PLUTO and HADES codes. The use of molecular dynamics techniques for the study of transport in solids is indeed the main concern of Chapters (5) and (18).

(iv) <u>Defect reactions in irradiation solids,</u> where application of the HADES code has made substantial contributions to our understanding of the defect process following the primary act of damage in alkali halide[9,10], and alkaline earth fluoride[11] crystals. For a discussion of the experimental problems we refer to Hobbs[12].

Finally, the simulation techniques have recently been extended to the modelling of minerals - a problem also considered by Jenkins in Chapter (16). The applications of the methods to this area of complex structural chemistry are very similar to those found in the modelling of disordered solids; we shall therefore discuss progress in this field below. This, together with our account of recent advances in the study of the anion deficient fluorites (topic (i) above) and clustering in superionic fluorites (topic (iii)) should provide a representative illustration of the present status of the field. This chapter thus concentrates on <u>applications</u> of the computer modelling methods. For accounts of the techniques we refer the reader to Chapter (1).

2. <u>Defect interactions and ion transport in anion deficient fluorites</u>

Interest in the fluorite structured oxides, e.g. CeO_2 and ThO_2, has been stimulated by the exceptionally high ionic conductivities of these compounds when they are doped with low valence cations, e.g. Y^{3+} and Ca^{2+}.* The defect structure of doped crystals is based on oxygen vacancy compensation for the low valence impurity which enters the lattice at cation sites.

---

*ZrO_2 also adopts the fluorite structure when heavily doped with low valence cations.

The exceptional conductivities of these materials are attributable to high oxygen mobility which in turn is due to two factors: first to the low activation energy for anion vacancy migration in fluorite structured materials[13,14], and secondly to the possibility of introducing very high dopant concentrations, (~ 10-20 mole %) and hence high vacancy populations. Recently, however, it has become clear that the behaviour of these systems is less simple than the above simple description might imply. In particular it is evident that vacancy-dopant interactions exert a critical influence on the rate of oxygen migration. Evidence comes from neutron scattering studies of Fender and co-workers[15] which suggest that vacancies are captured by the oppositely charged* dopant ions. In addition, the study of the variation of the electrical conductivity (σ) with dopant concentration (x) indicates that defect interactions have a decisive effect on the behaviour of these systems. In the absence of interactions the conductivity, σ, would vary with x according to the relationship; $\sigma \propto x(1-x)$. Spectacular deviations from this simple prediction are observed for e.g. Y^{3+} doped CeO_2, which shows a maximum in the plot of σ vs. x at a dopant concentration of ~ 8 mole % as illustrated in Figure 1; after the maximum the conductivity falls by several orders of magnitude. The remainder of this section concentrates on this observation since a satisfactory explanation of the phenomenon necessitates a radical revision in our theoretical approach to defect interactions in heavily defective solids.

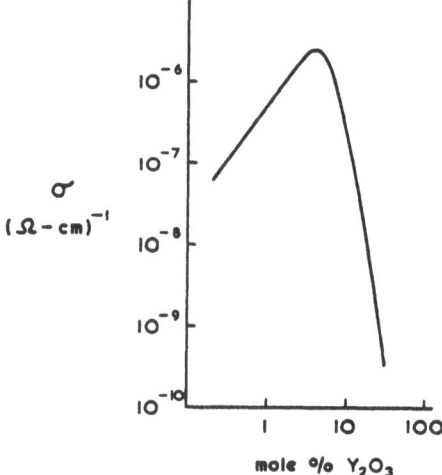

Figure 1   Plot of conductivity (σ) vs. dopant concentration (x) for Y^{3+} doped CeO_2
             - after Nowick et al[16]

---

*We are of course referring here to the <u>effective</u> charges of the vacancy and dopant ion which have opposite signs

Previous approaches to this problem have rested on two main assumptions: first, that the solid can be described in terms of a number of discrete localised clusters; and secondly that the relative concentrations of the different clusters can be treated by the methods of equilibrium thermodynamics usually based on a mass-action formalism[17] (although more sophisticated treatments of cluster concentrations are available[18]). A satisfactory treatment of the heavily defective fluorites requires revision of both these assumptions. First, we argue that at the defect concentrations with which we are concerned here, where ~ 1 in 5 of the lattice ions are replaced by dopant cations, it is no longer permissible to use the concept of discrete cluster models, which imply that indentifiable clusters are separated by regions of perfect lattice. Secondly, the maximum in the plot of σ vs. x cannot be explained by models based on a full thermodynamic equilibrium of the system; such models will predict a monotonic if non-linear increase of σ with x.

What constraint in the doped fluorite systems presents full attainment of thermodynamic equilibrium? We suggest that the low cation diffusion coefficients characteristic of the fluorite structured compounds are crucial, in that they lead to a non-equilibrium dopant ion distribution. Indeed we propose the simplest possible model for this distribution, namely a random dopant distribution frozen in from the high temperatures at which these solid solutions are prepared. Our theory could, however, be modified to allow for other dopant distributions which might be suggested by experimental data.

With this assumption we are now in a position to outline our treatment of the structure and transport in these materials. The main components are as follows:

(i) Discrete cluster models are abandoned. To describe the interactions of anion vacancies with the dopant ion, we classify anion sites according to the distribution of dopants in the surrounding cation shells. For the sake of simplicity we only consider the first cation shell in our classification; sites may be labelled by a single integral number, i, which can vary from 0 to 4, there being four cation sites immediately surrounding an anion in the fluorite structure. Refinements of the model are possible in which subsequent cation shells are used in the definition of the anion sites. Second neighbour cation shells have been considered in a theory of the thermodynamics of these systems[19].

(ii) The numbers, n_i, of anion sites with the dopant cation distribution, i, are calculated as a function of the dopant concentration, x. The calculations of the n_i will of course depend on the model assumed for the dopant distribution. A simple binomial expression may be used for the case of the random distribution which is at present assumed.

(iii) The binding energies, e_i, for the vacancies at the sites i are calculated using the HADES defect simulation program (see Chapter (1)) employing empirically developed potentials to describe the oxide crystal; further reference to these potentials is made below.

(iv) The distribution of vacancies between the available sites is calculated; a Fermi-Dirac function has been shown to be appropriate[20]. Thus we write for the factors x_i of the sites n_i that are occupied by a vacancy

$$x_i = \frac{1}{1 + \exp[(e_i - e_0)/kT]} \tag{1}$$

e_0 is the Fermi energy for the distribution, which is obtained from the condition

$$\sum_i n_i x_i = \tfrac{1}{2} Nx \tag{2}$$

where x is the total dopant concentrations* and N is the number of anion sites per mole of the crystal.

At this stage, the theory may be applied to the treatment of thermodynamic or structural properties. Extensions are required, however, in order to treat transport phenomena. Additional expressions are needed, which include terms representing the rate of oxygen migration between different sites in the crystal. Thus assuming conductivity to be effected entirely by oxygen vacancy jumps, we may write

$$\sigma = C \sum_i \sum_j n_i x_i p_{ij} f_{ij} \exp(-E_{ij}/kT) \tag{3}$$

where C is a constant. The sum over j refers to all the possible types of site that can surround site i; p_{ij} is the probability that site i is surrounded by site j. E_{ij} is the activation energy for the jump between sites i and j, and is calculated using the HADES code. The term f_{ij} is the probability that the jump i → j is not nullified subsequently by the jump j → i. A simple method is used to estimate these coefficients. We assume that the probability that the reverse jump would not occur is given by the probability that j is surrounded by another site that provided a deeper trap than i; it is assumed that a deeper trap requires more dopant ions in the surrounding cation shell.† This approach is very crude; but it reflects physical reality in that reversal of a jump is less likely to occur if an ion can subsequently move on to a deeper trap.

---

*The factor $\tfrac{1}{2}$ enters the formula as one vacancy is created for every two dopant ions, since the effective charge of the vacancy is twice that of the substitutional dopant.

†If j has an adjacent site with the same number of dopant ions as i, then we assume that the probability of the jump i → j being reversed is $\tfrac{1}{2}$.

Our theory thus has three essential components. First, we establish a simple general statistical mechanical formalism for treating the distribution and transport of vacancies. Secondly, we propose a model for the distribution of dopant ions on the cation sub-lattice. In the present case we have used the simplest possible model, namely that of a random distribution of dopant ions, but we should emphasise that more sophisticated models could be employed in our theory. Thirdly, we calculate the binding and activation energies (E_i and E_{ij}) using the HADES code. The computational methods thus play a vital role in quantifying our theory which at no stage, however, postulates distinct discrete cluster models.

To calculate E_i and E_{ij} we must, as noted above, specify interatomic potentials for the interactions between the host lattice ions and between host and dopant ions. We developed a standard shell model[*] potential for CeO_2 using empirical potential fitting of the type discussed in Chapter (10) and which have been successfully applied to UO_2[(2)]. In our treatment of the dopant ion, we neglected ion-size effects, i.e. we assumed that the short range potentials involving the dopant ion were the same as those involving the lattice cation. This approximation has important effects on the results of our calculations as discussed below.

Details of the potentials and of the calculated values of E_i and E_{ij} are presented elsewhere[(23)]. The main results are summarised in Figures 2 and 3. Figure 2 gives the calculated conductivity as a function of x and compares the theoretical curve with the experimental data of Nowick et al[(16)]. The results are encouraging in that the maximum in the plot is predicted at roughly the correct concentration. The theory, however, clearly fails at higher dopant concentrations, where a flattening off of the curve is calculated in contrast to the experimental finding of a continued decrease in σ with increasing x.

The reason for these inadequacies may be due to our neglect of ion size effects. Recent work on these[(24)] and other systems[(25,26)] has suggested that such effects may be substantial. Indeed, our results[(24)] find that the inclusion of potentials which incorporate the effects of the different ionic radius of Y^{3+} compared with Ce^{4+} could affect the calculated E_i and E_{ij} by a factor of ~ 2. Thus to obtain a crude estimate of the effects of ion size factors on our calculated plot of σ vs x, we applied a scaling factor of 2 to our values of E_i and E_{ij} and then recalculated the plot of σ vs x. The new results, given in Figure 3, accord more satisfactorily with experiment, especially at higher dopant concentration, where we now obtain a continued decrease of σ as x is increased.

---

[*]The advantages in defect simulations of Dick and Overhauser's [(21)] shell model treatment of ionic polarisation are discussed extensively in Chapters (10) and (12) and elsewhere[(1,2)].

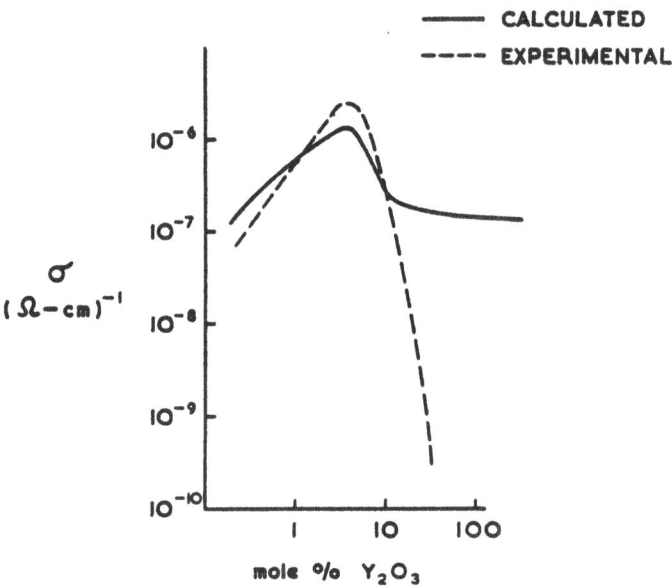

<u>Figure 2</u> Calculated and experimental plots of σ vs x for Y^{3+} doped CeO_2.

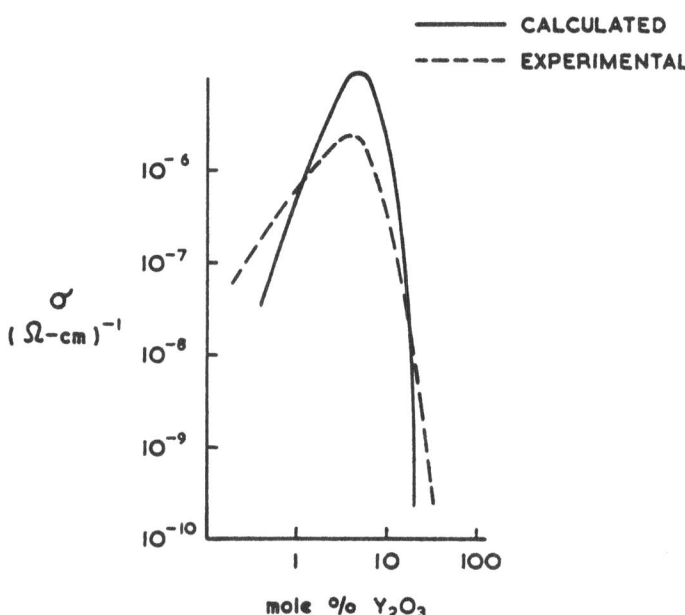

<u>Figure 3</u> Calculated and experimental plots of σ vs x for Y^{3+} doped CeO_2, after scaling binding energies and activation to allow for ion size effects.

The results in Figure 3 probably represent the best accord between theory and experiment that can be expected from such a simple theory. A detailed quantitative theory will require more sophisticated models of the dopant distribution, and a less simple calculation of the correlation factors f_{ij} and of the effect of ion size. However, the main aims of the theory at this stage are first to test whether the type of description of the disordered system that we have given, is capable of reproducing the remarkable qualitative features of the conductivity measurements on the anion deficient fluorite oxides, and secondly to identify the crucial factors controlling the magnitude of the conductivity. We find that our theory can generate the main features of the variation of σ with x for doped CeO_2; and we are able to identify ion-size factors as having a major effect on the conductivities of these systems. Moreover, analysis of our results yields insight into the physical basis of the unusual behaviour of these systems. The pronounced decrease of σ after the maximum at ~ 8 mole % dopant is due to the increasing number of deep traps-sites surrounding several dopant ions at these higher dopant concentrations, which effectively immobilise the vacancies.

This section has concentrated on the necessity of replacing simple concepts of defect clustering when treating heavily defective fluorite oxides. In the next section, we shall see how defect cluster models are, however, important in other highly conducting solids.

3. Defect clustering in superionic conductors

Defect aggregation is expected to be important in superionic materials owing to the high levels of disorder in these solids. Two special features of the disorder in superionics concern, first, the possibility of transient cluster structures and second, the occurrence of unusual modes of defect aggregation owing to the 'fluidity' of certain superionic systems. The first point is illustrated by the high temperature fluorites, and the second by our recent work on the widely studied βAl_2O_3 class of compounds.

3.1 Superionic fluorite structured halides

All fluorite structured halides show a diffuse phase transition within $100-200°$ of the melting point. Above this temperature the solids show exceptionally high ionic conductivities. The transition, which is manifested most obviously by a λ-type specific heat anomaly has been extensively studied by thermodynamic[27] transport[28] and neutron scattering techniques[30,31,35]. It is clear that the transition is associated with the generation of anion interstitial disorder, although it now appears[32] that the extent of this disorder is less than proposed by earlier workers. Neutron scattering studied[31] as well as the results of computer simulation[29,33,34] suggest interstitial concentrations of < 10 mole %.

However, even with interstitial concentrations of a few per cent, extensive interactions between vacancies and interstitials will occur.   Indeed, it is generally agreed that such interactions cause the sudden increase in the interstitial concentration that occurs at the phase transition temperature, T_c. Recently it has become clear that these interactions have a major effect on the structure of the interstitials in the high temperature phase.   Interstitials in lightly doped, low temperature fluorites are known to adopt the body centre position of the cubic interstitial site present in the fluorite structure.   However, recent quasi-elastic neutron scattering data obtained by Dickens et al[35] on the high temperature phase is incompatible with such structures.   The quasi-elastic scattering shows pronounced features for wave-vectors, Q of <1,2,2>, <2,1,1> and <1.5,1.5,1.5>.   Body centre interstitials could not give rise to such scattering.

Recently, calculations have confirmed that the modification of the interstitial structure indicated by the results of the quasi-elastic scattering experiments, is due to the interactions between vacancies and interstitials in these materials.

The type of cluster suggested by the calculations is shown in Figure 4. It contains a pair of interstitials stabilised by the relaxation of two neighbouring lattice ions - the so-called 2:2:2 cluster whose formation in fluorites doped with trivalent ions is well established. Recent calculations[36] have shown that such clusters may be stabilised by vacancies. Although our calculations find that there is a barrier with respect to recombination of this cluster, it is probable that their existence in the high temperature phase will be transient.   The mobility of the F^- ions will lead to rapid vacancy - interstitial recombination.   The life times of the clusters must, however, be several atomic vibrations in order to account for the pronounced peaks in the quasi-elastic scattering.

An explanation of the quasi-elastic data also requires that the clusters should be present in aggregates, otherwise the wave vector dependence would not show the peaks referred to above.   It is probable that clusters of the type shown in Figure 4 could indeed attract each other;   the complexity of the problem at present precludes accurate estimates of the interaction energy.

We have shown elsewhere[36] that the type of model discussed above accounts for the principal features of the quasi-elastic scattering data of Dickens et al[31,35]. The high temperature fluorites thus contain transient aggregates of short-lived clusters - a good example of the novel and intriguing defect structures that are found in superionic solids. Further illustrations of the unusual clustering in superionic solids follow below.   In concluding this section we should point out that the transience of the defect structure we have been discussing raises problems for 'static' simulation codes such as PLUTO and HADES.   More details can be

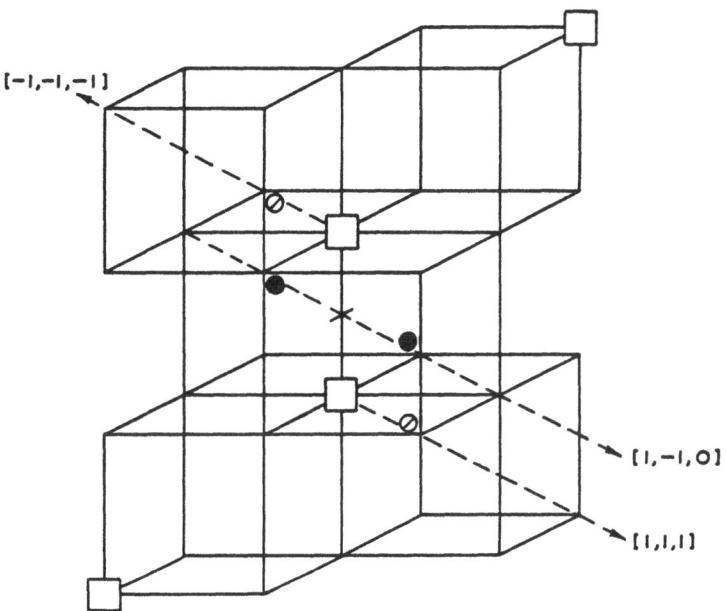

Figure 4   Vacancy stabilised 2:2:2 cluster proposed for superionic fluorites

obtained by the application of dynamical simulations which have been carried out for the high temperature fluorites by Dixon and Gillan[33,34] and which are discussed in greater detail in Chapter (18).   Their results support the type of defect models we have described in this section, but extend considerably our understanding of the transport mechanism in the high temperature fluorites.

3.2   Cluster structures in β-Al$_2$O$_3$

This widely studied superionic material has an unusual layer structure.   Spinel structured alumina blocks are separated by layers containing alkali metal cations and oxide anions;   the latter act as bridges.   The superionic properties are due to the high mobility of the alkali cations in the 'conduction plane' between the spinel blocks.   Details are given by Roth and Reidinger[39].

The stoichiometric β aluminas have the composition M$_2$O.11Al$_2$O$_3$ (M=alkali metal). However, most preparative methods result in a grossly non-stoichiometric material; a typical composition of sodium β-alumina is (Na$_2$O)$_{1.3}$ 11Al$_2$O$_3$.   The excess sodium and oxide ions are accommodated within the conduction plane.   The additional Na$^+$ ions are thought to be essential for the high conductivity of the material[11,44]. However, it is with the oxygen interstitial that we are mainly concerned here. Recent calculations[38] found that the interstitial was strongly stabilised by displacement of two neighbouring Al^{3+} ions from their lattice sites into interstitial positions in the conduction plane. The structure of the resulting complex is shown in Figure 5; evidence for the formation of such clusters was obtained previously by Roth et al[39], in an x-ray diffraction study of non-stoichiometric Na-βAl_2O_3.

Our calculations[38] have also shown that the feature exemplified by the interstitial cluster in Figure 5 - that is, the stabilisation of a conduction plane defect in βAl_2O_3 by large relaxations of surrounding lattice ions - is general to the material. This is a consequence of the fluidity of the lattice surrounding the conduction plane, which is, of course, the ultimate reason for the superionic properties of βAl_2O_3. As a result of these stabilising relaxations, defect formation is unusually favoured in this compound - a feature which may be responsible for the gross deviations from stoichiometry to which we have referred.

Very recent work of Walker and Catlow[40] has demonstrated the importance of a second type of clustering, which involves excess Na^+ ions and O^{2-} interstitials in the conduction plane. Their results support earlier suggestions of Wolf[41] that such clustering would strongly influence the transport properties of the material.

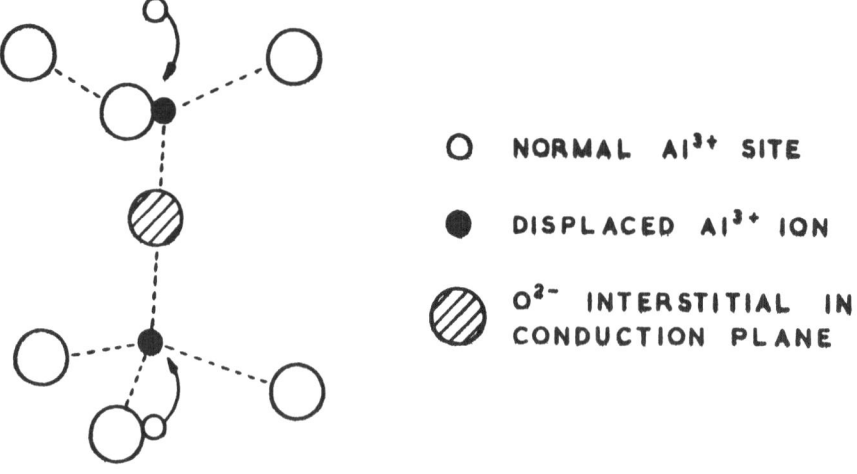

O NORMAL Al^{3+} SITE

● DISPLACED Al^{3+} ION

⊘ O^{2-} INTERSTITIAL IN CONDUCTION PLANE

Figure 5 Stabilised O^{2-} interstitials in β-Al_2O_3

Once more therefore we see unusual and complex modes of clustering in a superionic material. Structural complexity is a major feature of the systems examined in the next section, namely minerals which provide the most severe test of present potential models and computational techniques.

5. Modelling of mineral systems

In Chapter (16) Jenkins clearly shows that lattice energy calculations are a valuable tool in understanding complex problems in silicate chemistry. The work described in this section resembles that of Jenkins in that lattice energy techniques are used, but differs in that it is primarily concerned with structural rather than thermodynamic properties. We aim to predict the structures of minerals and to understand the factors controlling structural variations. It is therefore essential in our work to include a representation of short range interactions - a feature which was omitted in much of the earlier simulation studies on minerals which considered only electrostatic energies. The first step in our approach is thus to develop adequate potential models for silicates. We will describe below the extent to which this has been achieved, after which we illustrate the application first to structure prediction, taking the case of the olivine system, and secondly to advancing our understanding of structure variation by considering the factors controlling the nature of the silicate backbone in pyroxenoid minerals.

5.1 Potential model for silicates

Computer modelling has already been applied with success to simple minerals, e.g. Al_2O_3[42] and TiO_2[6]. Shell model potentials of the type described in Chapter (12) were developed for the oxides and successfully predict a wide range of perfect and defective lattice properties. In view of this success we were encouraged to apply this same method to the silicates. Covalence is unquestionably present in these systems; but covalent effects may at least in part be included in two body potentials, and such potentials provide the best starting point for our studies.

We recall from Chapter (10) that the usual procedure in parameterising potential models is to fit the potential parameters to experimental crystal data, including dielectric and elastic constants. Unfortunately, although there is a great diversity of silicate minerals there are few examples of accurately determined structures for which there are a wide range of crystal data. As a consequence, we have developed an alternative approach which takes advantage of the complexity of silicate mineral structures. In low symmetry structures, for a given pair of ions there may be a large number of separations for which there is an appreciable short range interaction. In contrast for high symmetry crystals short range potentials are sampled generally over only one or two separations. For this reason, in crystals with low symmetry structures, a considerable amount of information is available on potentials <u>from the structure itself</u>.

Structural data can therefore be used via the equilibrium condition to determine potential parameters. This may be achieved by minimising the calculated forces acting on the atoms in the unit cell and the unit cell as a whole, using the PLUTO program as described in Chapter (1). In our work on silicate systems, parameters, other than those for the Si-O interaction were held at values calculated using electron gas methods for the simple oxides[42]; the assumption here is that potential parameters can be transferred provided that the ionic environments are similar.

The procedure worked well in systems containing isolated SiO_4 groups as well as for chain and ring silicates; Table 1 gives a list of structures used in obtaining potential parameters. Moreover the Si...O potentials obtained by fitting to each individual crystal of a given structural type, were found to be similar. It was therefore possible to develop a single set of parameters for all silicates containing isolated SiO_4 groups and a second set applying to all ring and chain structural systems. Parameters are given in Table 2.

Table 1

Mineral structures used in calculating
short range parameters for silicon-oxygen interaction

Name	Formula unit
Monticellite	$CaMg\ SiO_4$
Sillimanite	$Al_2\ SiO_5$
Titanite	$Ca(TiO)\ SiO_4$
Benitoite	$BaTi(SiO_3)_3$
Diopside	$CaMg(SiO_3)_2$
Jadeite	$NaAl(SiO_3)_2$
β-Wollastonite	$CaSiO_3$

Table 2
Short range parameters for silicon-
oxygen interaction

Structural type	$V(R)=A \exp(-R/\rho$	
	$A(eV)$	$\rho(\overset{\circ}{A})$
Isolated SiO_4	471.19	0.4297
Ring and chain silicates	998.98	0.3455

To test these potentials we equilibrated the structures referred to in Table 1, i.e. we adjusted all ion coordinates until there are no net forces acting on any of the species in the unit cell.   The potential was considered satisfactory when the maximum displacement on equilibration was less than $0.25\overset{\circ}{A}$ - a value which was chosen as it is roughly twice the root mean square amplitude of the thermal motions obtained from the temperature factors reported in crystallographic studies of the ambient temperature structure of the minerals referred to in Table 1.   It is unlikely that a <u>static</u> simulation of the type discussed here could achieve agreement with experimentally quoted coordinates by an amount significantly better than this factor. All the calculated atomic positions of the structures in Table 1 accord with experiment to within this criterion. And we should emphasize that we predicted structures for a number of minerals not used in deriving the potential parameters given in Table 2.

For systems containing layers and networks of silicate tetrahedra we have not, as yet, successfully derived potentials. To achieve this, the models will need to be extended, to include angle dependent terms.   However, potential models, based on the parameters reported in Table 2 have been successfully applied to a study of aluminium ordering in the framework silicate, Na-zeolite A[43].

It will be noted that our potential model does not include short range Si-Si interactions. A number of earlier studies[44,45] suggested that short range inter-actions between Si atoms might play an important role in determining the crystal structure of certain silicates. Indeed, it is a curious feature of silicate structural chemistry that the Si-Si separation is virtually invariant, approximately $3.06\overset{\circ}{A}$. We initially included Si-Si short range parameters when trying to fit to the structures, but found that it gave only an insignificant contribution to both forces and energies. We have, more recently, calculated a silicon atom-silicon atom potential using the electron-gas methods discussed in Chapter (10); this should overestimate the short range potential between these atoms in silicates.   We then included the resulting terms in the potential model for several structures for which the Si-Si short range interactions had been suggested to be important in determining the structure. We found that inclusion of the Si...Si interactions had a negligible effect on the lattice energy and that equilibration of a structure which omitted these interactions gave a calculated structure which was close to the experimentally determined one. Moreover, inclusion of the Si-Si interaction did not affect the final positions after equilibration. We therefore conclude that the Si-Si interaction is unimportant and that the rough consistancy of the Si-Si separation is due to other factors.

Having developed our potential models, we may now apply them to structural problems in silicate chemistry. The first problem we considered concerns the prediction of structures, in particular the extent to which our potentials can reliably reproduce the distortions of real structures from ideal models. This application is illustrated by a study of the olivine class of minerals. A second illustration is then presented of a rather different type of application to estimating the stabilities of different silicate chain structures in pyroxenoids.

5.2 Prediction of the structure of Olivine

The structure of olivine can be described in terms of a hexagonal close packed oxygen sublattice with the divalent cation occupying the octahedral holes and silicon the tetrahedral holes. The presence of the cations causes distortions of the oxygen sublattice from the ideal close packed structure. We investigated these distortions for three systems: magnesium, iron and manganese olivine. It must be noted that none of these was used in calculating the potential parameters reported in Table 2.

We calculated the final structures by starting from the idealised olivine structures which were then equilibrated. We compared the final atomic positions with those from a recent structure determinations[46], and found that the discrepancies were within our limit of 0.25Å, as shown in Table 3. These results demonstrate the predictive capacity of our methods and suggest that the potentials we have developed could be applied to calculating mineral structures where the details of atomic positions are unknown.

5.3 Pyroxenoid structures

The pyroxenoid class of minerals have chain structures; different members of the class have different stacking modes for the corner linked tetrahedra, as shown in Figure 6. Ion size effects have been considered as the major factor controlling the stacking mode, although no definite evidence was available to support of this hypothesis.

We investigated this problem by comparing the energies of each of the pyroxenoid structures in Figure 6 after equilibration; in each case the calculations were carried out for the four cations: Ca^{2+}, Mg^{2+}, Fe^{2+} and Mn^{2+}. We found that in such calculations it was not sufficient simply to adjust the coordinate

Table 3
Comparison of experimental (46) and
calculated atomic coordinates for Mg, Fe and Mn Olivine

Magnesium Olivine $MgSiO_4$
a=4.7534Å, b=10.1902Å, c=5.9783Å, space group Pbnm

	Experimental			Calculated			Displacement 1(Å)
	x/a	y/b	z/c	x/a	y/b	z/c	
M(1)	0.0	0.0	0.0	0.0	0.0	0.0	0.0
M(2)	0.99169	0.27739	0.25	0.99323	0.28481	0.25	0.076
Si	0.42645	0.09403	0.25	0.45655	0.10909	0.25	0.210
O(1)	0.76594	0.09156	0.25	0.76914	0.09442	0.25	0.033
O(2)	0.22164	0.44705	0.25	0.20747	0.44018	0.25	0.150
O(3)	0.27751	0.16310	0.03304	0.29382	0.16270	0.04429	0.103

Iron Olivine $FeSiO_4$

a=4.8195, b=10.4788, c=6.0873, space group Pbnm

	Experimental			Calculated			Displacement 1(Å)
	x/a	y/b	z/c	x/a	y/b	z/c	
M(1)	0.0	0.0	0.0	0.0	0.0	0.0	0.0
M(2)	0.98598	0.28026	0.25	0.98938	0.28562	0.25	0.059
Si	0.43122	0.09765	0.25	0.45437	0.11086	0.25	0.178
O(1)	0.76814	0.09217	0.25	0.74837	0.09888	0.25	0.118
O(2)	0.20895	0.45365	0.25	0.17983	0.47113	0.25	0.230
O(3)	0.28897	0.16563	0.03643	0.29781	0.16626	0.05581	0.126

Manganese Olivine
a=4.9023Å, b=10.59640Å, c=6.2567, space group Pbnm

	Experimental			Calculated			Displacement 1(Å)
	x/a	y/b	z/c	x/a	y/b	z/c	
M(1)	0.0	0.0	0.0	0.0	0.0	0.0	0.0
M(2)	0.98792	0.28041	0.25	0.98865	0.28580	0.25	0.057
Si	0.42755	0.09643	0.25	0.45359	0.10952	0.25	0.211
O(1)	0.75776	0.09363	0.25	0.74741	0.09731	0.25	0.064
O(2)	0.21088	0.45369	0.25	0.18338	0.47166	0.25	0.2333
O(3)	0.28706	0.16384	0.04140	0.30436	0.16579	0.05827	0.137

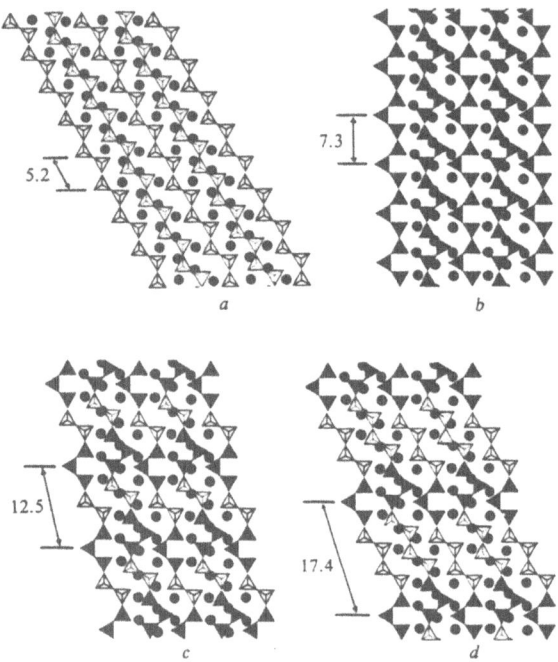

Figure 6   Chain structures in Pyroxenoid minerals

positions, but that it was necessary to allow the unit cell paramaters to vary. These dilations (or contractions) were often strongly anisotropic and their inclusion was essential, if consistent results were to be obtained.   The most efficient approach was to adjust the cell parameters iteratively with minimisation of the coordinate positions after each iteration.   This procedure was repeated until both internal and bulk strains were minimised.   The same procedure was used in the calculations on shear planes discussed by Cormack in Chapter (20).

The results of the calculation are displayed graphically in Figure 4, which for each cation gives the calculated lattice energy per Si atom, with respect to the value calculated for the diopside structure.

Our results show first that the wollastonite structure is always preferred, on energetic grounds, over the diopside structure when the structure contains only one type of cation.   However, we found that the diopside structure could become energetically preferred if both Ca and Mg were present.   This accords with observation, as the diopside structure is observed for $CaMg(SiO_3)_2$ but not for structures containing a single type of alkaline earth cation.

The calculations suggest that there is a strong 'mixed cation' effect in pyroxenoid structure chemistry.   This point was further emphasised by the calculation on the rhodonite and pyroxmangite structures which again were

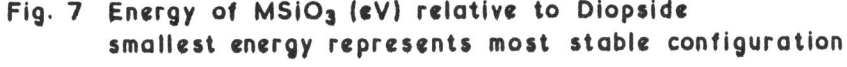

Fig. 7 Energy of MSiO₃ (eV) relative to Diopside
smallest energy represents most stable configuration

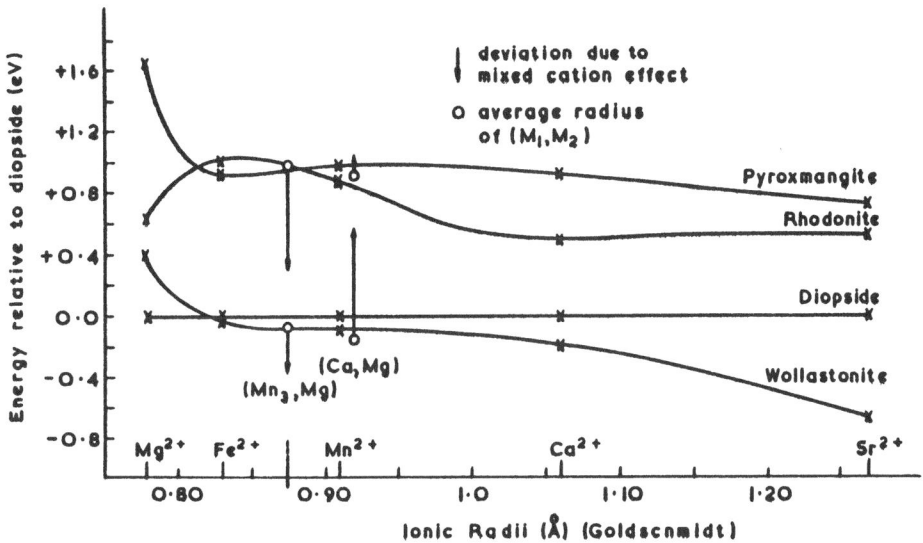

Figure 7 Calculated lattice energies (per $MSiO_3$ unit) of pyroxenoids. Values are given for different cations and are expressed relative to the value calculated for the diopside structure.

predicted to be less energetically favoured than the wollastonite structure, but which were stabilized by including a mixture of cation types, as shown, for example in our results on the synthetic pyroxenoid $Mn_3Mg(SiO_3)_3$. However, for the rhodonite and pyroxenoid structures we were unable to find mixed cation configurations which showed greater stability than that of the wollastonite modification. These structures may be metastable.

5.4 Further developments

Simulation of mineral systems may, we believe, be a most rewarding area of computational chemistry in the near future. The work of Jenkins and co-workers, discussed in Chapter (16) clearly shows the value of calculations in extremely complex problems such as hydration and intercalation. The work outlined here shows that, at least for certain classes of mineral, the techniques of structure prediction can allow a surprisingly large degree of precision. Future work must, we believe, concentrate on extending the sophistication of the potential models so that layer and framework structures can be handled. Some type of many body potential, as described by Stoneham and Harding in Chapter (11), will probably be required.

6. Conclusions

Computer modelling of solids can now clearly handle complex problems; examples are given by the superionic and mineral systems described earlier in this paper. Complex systems can also be tackled, of which the minerals discussed above provide a good illustration. Future progress will depend first on improvements in potential models to include more sophisticated features, e.g. three-body terms, in a computationally viable manner, secondly on advances in numerical techniques - for example developments in minimisation techniques to permit a large number of variables to be handled - an important factor if very large unit cells are to be simulated - and thirdly on greater integration with other computational methods, especially molecular dynamics simulations. The success summarised in this paper we believe encourages such developments.

References

1. Catlow, C.R.A., Chadwick, A.V. and Corish, J., J. Solid State Chem. In press.

2. Bendall, P.J., Catlow, C.R.A. and Fender, B.E.F., J. Phys. C. - to be published.

3. Bendall, P.J., D. Phil. Thesis, University of Oxford, 1980.

4. For an account of shear plane structures we refer to Bursill, L. and Hyde, B.G. Prog. in Solid State Chem. $\underline{7}$, 177 (1972).

5. Catlow, C.R.A. and James, R., Nature $\underline{272}$, 602.

6. James, R., Ph.D. Thesis, University of London, 1979. Also available as UKAEA Report, AERE-TP.814).

7. Catlow, C.R.A., Proc. Int. Conf. on Modulated Structures (Kailua-Kona, Hawaii, 1979), AIP Conference Proceedings, No.53, p.149.

8. Catlow, C.R.A., Proc. Nato School on Refractory Oxides, Corsica. Plenum Press, 1982.

9. Catlow, C.R.A., Diller, K.M. and Norgett, M.J., J. Phys. C$\underline{8}$, L34 (1975).

10. Catlow, C.R.A., Diller, K.M. and Hobbs, L.W., Phil. Mag.

11. Catlow, C.R.A., J. Phys. C $\underline{12}$, 969 (1979).

12. Hobbs, L.W., in Surface and Defect Properties. Eds. M.W. Roberts and J.M. Thomas, Chem. Soc. Spec. Publication (197).

13. Catlow, C.R.A. and Norgett, M.J., J. Phys. C. $\underline{6}$, 1393 (1973).

14. Catlow, C.R.A., Lidiard, A.B. and Norgett, M.J., J. Phys. C $\underline{8}$, L35 (1975).

15. Fender, B.E.F. and Steele, D., J. Phys. C $\underline{7}$, 1 (1974).

16. Nowick, A.S., Wang, Da Yu., Park, D.S. and Griffith, J., in Fast Ion Transport in Solids (Eds P. Vashishta et al), North Holland, 1979.

17. Lidiard, A.B. In Handbuck der Physik, Vol.20, (Ed. S. Flügge), Springer Verlag, Berlin 1957.

18. Allnatt, A.R. and Rowley, L.A., J. Chem. Phys. 53, 3217 (1970).

19. Catlow, C.R.A. and Tasker, P.W., Phil.Mag. - to be published.

20. Catlow, C.R.A., Phys. Stat. Sol. 38A, 191 (1978).

21. Dick, B.G. and Overhauser, A.W. Phys. Rev. B, 112, 90 (1958).

22. Catlow, C.R.A., Proc. Roy. Soc. A333, 533 (1977).

23. Catlow, C.R.A. To be published.

24. Butler, V., Catlow, C.R.A. and Fender, B.E.F., Solid State Ionics 5, 539 (1981).

25. Catlow, C.R.A. and Wapenar, K., Solid State Ionics. 2, 245 (1981).

26. Corish, J., Quigley, J.M., Jacobs, P.W.M. and Catlow, C.R.A., Phil. Mag. A44, 13 (1981).

27. Nolting , J. and Schroter, W., J. de Physique 41 C6-20 (1980).

28. Carr, V.M., Chadwick, A.V. and Saghafian, R., J. Phys. C 11, L637 (1978).

29. Catlow, C.R.A., Comins, J.D., Germano, F.A., Harley, R.T. and Hayes, W. J. Phys. C. 11, 3197 (1978).

30. Dickens, M.H., Hayes, W., Hutchings, M.T. and Smith, C., J. Phys. C 12 L97 (1979).

31. Dickens, M.H., Hutchings, M.T., Kjems, J. and Lechner, R.E., J. Phys. C 11, L538 (1978).

32. Catlow, C.R.A. In Comments on Solid State Physics, 9, 157 (1980).

33. Dixon, M. and Gillan, M.J., J. Phys. C. 13, 1901 (1980).

34. Dixon, M. and Gillan, M.J., J. Phys. C. 13, 1919 (1980).

35. Dickens, M.H. and Hutchings, M.T. To be published.

36. Catlow, C.R.A. and Hayes, W. J. Phys. C. 15, L9 (1981).

37. Perram, J. and Leeuw, S.W. In Fast Ion Transport in Solids. (Eds. P. Vashishta et al), North Holland, 1979.

38. Walker, J.R. and Catlow, C.R.A. Nature. 287, 186 (1980).

39. Roth, W.L., Reidinger, F. and La Placa, S., Proc. Int. Conf. in Superionic Conductors. (Eds. G.D. Mahon and W.L. Roth), Schenectady, New York, Plenum Press (1976).

40. Walker, J.R. and Catlow, C.R.A., J. Phys. C. J. Phys. C. - in press.

41. Wolf, D., J. Phys. Chem. Solids 40, 729 (1979).

42. Mackrodt, W.C. and Stewart, R.F., J. Phys. C 12, 431 (1979).

43. Catlow, C.R.A., Parker, S.C. and Sanders, M. To be published.

44. O'Keefe, M. and Hyde, B.G., Acta Cryst B32, 2923 (1976).

45. O'Keefe, M. and Hyde, B.G., Acta Cryst B34, 27 (1978).

46. Fujino, K., Suslaki, S., Takeuchi, Y. and Sudanaga, R., Acta Cryst B37, 513 (1981).

CHAPTER 16   ASPECTS OF THE CHEMISTRY OF

PHYLLOSILICATES AND INTERCALATION IN VERMICULITES

by

H.D.B. Jenkins
Department of Chemistry and Molecular Sciences,
University of Warwick, Coventry CV4 7AL, West Midlands

1. Introduction

The existence of covalent contributions to the bonding in silicates is un-
disputed.   Nevertheless, studies which consider only electrostatic forces have led
to valuable insights into the relative stabilities and energetics of these systems.
The usefulness of purely electrostatic models has already been demonstrated in a
number of studies.   These include, first, predictions of the oxygen atom positions
in garnets[1] and M(2) positions in olivines[2], second, calculations of the
relative stabilities of Mg-, Fe- and Ni-olivines and spinels[3], third, predictions
of the cation site preferences in amphiboles and fourth,   studies of the hydroxyl
orientations in muscovite[5] and kaolin[6].

The advent of high speed computers has enabled electrostatic contributions
to be calculated with relative ease.   Indeed we have reached the stage now where
such calculations can be refined and we may attempt to study the secondary effects
of repulsion and dispersion energies.   Studies of the latter type yield lattice
potential energies and in turn, via thermochemical cycles, thermodynamic data which
cannot be generated by the purely electrostatic approach.

Recently we have embarked on a programme of study of silicates with the
following objectives:

(i) Development of computational aspects of the theory involving the
treatment of the electrostatic contribution to the energies of large structures.

(ii) Examination of the possibility of calculating the energetics of certain
important minerals of current interest which occur, however, as extremely small
particles so that single crystal x-ray diffraction data cannot be obtained.

(iii) Determination of the energy change on expansion of silicates . This work
is motivated by the growing interest in intercalation phenomena whereby small
molecules and ions are incorporated within the layers of expanded silicates

(iv) Development of models for vermiculites in which water molecules are
incorporated within the layers

(v) Examination of the feasibility of obtaining total lattice potential
energies of large structures, thus providing thermodynamic information on the
relative magnitudes of various substitution, expansion and intercalation processes.

The present chapter considers each of these areas and reports on studies
currently being undertaken.

2. Computational aspects

We shall consider the example of the phyllosilicate systems which are of general formula $KX_2X'T_4O_{10}(OH)_2$ where X' corresponds to a 'trioctahedrally' occupied site (i.e. an extra site occupied by an atom in a trioctahedral mica but vacant in the dioctahedral case), X corresponds to a dioctahedrally occupied site and T refers to the tetrahedral site (T_4 is for example Si_4 or Si_3Al). The basic structural framework of the monoclinic cell is shown in Figure 1.

A conventional computation of the electrostatic energy of this mineral would input the cell parameters and atomic co-ordinates into a suitable program (e.g. MADPROG(7) , LATEN(8) or the PLUTO program discussed in Chapter (1)). The assigned charges usually have formal oxidation state charges (q_K=+1, $q_X = q_{Mg}$=+2, $q_{X'} = q_{Mg}$=+2, q_T=+4, $q'_0 = q_0$=-2, q_H=+1) where q_0 and $q_{0'}$ are the charges on the oxygen of the silicate layer and of the hydroxyl oxygen respectively; q_H is the charge of the hydroxyl hydrogen. Using the crystal structure reported for this mineral[9] an electrostatic energy 74915 kJ mol^{-1} is obtained. Recently however, Jenkins and Hartman[10], recognised that the electrostatic energy for $KX_2X'T_4O_{10}(OH)_2$ can be written in the following expansion

$$U_{elec} = \sum_{i=0}^{2} \sum_{j=0}^{(2-i)} \sum_{k=0}^{(2-j)} \sum_{l=0}^{(2-k)} \sum_{m=0}^{(2-l)} \sum_{n=0}^{(2-m)} \sum_{p=0}^{(2-n)} A_{ijklmnp} \; q_K^i q_X^j q_{X'}^k q_T^l q_0^m q_{0'}^n q_H^p \quad . \quad (1)$$
$$i+j+k+l+m+n+p=2$$

They then considered the 1-M Al mica structure reported by Siderenko et al[11] which has 2C symmetry. This was then modified by the introduction of a mirror plane and addition of two atoms X' in positions $(0,\frac{1}{2},\frac{1}{2})$ and $(\frac{1}{2},0,\frac{1}{2})$ (the trioctahedral site positions) to generate a 'generic' hypothetical mica structure, for which converged values of the coefficients $A_{ijklmnp}$ were calculated.

The advantage of this approach is that by using the values obtained for these coefficients, as given in Table 1, we can obtain from this <u>single</u> calculation the lattice electrostatic energies of all the minerals shown in the scheme given in Table 2, merely by using the charges appropriate to the specific system, as indicated in Table 2.

The separation of structurally dependent coefficients, $A_{ijklmnp}$ from the product of the magnitudes of the interacting charges enables us to examine the change in electrostatic energy on substitution of one atom for another without corresponding relaxation of the positions of the surrounding atoms. Comparison of the energies of the various phyllosilicates, calculated using precise crystal structures in each case, with the energies of the same phyllosilicates calculated using the scheme of Table 2 have enabled Jenkins and Hartman[10] to evaluate, first, energies of relaxation of the structures and second the energy changes caused by

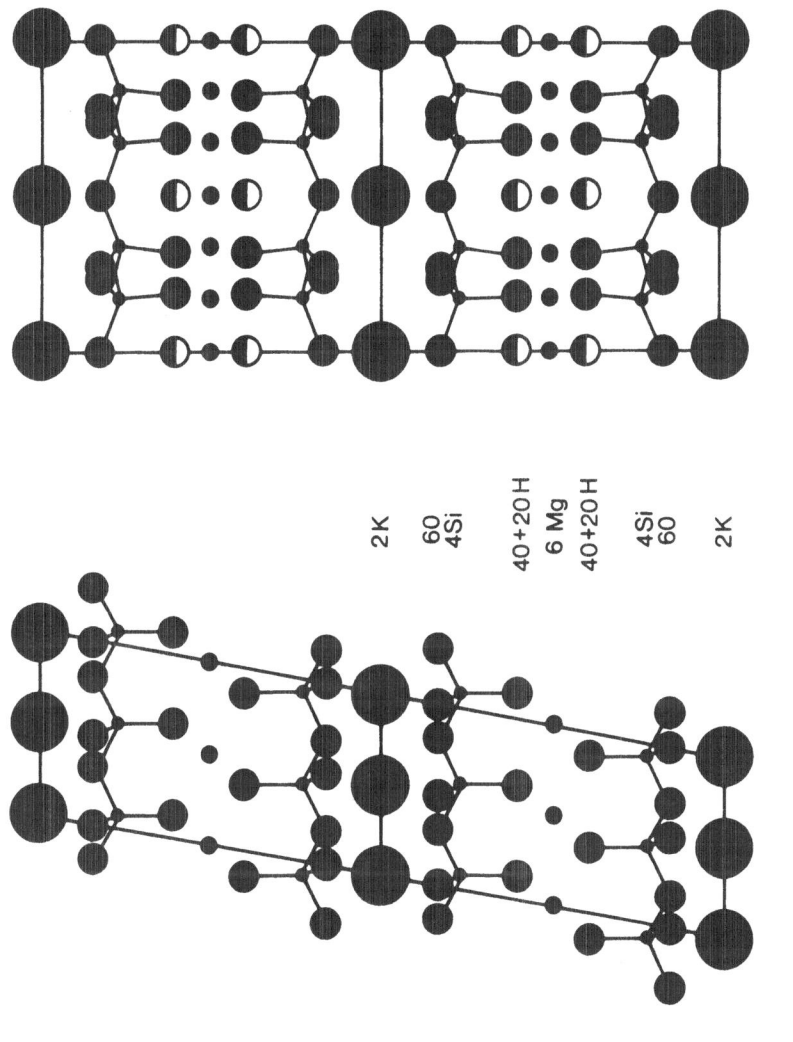

2K

6O
4Si

4O+2OH
6 Mg
4O+2OH

4Si
6O

2K

Figure 1   Crystal structure of a general phyllosilicate showing two aspects of the unit cell and the positions of the various atoms. (Note: Octahedral Mg sites can be substituted by two Al atoms thus leaving a vacant site. Tetrahedral Si_4 is replaced by Si_3Al in presence of interlayer K^+ ions)

Table 1

$A_{ijklmnp}/(kJ\ mol^{-1}e^{-2})$ coefficients for hypothetical mica, $KX_2X'T_4O_{10}(OH)_2$

ijklmnp	Interaction	$A_{ijklmnp}/(kJ\ mol^{-1})$
2000000	KK	254.5
1100000	KX	619.2
1010000	KX'	309.0
1001000	KT	276.2
1000100	KO	795.8
1000010	KO'	524.2
1000001	KH	486.5
0200000	XX	300.8
0110000	XX'	-416.0
0101000	XT	884.4
0100100	XO	319.8
0100010	XO'	-1008.2
0100001	XH	-515.8
0020000	X'X'	254.4
0011000	X'T	455.9
0010100	X'O	423.4
0010010	X'O'	-421.0
0010001	X'H	-772.3
0002000	TT	1814.1
0001100	TO	-980.8
0001010	TO'	833.9
0001001	TH	716.7
0000200	OO	6290.2
0000110	OO'	1228.5
0000101	OH	1091.0
0000020	O'O'	486.3
0000011	O'H	-1695.5
0000002	HH	592.7

Table 2

Scheme showing necessary substitution of charges, q, in equation (1) in order to give energies of related silicates

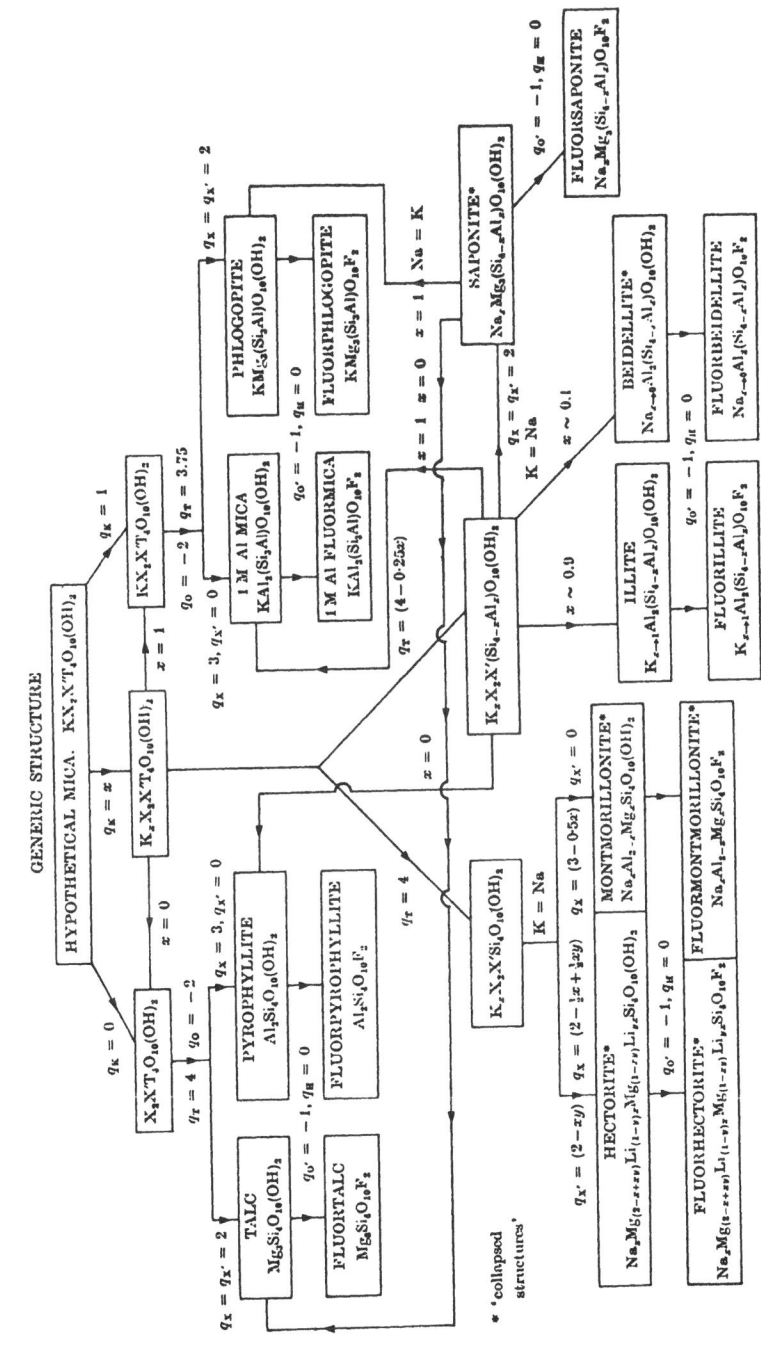

changes in angle between hydroxyl groups and the mica layer. In addition they have been able to calculate the energies of transition of trioctahedral to dioctahedral structures. The following important conclusions were obtained in these studies[10].

(i) Insertion of an OH group in place of an F atom in a trioctahedral mica requires less energy than in a dioctahedral mica.

(ii) The energy of replacement of Al in the tetrahedral layer by Si to convert $Si_3Al \rightarrow Si_4$ and consequent removal of the interlayer K^+ ion is similar in both di and trioctahedral micas.

(iii) The dioctahedral to trioctahedral conversion needs less energy in a fluormica than in a hydroxymica.

(iv) The presence or absence of K^+ ions in the phyllosilicate has only a small influence on the OH orientation with respect to the mica layer.

(v) The surface energies of hydroxy and fluormicas are similar.

(vi) The surface energies of K^+ containing micas are almost the same irrespective of whether they are di or trioctahedral.

In view of the success of the work, it clearly becomes an attractive prospect to apply the same method to the types of problem which were cited in the Introduction as illustrating the success of the electrostatic method. For example, let us consider the problem of predicting the most likely position of a cation M^{Z+} within a given structural framework containing n different atoms whose atomic positions are known precisely. Selecting m possible positional co-ordinates for a given M^{Z+} cation (labelled $M_1, M_2, \ldots, M_m$, to distinguish them) we can then carry out a <u>single</u> calculation determining the (m+n)! Jenkins-Hartman coefficients, A, for the (m+n)! interactions. If we then calculate the energy of the m cases taking $q_{M_i} = z(i=1,2,\ldots,m)$ while taking $q_{M_j} = 0$ (for all $j \neq i$) we can determine the electrostatic energy as a function of M^{Z+} ion position and hence assign the position corresponding to the most stable configuration. This type of approach is being adopted for the study of interlayer cation positions in the model potassium vermiculite calculation discussed in section 5. The advantage lies in the fact that the basic fixed framework interactions need be computed only once.

3.   Energies of silicates whose full crystal structures are not determined

A detailed crystal structure determination has not been reported for many of the minerals included in Table 2. An example is provided by the beidellite-illite series, $A_xAl_2(Si_{4-x}Al_x)O_{10}(OH)_2$ which for small values of $x \simeq 0.1$ and for A = Na corresponds to the mineral beidellite; no accurate determination of its structure has been made. For $x \simeq 0.9$ and A = K we have the mineral illite for which there is a similar lack of detailed structural information. The approach described above has enabled us to estimate both surface, and lattice energies of these structures and to conclude that the hydroxyl group is not perpendicular to the mica layer[10]. In montmorillonites, $Na_xAl_{2-x}Mg_xSi_4O_{10}(OH)_2$, for which again full crystal data are not available, we have been able to estimate lattice and surface energies and to show that, when x = 1, substitution in the octahedral layer increases the surface energy with respect to substitution in the tetrahedral layer. Conclusions of this kind can emerge easily by following the scheme outlined in Table 2, while calculations on the individual minerals involve a large amount of computer time.

4. Expansion of phyllosilicates

The first calculation of the electrostatic contribution to interlayer bonding was made by Giese[12] for the dioctahedral mica 2M-muscovite. His method is based on the comparison of the energies of a normal and an expanded mica. The expansion is carried out in such a manner as to move the layers apart and is continued until additional expansion effects no further change in the energy. Any interlayer K^+ ions are divided between the two separating layers in an ordered fashion as illustrated schematically in Figure 2. The surface energy is then taken as the difference between the fully expanded and normal mica lattice energy.

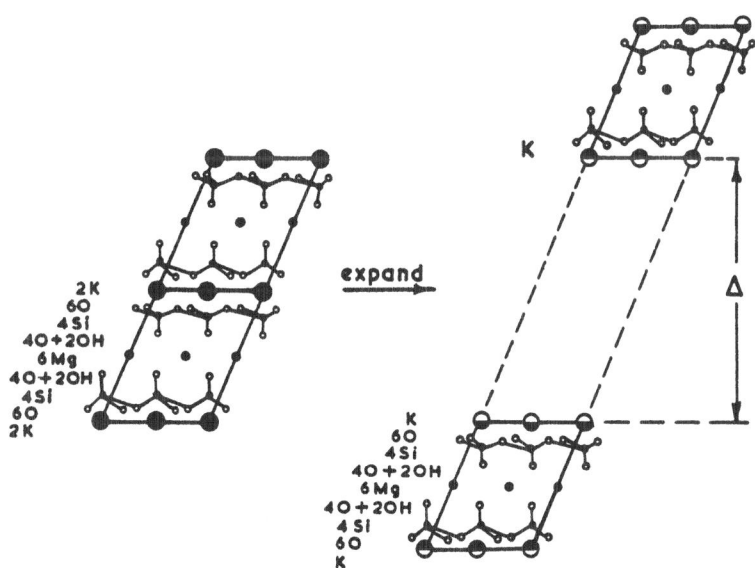

Figure 2   Expansion of a phyllosilicate with division of the potassium ions in the interlayer such that half move with the upper silicate unit and half with the lower. Δ = interlayer separation on expansion, Δ = 0 corresponds to normal phyllosilicate.

Using our hypothetical structure, $KX_2X'T_4O_{10}(OH)_2$ at the head of Table 2 we have obtained the co-ordinates of the expanded structures by separating the layers along the direction of the c axis. For the expanded calculations we can operate the basic scheme of Table 2 to investigate the effects of octahedral and tetrahedral substitution on expansion energies. For an expansion in which interlayer K^+ ions are divided between the separating layers as shown in Figure 2, we considered interlayer separations of 2.5Å, 4.5Å and 10.0Å. We can consider of course an alternative expansion process, that of Figure 3, in which expansion takes place leaving the interlayer cations in the centre of the expanded interlayer gap.

The influence of the interlayer charge on the expandability of micas has been the subject of several studies. Appelo[13,14] considered the expansion of the di- and trioctahedral micas for the expansion mode shown in Figure 3. In addition to considering the mode illustrated in Figure 2, Giese[5] also examined the mode in which all the potassium ions remain on one side of the separating layers. The importance of this topic in relation to intercalation properties of micas has led us to examine this question as a first step for an extended study on intercalation of micas.

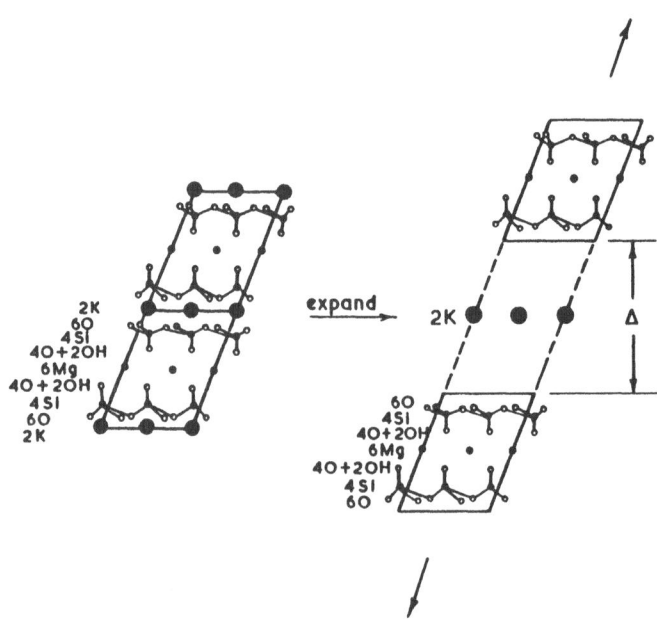

Figure 3 Expansion of a phyllosilicate while retaining the potassium ions in the interlayer mid-way between the separating silicate units. Δ = interlayer separation on expansion, Δ = 0 corresponds to normal phyllosilicate.

In this work we have considered the cases in which the layer charge is either in the tetrahedral sites, corresponding to the formula $K_xX_2X'(Si_{4-x}Al_x)O_{10}(OH)_2$, or in the octahedral sites, corresponding to $K_xAl_{2-x}Mg_xSi_4O_{10}(OH)_2$. Both formulae are derived from the 'generic' $K_xX_2X'T_4O_{10}(OH)_2$, where $T_4 = (Si_{4-x}Al_x)$ and where X is the octahedral cation present in both di- and trioctahedral micas; X' is the octahedral cation present only in trioctahedral micas. Calculations have been performed[10,15] for the unexpanded micas and the micas expanded by 1.0Å, 2.5Å and 4.5Å. The mode of expansion adopted in this paper is such that the silicate layers move apart in the direction of c*, while the K^+ions remain exactly midway between the layers as shown in Figure 3.

For an expansion by ΔÅ we define the expansion energy ΔU^Δ_{elec} as:

$$\Delta U^\Delta_{elec} = U_{elec} \text{ (unexpanded)} - U_{elec} \text{ (expanded)} \qquad (2)$$

which can be written in the general form

$$\Delta U^\Delta_{elec} = D_1x^2 + D_2x + D_3 \qquad (3)$$

where x is the interlayer charge in the micas.

The conclusions of the calculations are as follows:

1. More energy is required to expand a mica when the layer charge is in the octahedral sites; the energy difference does not generally exceed 27kJ mol^{-1}.

2. The expansion energy of the fluormicas is larger than that of the OH-micas by a few kJ mol^{-1}.

3. When the substitution is tetrahedral, the dioctahedral micas have a slightly larger expansion energy than the trioctahedral micas, the difference being about 1-2 kJ mol^{-1}.

We now turn to the mode of expansion. Giese[12,16] and Jenkins and Hartman[10] took the mode of expansion of Figure 2. Figure 4 shows the effect of the two different modes of expansion. The expansion mode of the type shown in Figure 2 gives the more stable configuration for the interlayer, although at low interlayer separations the expansion energies are almost identical. At 2.5Å separation the gap is about 60 kJ mol^{-1}, suggesting that intercalation of water might, owing to the hydration energy, stabilise the structure in which the potassium ions are in the middle. Work is in progress on such a model of a 12.5Å vermiculite.

As a final illustration of this type of study we present in Figure 5 a plot of ΔU^Δ_{elec} as a function of interlayer charge x for three chosen values of Δ in the cases of tetrahedral and octahedral substitution.

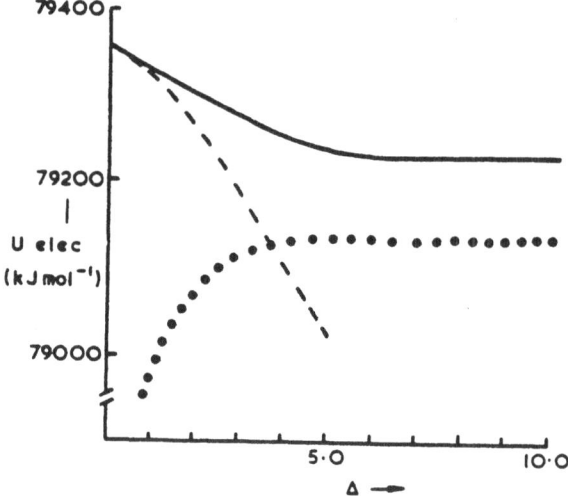

<u>Figure 4</u> U_{elec} (kJ mol^{-1}) as a function of the separation distance Δ of the layers.   <u>Full line</u>: K$^+$ divided between the layers (Figure 2).   <u>Broken line</u>: K$^+$ ions midway between the layers (Figure 3).   <u>Dotted line</u>: K$^+$ ions divided between layers having a charge $+\frac{1}{2}$.

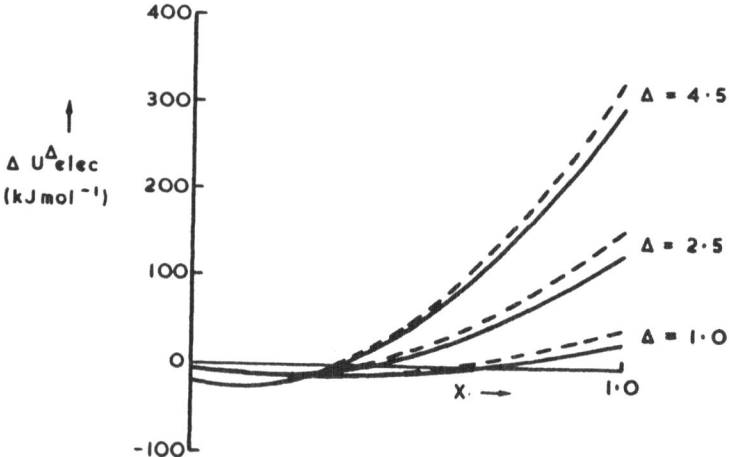

<u>Figure 5</u>   Expansion energy ΔU^{Δ}_{elec} as a function of layer charge x.   <u>Full lines</u>: tetrahedral substitution.   <u>Broken lines</u>: octahedral substitution.

5. Calculations on a model potassium vermiculite

Telluria, Slade and Radoslovich[17] have reported an x-ray diffraction study of a barium vermiculite which has a triclinic unit cell with parameters $a = 5.33\text{Å}$, $b = 9.26\text{Å}$, $c = 12.47\text{Å}$, $\alpha = 100.75^\circ$ and $\beta = 93.5^\circ$. The interlayer contains a single layer of water molecules; the following structural features are of note:

(i) In each unit cell four water molecules are arranged in a hexagonal pattern (similar to the arrangement of carbon atoms in a graphite layer).

(ii) The interlayer Ba^{2+} ions are located above and below these water molecules.

(iii) The layers of Ba^{2+} ions lie upon the silicate layer so that the Ba^{2+} ions fit, approximately, into the ditrigonal holes caused by the arrangement of oxygen atoms.

Adopting these features of the intercalated water geometry we can model a potassium vermiculite, $K_{2x}Mg_6(Si_{4-x}Al_x)_2O_{20}(OH)_4$ by taking phlogopite based on McCauley, Newnham and Gibbs' structure[9] and expanding this structure in a direction perpendicular to the (001) plane such that $a = 5.308\text{Å}$, $b = 9.183\text{Å}$, $c = 12.608\text{Å}$, $\beta = 98.08^\circ$ and incorporating the H_2O in the interlayer (Figure 6).

For the case where $x = 1$ the calculation is straightforward and is based on the approach discussed in section 2 of this chapter. We can consider variable positions of the K^+ interlayer ions above and below the H_2O molecules by performing the calculation including all the positions $K_A, K_B, K_C, \ldots$, etc. and then employing stepwise elimination of the coefficients in the final calculation.

For vermiculites in which $x < 1$ specific consideration must be given to the arrangement of the interlayer cations and the water molecules; these interactions are then calculated separately. When x is low we have isolated K^+ ions surrounded by larger numbers of water molecules as shown in Figure 7.

In each case we have to give careful consideration to the arrangement of the water dipoles. Figures 8,9,10 and 11 show this explicit consideration of the interlayer geometry for the cases where $x = 1$, thus corresponding to $K_2Mg_6(Si_3Al)_2O_{20}(OH)_4 \cdot (H_2O)_4$, where $x = 15/16$ corresponding to $K_{1.88}Mg_6(Si_{3.06}Al_{0.94})_2O_{20}(OH)_4 \cdot (H_2O)_4$, where $x=\frac{3}{4}$ corresponding to $K_{1.50}Mg_6(Si_{3.25}Al_{0.75})_2O_{20}(OH)_4$ and where $x=1/3$ corresponding to the case of $K_{0.67}Mg_6(Si_{3.67}Al_{0.33})_2O_{20}(OH)_4 \cdot (H_2O)_4$. The arrows point from the oxygen atoms towards the hydrogen atoms.

As the interlayer charges, $q_K = x$, are decreased from unity the vermiculite is correspondingly modified, first in the interlayer by geometrical changes due to creation of vacant cation sites and the corresponding adjustments, shown in Figures 9,10 and 11, in the orientations of the water molecule dipoles; and secondly in the silicate framework by the occupational changes in the tetrahedral layer sites

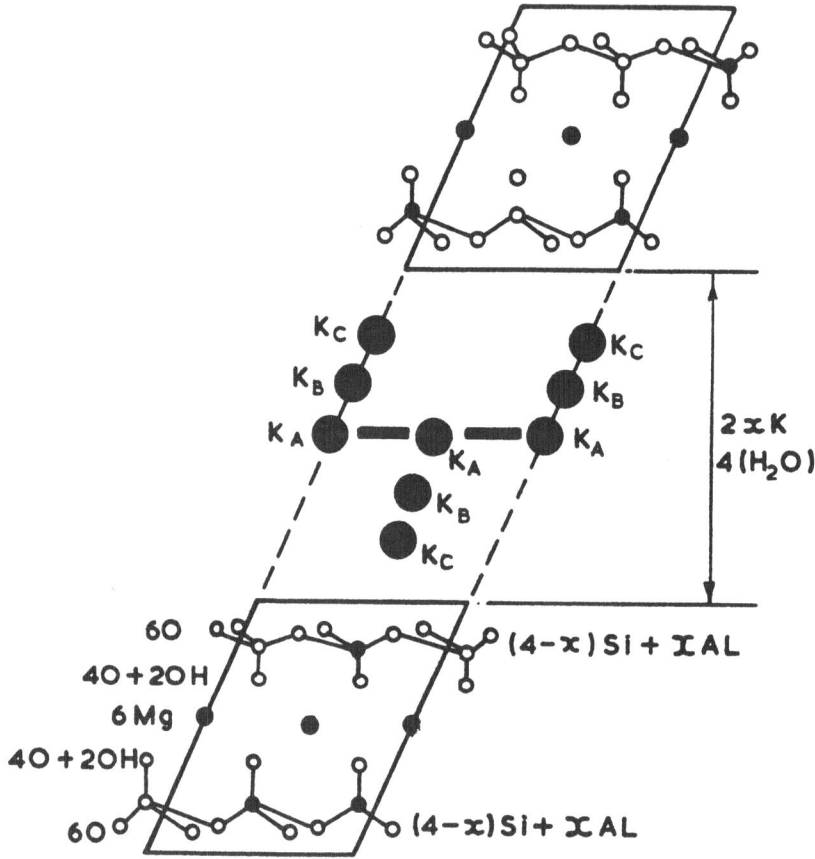

<u>Figure 6</u> Potassium vermiculite showing the interlayer arrangement taken
for the K[+] ions and for the water molecules (illustrated by
solid horizontal lines) moving between the layers.

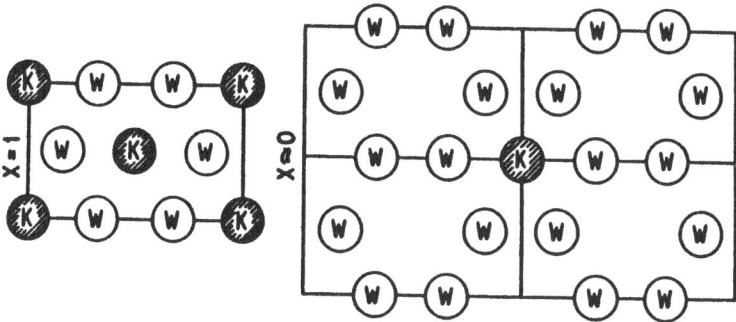

<u>Figure 7</u> For x = 1, the vermiculite $K_2Mg_6(Si_3Al)_2O_{20}(OH)_4 \cdot (H_2)_4$ has the hexagonal arrangement of $H_2O(W)$ molecules around each K^+ ion and a full complement of K^+ ions in the interlayer. For intermediate values of x, each situation involves a specific arrangement. When x → 0, we have few K^+ ions surrounded by large numbers of $H_2O(W)$ molecules. The figure shows the two extreme arrangements.

that are required to maintain electroneutrality. The latter modifications are readily incorporated using our 'generic' approach; the former changes affect the $K-H_2O$, H_2O-H_2O and K—K interaction energies within the interlayer. For the purposes cf calculations when x < 1 the interlayer cations are assumed to be located in the central position within the interlayer (position K_A in Figure 6) throughout.

6.   Intercalation energies [26]

The intercalation energy, $\Delta U_{inter}(g)$, of the process

$$2K_xMg_3(Si_{4-x}Al_x)O_{10}(OH)_4(c) + 4H_2O(g) \xrightarrow{\Delta U_{inter}(g)} K_{2x}Mg_6(Si_{4-x}Al_x)_2O_{20}(OH)_8 \cdot (H_2O)_4(c)$$

PHLOGOPITE   VERMICULITE INTERCALATE

can be written as

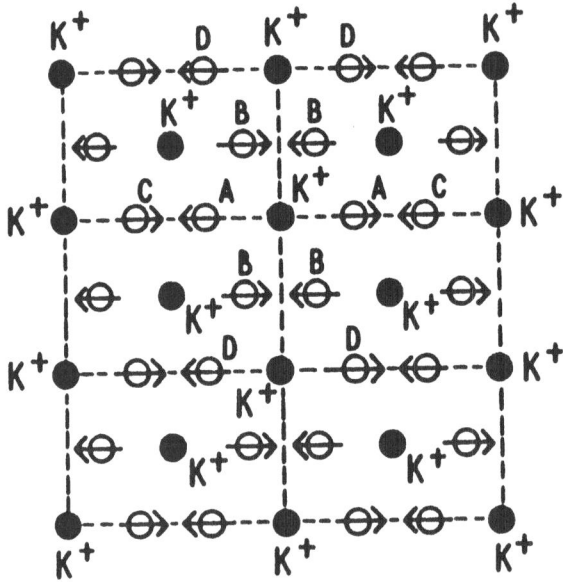

Figure 8   The diagram illustrates explicit consideration of the interlayer arrangement of water dipoles and K^+ ions for $x = 1$ and hence for $K_2Mg_6(Si_3Al)_2O_{20}(OH)_4 \cdot (H_2O)_4$.   Arrows representing water molecules point from oxygen atoms in the direction of the hydrogen atoms.

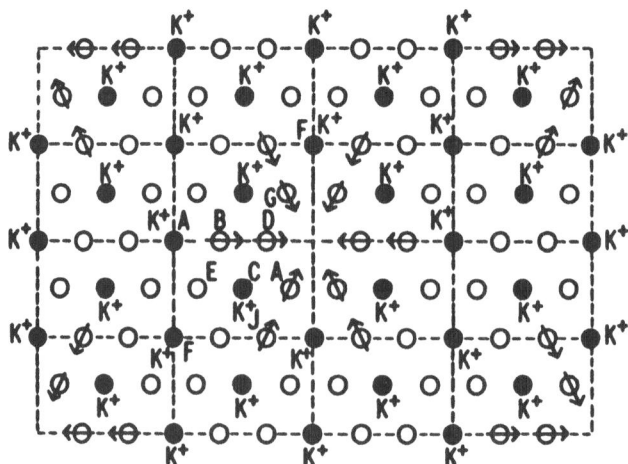

Figure 9 The diagram illustrates explicit consideration of the interlayer arrangement of water dipoles and K^+ ions for $x = 15/16$ and hence for $K_{1.88}Mg_6(Si_{3.06}Al_{0.54})_2O_{20}(OH)_4 \cdot (H_2O)_4$.   Arrows representing water molecules point from oxygen atoms in the direction of the hydrogen atoms.

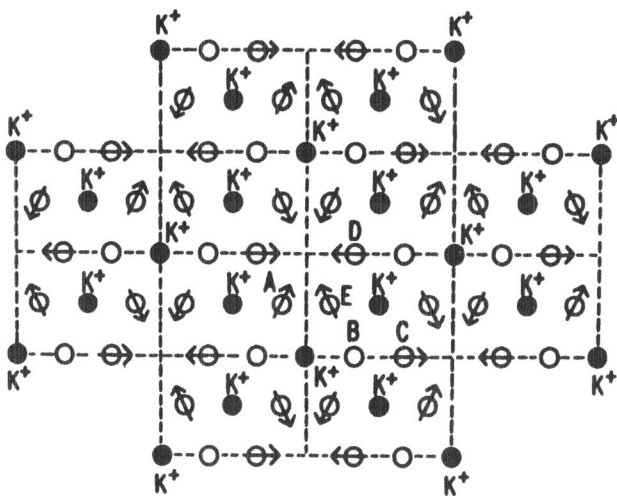

Figure 10   The diagram illustrates explicit consideration and the interlayer
arrangement of water dipoles and K^+ ions for $x = \frac{3}{4}$, $K_{1.50}Mg_6(Si_{3.25}Al_{0.75})_2O_{20}(OH)_4 \cdot (H_2O)_4$

Arrows representing water molecules point from oxygen atoms in the direction of the
hydrogen atoms.

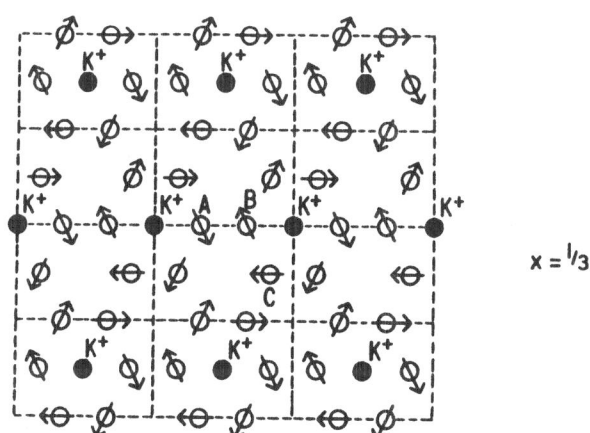

$x = \frac{1}{3}$

Figure 11   The diagram illustrates explicit consideration and the interlayer
arrangement of water dipoles and K^+ ions for $x = 1/3$.
$K_{0.67}Mg_6(Si_{3.67}Al_{0.33})_2O_{20}(OH)_4 \cdot (H_2O)_4$.   Arrows representing water
molecules point from oxygen atoms in the direction of the hydrogen atoms.

$$\Delta U_{inter}(g) = U_w + \Delta U_{exp}^{\Delta} \qquad (4)$$

where ΔU_{exp}^{Δ} is the expansion energy of phlogopite as given by equation (2) and U_w is given by

$$U_w = - E(H_2O-H_2O)-E(K-H_2O)-E(H_2O-silicate\ layer) \qquad (5)$$

in which $E(H_2O-H_2O)$, $E(K-H_2O)$ and $E(H_2O-silicate)$ are the H_2O-H_2O and $K-H_2O$ interlayer interactions and $E(H_2O-silicate)$ is the interaction energy of the water molecules with the silicate layer.

For the same process in which liquid water rather than water from the gas phase, is taken up by the expanded phlogopite lattice, we have for the interaction energy $\Delta U_{inter}(\ell)$

$$\Delta U_{inter}(\ell) = \Delta U_{inter}(g) + 4\Delta H_{vap}(H_2O)(\ell) \qquad (6)$$

where $\Delta H_{vap}(H_2O)(\ell)$ is the enthalpy of vapourisation of water, which has a value of 44 kJ mol^{-1}.

Figure 12 shows the variations of $\Delta U_{inter}(g)$ with x for various choices of the charge $q_{H''}$ on the hydrogen atoms of the intercalated water molecules. We see that for $q_{H''} \geqslant 1/3$ no uptake of water is predicted as possible and that for for $q_{H''} = \frac{1}{2}$ (a value commonly chosen to model water in work on crystalline hydrates[18]) no intercalation is possible when $0.1 \leqslant x$ or $x \geqslant 0.8$. Figure 13 shows the behaviour of the individual contributions, U_w and ΔU_{exp}^{Δ} of equation (4) for the case where $q_{H''} = 1/3$. The behaviour of ΔU_{exp}^{Δ} is of course similar to that illustrated in Figure 5 while that for U_w is a complex and irregular, due to the variations in the configurations adopted by the water dipoles. $\Delta U_{inter}(g)$, however, is found to be a smooth function of x. The study therefore establishes the following points

(1) At large values of x, the magnitude of the expansion energy, ΔU_{exp}^{Δ}, that is required to separate the interlayer of the phlogopite is greater than the stability provided by U_w. Thus $\Delta U_{inter}(g)$, and hence $\Delta U_{inter}(\ell)$, rise above the intercalation limit and no uptake of water is possible.

(2) At low values of x, the expansion energy is very small and $\Delta U_{inter}(g) \simeq U_w$. The predominant term in U_w is the $K-H_2O$ interaction energy while the interaction of the water molecules with the silicate layer is vanishingly small.

(3) The interaction of water molecules and the silicate layer becomes an important factor when $x > 0.5$.

(4) For intermediate values of x intercalation occurs due to the large $K-H_2O$ interaction layer.

There remains work to be done on these systems and we are currently examining a number of further questions. These include the effect on these calculations of

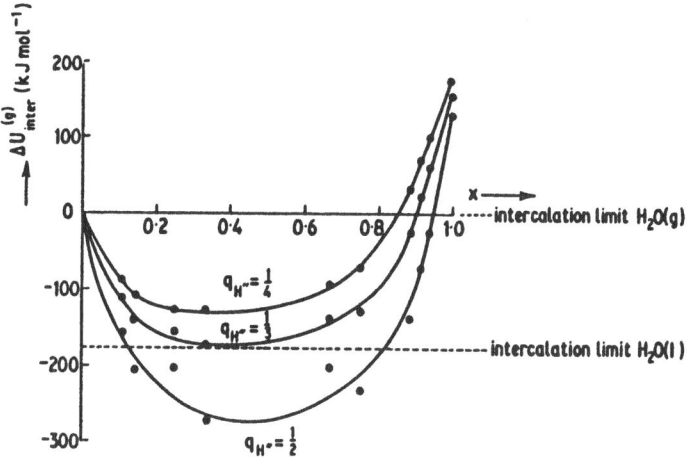

<u>Figure 12</u>   Variation of $\Delta U_{inter}(g)$ and $\Delta U_{inter}(\ell)$ as functions of x for hydrogen atom charges $q_{H''}$ of $\frac{1}{4}$, 1/3 and $\frac{1}{2}$.

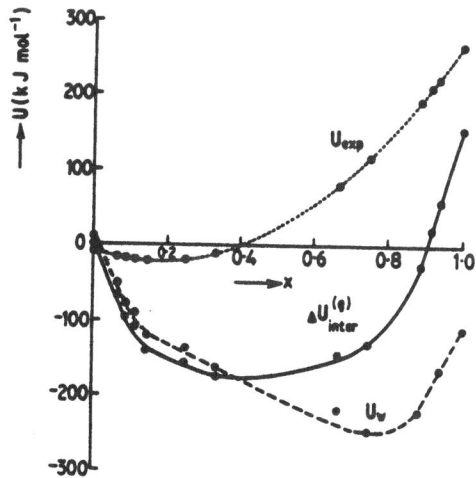

<u>Figure 13</u>   Variation of $\Delta U_{inter}(g)$ and the component terms U_w and U_{exp} as a function of interlayer charge, $q_K = x$ for $q_{H''} = 1/3$.

including repulsion and dispersion energies, and the more accurate estimation of $q_{H''}$ (the hydrogen charge on the water molecules). Experimental studies have already recognised that layer charge is important in determining whether intercalation takes place. The present work[27] underlines the importance of this factor and shows that other factors arising from the difference between tetrahedral and octahedral substitution are much less significant.

7. Full lattice potential energy calculations on silicates

Extensive studies have been reported on the calculation of total lattice potential energies of complex inorganic salts[20,21,22]; these have included the extension of the Huggins and Mayer[23] approach for the calculation of repulsion energies to encompass a minimisation procedure[22,23]; their success encourages the extension of this approach to the field of silicate chemistry.

Following the Huggins-Mayer approach, we will assume that the repulsion energy between two ions i and j is given by the equation:

$$U_R(i,j) = bC_{i,j} \exp\left[\frac{\bar{r}_i + \bar{r}_j}{\rho}\right] \sum_j^{ions} \exp\left(-\frac{R_{ij}}{\rho}\right) \tag{7}$$

where b and ρ are constants, and $\bar{r}_i$ and $\bar{r}_j$ are the 'basic' radii of the ions i and j, and $C_{i,j}$ is defined by the equation

$$C_{i,j} = (1 + q_i/n_i + q_j/n_j) \tag{8}$$

where n_i and n_j are the number of valence electrons in the ions i and j having charges q_i and q_j. The summation over the ions j is continued out to distances at which the interactions become negligible.

We can calculate pairwise repulsion energies for the phyllosilicates using parameters derived from our study on complex salts. The dipole-dipole (R^{-6}) and the quadrupole-dipole (R^{-8}) dispersion energies U_{dd} and U_{qd}, are given by equations (9-12) below. For the interaction between ions i and j we have

$$U_{dd}(i,j) = bc_{i,j} \sum_j^{ions} (R_{i,j})^{-6} \tag{9}$$

$$U_{qd}(i,j) = bd_{i,j} \sum_j^{ions} (R_{i,j})^{-8} \tag{10}$$

in which

$$c_{i,j} = \frac{3\alpha_i\alpha_j E_i E_j}{2(E_i + E_j)} \tag{11}$$

$$d_{i,j} = \frac{9c_{i,j}}{32e^2} (\alpha_i E_i + \alpha_j E_j) \tag{12}$$

where α_i and α_j are the polarisabilities of the ions i and j, E_i and E_j are the characteristic energies (usually taken to be 3/4 of the ionisation potentials) and e is the electronic charge.

The total lattice energy is now given by

$$U_{pot} = U_{elec} - U_R + U_D \qquad (13)$$

where U_R and U_D are the total repulsion and dispersion energies obtained by pairwise interaction summation

$$U_D = U_{dd} + U_{qd} \quad . \qquad (14)$$

Jenkins and Thakur[24] have studied the application of the above equations to a range of silicates. The electrostatic calculations on the model vermiculite are also being augmented by studies of repulsion and dispersion contributions.

Starting from basic crystal data we have calculated the total lattice energies of the silicates listed in Table 3 where the calculated values are compared with the predicted lattice energies obtained using the experimental enthalpies of formation of the silicates. The differences are undoubtedly due mainly to covalence, but the measure of agreement between the independently calculated cycle value is certainly encouraging.

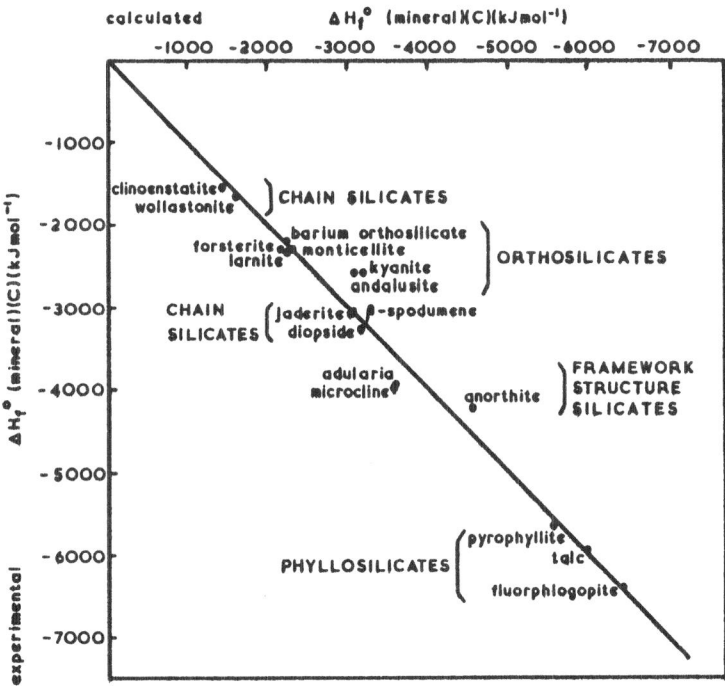

Figure 14 Predicted and experimental standard enthalpies of formation, ΔH_f(silicate)(c)(kJ mol^{-1}) for various silicate types. Predicted values are obtained from complete lattice energy calculations involving considerations of electrostatic, repulsion and dispersion contributions.

An alternative way of looking at the results is to use the calculated lattice energies in a thermochemical cycle and compute the standard enthalpies of formation of the minerals listed from their elements in their standard states. Figure 14 shows a plot of the calculated enthalpies of formation against the experimental

Table 3

Comparison of thermochemical cycle and calculated lattice potential energies of silicates

Silicate	$U_{pot}/(kJ\ mol^{-1})$	
	Calculated	Experimental (via cycle)
Phyllosilicates		
fluorphlogopite, $KMg_3Si_3AlO_{10}F_2$	58609	58566
talc, $Mg_3Si_4O_{10}(OH)_2$	67903	67829
pyrophyllite, $Al_2Si_4O_{10}(OH)_2$	71359	71460
Framework structures		
low sanidine, $KAlSi_3O_8$	47746	-
microcline, $KAlSi_3O_8$	47583	47963
adularia, $KAlSi_3O_8$	47626	47958
anorthite, $CaAl_2Si_2O_8$	45005	44715
Chain structures		
diopside, $CaMgSi_2O_6$	33324	33378
clinoenstatite, $MgSiO_3$	16725	16846
spodumene, $LiAlSi_2O_6$	34815	35108
-spodumene, $LiAlSi_2O_6$	35097	35088
jadeite, $NaAlSi_2O_6$	35034	35032
wollastonite, $CaSiO_3$	16439	16509
Orthosilicates		
kyanite, Al_2SiO_5	28687	28342
forsterite, Mg_2SiO_4	20697	20664
barium orthosilicate, Ba_2SiO_4	19300	19403
monticellite, $CaMgSiO_4$	20364	20327
larnite, Ca_2SiO_4	19831	19949
andalusite, Al_2SiO_5	28807	28188

values taken from standard tabulations[25]. The trends shown by this comparison encouraged our belief that meaningful calculations are beginning to be made of the non-electrostatic contributions within these large structures.

8. Conclusions

We believe that the calculations summarised in this chapter show that a considerable amount of information emerges from the new approach which we have developed for the treatment of silicates. The more extended calculations are at a preliminary stage, but the initial studies are encouraging and the comparisons that we have made suggest that we are able to calculate repulsion energies using a parameter scheme that is transferable between different minerals reasonably. Our work on these problems continues.

Acknowledgements

The work described was carried out by the author and his colleague Professor Piet Hartman of the University of Utrecht whose contribution to this work is here acknowledged. The provision of a six month visiting fellowship by the ZWO (Netherlands Organisation for Pure Science) to carry out much of this work at the University of Leiden, The Netherlands and the assistance of the Science Research Council is also acknowledged.

References

1. Born, L. and Zemann, J., Beitr Mineral Petrog, 10, 2 (1964).

2. Born, L., N Jb Mineral Monat. 1964, 81 (1964).

3. Gaffney, E.S. and Ahrens, T.J., Phys. Earth Planet Inter. 3, 205 (1970).

4. Whittaker, E.J.W., Amer. Mineral, 56, 980 (1971).

5. Giese, R.F., Jnr. Science, 172, 263 (1971).

6. Giese, R.F. Jnr. and Datta, P., Amer. Mineral, 58, 471 (1973).

7. Jenkins, H.D.B. and Pratt, K.F., Comp. Phys. Commun. 13, 341 (1978).

8. Jenkins, H.D.B. and Pratt, K.F., Comp, Phys. Commun. 21, 257 (1980).

9. McCauley, J.W., Newham, R.E. and Gibbs, G.V., Amer. Mineral, 58, 249 (1973).

10. Jenkins, H.D.B. and Hartman, P., Phil. Trans. Roy. Soc. A293, 169 (1979).

11. Siderenko, O.V., Zvyagin, B.B. and Sobolera, S.V., Soc. Phys. Crystl. 20, 332 (1975).

12. Giese, R.F. Jnr., Nature (Phys. Sci.), 248, 580 (1974).

13. Appelo, C.A.J., Amer. Mineral, 63, 782 (1978).

14. Appelo, C.A.J., "Aspects of mica-related clay minerals in hydrogeochemistry". Thesis, Vrije Universitat, Amsterdam (1978).

15. Jenkins, H.D.B. and Hartman, P., Phys. Chem. Minerals, 6, 313 (1981).

16. Giese, R.F. Jnr., Z. Kristallogr. 141, 138 (1975).

17. Telluria, M.I., Slade, P.G. and Radoslovich, E.W., Clays and Clay Minerals, 25, 119 (1977).

18. Baur, W.H., Acta. Cryst. 14, 209 (1961).

19. Schulz, L.G., Clays and Clay Minerals, 17, 115 (1969).

20. Jenkins, H.D.B. and Pratt, K.F., Proc. Roy. Soc. A356, 115 (1977).

21. Jenkins, H.D.B. and Pratt. K.F., J. Chem. Soc. Faraday II, 74, 968 (1978).

22. Jenkins, H.D.B. and Pratt, K.F., Adv. Inorg. Chem. Radiochem. 22, 1 (1979).

23. Huggins, M.L. and Mayer, J.E., J. Chem. Phys. 1, 643 (1933).

24. Jenkins, H.D.B. and Thakur, K.P. To be published (1982).

25. Robie, R.A., Hemmingway, B.S. and Fisher, J.R., "Thermodynamic properties of minerals and related substances at 298.15K and 1 Bar (10^5 P) pressure and at higher temperatures". Geological Survey Bulletin 1452, U.S. Govt. Printing Office, Washington 1978.

26. Jenkins, H.D.B. and Hartmann, P., Phil. Trans. Roy. Soc., A304, in the press (1982).

27. Jenkins, H.D.B. and Hartman, P., Phil. Trans. Roy. Soc., A304, 397 (1982).

CHAPTER 17

AGGREGATION AND PRECIPITATION IN ALKALI HALIDES

by

J. Corish
Department of Chemistry, University College Dublin,
Dublin 4, Ireland

and

J.M. Quigley
School of Pharmacy, Trinity College,
Dublin 4, Ireland

1. Introduction

Thermal depolarization (1-3), dielectric loss (4-8) and low temperature ionic conductivity data (9,10) in the alkali halides are greatly influenced by the association of divalent impurity ions with their charge compensating vacancies and the subsequent aggregation of the resulting dipoles. There is also considerable interest in the kinetics of the decay in dipole concentration as dimers, trimers and higher aggregates form. In some systems, e.g. $NaCl:Cd^{2+}$, the aggregation process results in the eventual formation of the metastable Suzuki phase, the structure of which has been determined by X-ray diffraction measurements (11). In most cases, however, the structures of intermediate dipole clusters cannot be identified by physical methods and there is an obvious need for reliable theoretical estimates of the relative stabilities of various configurations. Indeed their study provides a good example of the valuable rôle which can be played by theory in advancing our understanding of complex phenomena in defective solids.

We shall denote clusters of nearest neighbour dipoles by A_n and of next-nearest neighbour dipoles by B_n where n is the number of impurity ions which they contain. Thus the equation

$$A_n(x) + A_1 \rightleftharpoons A_{n+1}(y); \quad n=1,2,\ldots \tag{1}$$

describes the addition of an nn dipole to a cluster containing n impurity ions. Several different configurations of dimers (n=2), trimers (n=3), etc. are shown in Figure 1 and identified by the symbols x,y=a,b,...,j. Dielectric loss studies have been interpreted as a dipole aggregation process exhibiting third order kinetics in the initial and final stages, separated by a dipole-trimer equilibration stage (4,5,7,8,12,13). For nearest neighbour dipoles the mechanism

$$A_{2n+1}(x) + 2A_1 \underset{k_{2n+3}}{\overset{k_{2n+1}}{\rightleftharpoons}} A_{2n+3}(y); \quad n=0,1,2,\ldots \tag{2}$$

where $k_{2n+3}=0$ except when n=0. An alternative to this improbable third-order elementary kinetic process is a two-step model involving the formation of an intermediate dimer (14,15). When such an intermediate is loosely bound the observed

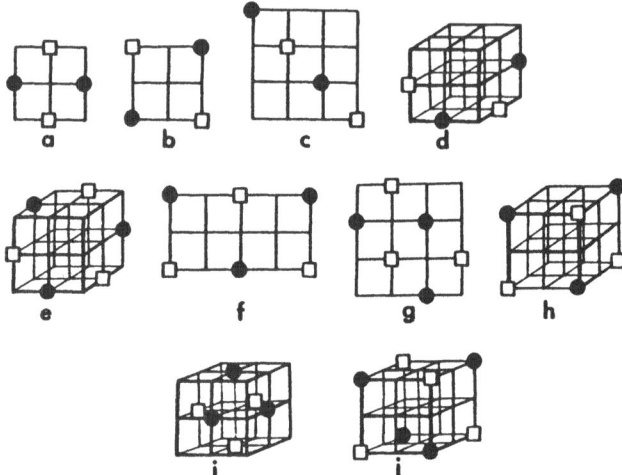

<u>Figure 1</u>   Dipole aggregates for which calculations have been made: ● is an
impurity ion and □ a cation vacancy

reaction rate will also be proportional to the third power of the dipole
concentration. Some experimental data has been found to conform to second-order
kinetics and recent analyses (16-18) of this problem indicate that the kinetics
observed may depend on the particular experimental conditions. Although it is
impossible to deduce detailed structural information from these kinetic experiments
it appears probable that a stable trimeric species is formed by a consecutive
reaction involving an intermediate dimer.

   This chapter illustrates the application of theoretical defect simulation methods
to the determination of the relative stabilities of dipole clusters containing two
or more impurity ions. This enables us to assess the relative importance of these
clusters as participants in an aggregation process.   Previous calculations, relying
on the Point Ion approximation, overestimated association energies of aggregates due
to the inherent inability of the model to account adequately for polarization in the
crystal, as discussed in Chapter (10).   This difficulty is overcome in the present
calculations by the use of the shell model.   In addition, since specific interionic
potentials are derived for each impurity ion, the calculations reveal differences
in the behaviour of the various dopants in the host crystals. Finally we combine
these defect energies with the results of lattice energy calculations to estimate the
relative stability of the Suzuki phase.   These latter calculations are useful in
that they elucidate those particular features which influence the formation of this
superstructure.

2. Methods and Potentials

In this study we used the two-body central-force interionic potentials derived by Catlow et al[24] for the host ion interactions: the dopant ion-host ion potential parameters were obtained using electron-gas techniques and have been quoted previously[25]. The defect energies were calculated using the HADES program discussed in Chapter (1) and, because of the size of the larger clusters, were based on an inner explicitly-relaxed region of approximately 170 sites. The lattice energies of the Suzuki phase were obtained using a version of the PLUTO program, modified by the inclusion of a Newton-Raphson minimization routine which enables the lowest energy configuration of the ions in the structure to be determined.

3. Defect Energies

The association energies of impurity-vacancy dipoles are given by

$$\Delta u_{ai} = u_{ai} - [u_+^\infty + u(M^{2+})] \tag{3}$$

where i=1,2, for nearest neighbour and next-nearest neighbour configurations respectively. The term u_+^∞ is the energy of formation of a cation vacancy and $u(M^{2+})$ is that for the substitution of a host cation by a divalent impurity ion. The defect energy of this substitutional ion with an associated vacancy in the appropriate position is given by u_{ai} and the resultant association energies are listed in Table 1.

Table 1

System	Calculated dipole association energies	
	$\Delta u_{a1}/eV$	$\Delta u_{a2}/eV$
NaCl:Mg^{2+}	-0.556	-0.563
NaCl:Ca^{2+}	-0.601	-0.521
NaCl:Sr^{2+}	-0.641	-0.483
NaCl:Ba^{2+}	-0.681	-0.461
KCl:Mg^{2+}	-0.597	-0.663
KCl:Ca^{2+}	-0.626	-0.636
KCl:Sr^{2+}	-0.656	-0.610
KCl:Ba^{2+}	-0.688	-0.600
KBr:Mg^{2+}	-0.565	-0.623
KBr:Ca^{2+}	-0.586	-0.598
KBr:Sr^{2+}	-0.608	-0.575
KBr:Ba^{2+}	-0.633	-0.566

An important conclusion from these data which is relevant to the aggregation processes is that the stabilities of these dipoles depend on the size of the impurity ion. The stability of the nn configuration increases while that of the next-nearest neighbour (nnn) configuration decreases with increasing size of the dopant ion. For larger substitutional ions the strain energy is substantially relieved by proximity to a nearest neighbour vacancy while the nnn complex is stabilized by relaxation of the intervening anion towards the dopant ion (25).

Table 2 contains the calculated defect energies of the aggregates of Figure 1. The energies of formation of these clusters from their isolated constituents are given in Table 3. They are expressed as the energy of formation per impurity ion, $\Delta u(x)/n$, where

$$\Delta u(x) = u(x) - n\{u_+^\infty + u(M'^{2+})\} \tag{4}$$

and n is the number of impurity ions in the cluster.

4. Lattice energies of Suzuki phases

The Susuki phase is a superlattice of impurity ions and vacancies in next-nearest positions with respect to each other in the host crystal. The unit cell which comprises 8 of the units j shown in Figure 1 contains 4 divalent impurity ions, 4 cation vacancies, 24 host cations and 32 host cations. Of these anions, 8 retain the six-fold coordination by host cations characteristic of the host lattice while two of the coordination sites of each of the remaining anions are replaced by a divalent impurity ion and a cation vacancy. The formation of this unit cell may be visualized as the combination of one impurity ion and one cation vacancy with six host cations and eight host anions on the sixteen lattice sites normally occupied by eight ion pairs. Thus, the stabilization energy of the Suzuki phase per divalent ion as compared to isolated dopant ions and vacancies is given by

$$\Delta u_{S1} = \{u_S - 8u_L\} - \{u_+^\infty + u(M'^{2+})\} \tag{5}$$

where u_L and u_S are respectively the lattice energies of the host crystals and Suzuki phases (per $6MX.M'X_2$ unit). The stabilization energy of the Suzuki phase relative to nnn dipoles, of which it is composed, is given by

$$\Delta u_{S2} = (u_S - 8u_L) - u_{a2} . \tag{6}$$

These values are listed in Table 4.

5. Discussion

5.1 Stability of aggregates

The most significant feature of the aggregation energies which we have calculated is that the trends evident in the relative stabilities of nn and nnn configurations become more pronounced with increasing aggregate size. This suggests that separate aggregation pathways exist for each kind of dipole. Nearest neighbour

Table 2

Calculated defect energies u(x) in eV for the aggregates a-j illustrated in Figure 1

System	$\Delta u(x)$									
	a	b	c	d^+	e	f	g	h	i	j
NaCl:Mg^{2+}	-17.562	-18.171	-17.717	-17.828	-27.025	-27.397	-26.158	-27.308	-25.940	-36.962
NaCl:Ca^{2+}	-13.782	-13.861	-13.721	-13.823	-21.082	-20.878	-20.616	-20.805	-19.879	-28.188
NaCl:Sr^{2+}	-11.202	-10.944	-11.060	-11.162	-17.128	-16.444	-16.812	-16.384	-15.806	-22.196
NaCl:Ba^{2+}	- 9.015	- 8.556	- 8.871	-8.973	-13.871	-12.800	-13.562	-12.761	*	-17.285
KCl:Mg^{2+}	-19.227	-20.463	-19.629	-19.788	-29.992	-30.842	-29.014	-30.778	-28.786	-41.741
KCl:Ca^{2+}	-15.840	-16.591	-16.062	-16.141	-24.569	-25.022	-23.534	-24.962	-23.390	-33.916
KCl:Sr^{2+}	-13.705	-14.056	-13.734	-13.809	-21.107	-21.194	-20.421	-21.141	-19.905	-28.757
KCl:Ba^{2+}	-11.960	-12.035	-11.867	-11.940	-18.328	-18.126	-17.891	-18.096	-17.120	-24.640
KBr:Mg^{2+}	-18.045	-19.161	-18.460	-18.551	-28.116	-28.889	-27.150	-28.823	-27.046	-39.088
KBr:Ca^{2+}	-15.208	-15.932	-15.429	-15.507	-23.584	-24.033	-22.601	-23.970	-22.508	-32.560
KBr:Sr^{2+}	-13.351	-13.741	-13.410	-13.484	-20.579	-20.725	-19.889	-20.667	-19.465	-28.101
KBr:Ba^{2+}	-11.982	-12.148	-11.938	-12.011	-18.391	-18.310	-17.909	-18.269	-17.267	-24.860

*invalid minimization due to excessive displacement.

$^+$u(d) - 123 sites

Table 3

Calculated association energies in eV for the aggregates shown in Figure 1. The energies are expressed per impurity-vacancy ion pair and are relative to the same number of isolated vacancies and impurities

System	$\Delta u(x)/n$									
	a	b	c	d+	e	f	g	h	i	j
NaCl:Mg^{2+}	-0.500	-0.805	-0.577	-0.632	-0.727	-0.851	-0.438	-0.821	-0.366	-0.960
NaCl:Ca^{2+}	-0.682	-0.722	-0.651	-0.702	-0.819	-0.751	-0.663	-0.726	-0.418	-0.838
NaCl:Sr^{2+}	-0.775	-0.646	-0.704	-0.757	-0.884	-0.656	-0.778	-0.636	-0.443	-0.723
NaCl:Ba^{2+}	-0.824	-0.594	-0.751	-0.808	-0.940	-0.583	-0.837	-0.570	*	-0.637
KCl:Mg^{2+}	-0.345	-0.963	-0.577	-0.624	-0.729	-1.012	-0.403	-0.991	-0.327	-1.167
KCl:Ca^{2+}	-0.536	-0.911	-0.647	-0.681	-0.805	-0.956	-0.460	-0.936	-0.412	-1.095
KCl:Sr^{2+}	-0.685	-0.860	-0.699	-0.731	-0.868	-0.897	-0.639	-0.879	-0.467	-1.021
KCl:Ba^{2+}	-0.794	-0.832	-0.748	-0.778	-0.924	-0.856	-0.778	-0.846	-0.521	-0.974
KBr:Mg^{2+}	-0.347	-0.905	-0.554	-0.598	-0.696	-0.954	-0.374	-0.932	-0.339	-1.096
KBr:Ca^{2+}	-0.495	-0.858	-0.606	-0.641	-0.753	-0.903	-0.425	-0.882	-0.394	-1.032
KBr:Sr^{2+}	-0.618	-0.813	-0.648	-0.680	-0.802	-0.851	-0.572	-0.832	-0.431	-0.968
KBr:Ba^{2+}	-0.706	-0.789	-0.684	-0.717	-0.845	-0.819	-0.685	-0.805	-0.471	-0.930

+u(d) - 123 sites.   *invalid minimization due to excessive displacement

Table 4

Calculated cohesive energies for Suzuki phases and host crystals and energy changes which occur when the Suzuki phase forms from isolated defects: the symbols are defined in the text

System	u_S/eV	u_L/eV	Δu_{S1}/eV	Δu_{S2}/eV
NaCl:Mg^{2+}	-73.693	-8.042	-1.073	-0.509
NaCl:Ca^{2+}	-71.290		-0.742	-0.221
NaCl:Sr^{2+}	-69.973		-0.808	-0.325
NaCl:Ba^{2+}	-68.614		-0.591	-0.130
KCl:Mg^{2+}	-69.835	-7.370	-1.604	-0.940
KCl:Ca^{2+}	-67.792		-1.445	-0.808
KCl:Sr^{2+}	-66.398		-1.267	-0.656
KCl:Ba^{2+}	-65.230		-1.081	-0.481
KBr:Mg^{2+}	-66.528	-7.048	-1.470	-0.847
KBr:Ca^{2+}	-64.816		-1.326	-0.728
KBr:Sr^{2+}	-63.609		-1.170	-0.595
KBr:Ba^{2+}	-62.692		-1.026	-0.460

dipoles can form dimers with structures a, c or d and of these d is found to be most stable. It is also the natural precursor of the trimer e which is oriented in the {111} plane. This trimer is consistently the most stable of the nn trimers for which we have calculated defect energies. Thus, for systems containing A_1 dipoles, the aggregation path which is favoured energetically is

$$A_1 \rightleftharpoons A_2(d) \rightarrow A_3(e) . \qquad (7)$$

In the case of large dopant ions the formation of the trimeric species g becomes more competitive.

The pronounced stability of the {111} trimer, as indicated by our calculations, is in agreement with experimental studies on a number of systems. The species was first proposed by Cook and Dryden[5] and its presence was later found to be consistent with optical (8,26) and epr studies (27). The results of a limited number of calculations for tetramers formed by adding a further A_1 dipole to defects e and g indicate that none of these is sufficiently stable for continuation of the aggregation process in this way. The orientation of this dipolar addition to the g trimer is especially sensitive to inter-dopant and inter-vacancy electrostatic interactions. Marginally favourable orientations are observed for some systems but the aggregation to the {111} trimer remains the clustering process of primary importance.

Next-nearest neighbour dipoles are predicted to form dimer b which is always stable with respect to two isolated B_1 dipoles. This stability decreases monotonically with increasing dopant ion size in all three host crystals with the trend being most obvious in NaCl. For systems with B_1 dipoles the energetically favoured aggregation pathway is

$$B_1 \rightarrow B_2(b) \rightarrow B_3(h) \rightarrow B_4(j) \quad . \tag{8}$$

Greater stabilization is evident for smaller impurity ions and may be expected to lead eventually to a Suzuki phase.

5.2 The Suzuki phase

The calculated energy changes given in Table 4 for the formation of Suzuki phase again emphasise that the $6MX.M'X_2$ structure is more effectively stabilized with smaller divalent ions. This is in agreement with available experimental evidence (28,29). The values for Δu_{S1} (Table 4) may be compared directly with those of $\Delta u(j)/4$ (Table 3) and indicate that, in the absence of any other effect, the formation of the Suzuki phase would result in a further lowering of the energy over that observed in the sequence represented by equation (8). An important structural feature of this phase, which influences its stability, is the position of the host anion between the impurity ion and cation vacancy. Displacement of these anions towards the dopant ion, shown in Table 5, stabilizes the structure and are largest for the smallest dopant ions. In these systems the elastic forces counteracting the Coulombic interactions are relatively unimportant.

Table 5

Calculated anion displacements, d, towards adjacent impurity ions. a_0 is the relative anion-cation separation and positive d values indicate relaxation towards the impurity.

System	d/a_0
$NaCl:Mg^{2+}$	0.095
$NaCl:Ca^{2+}$	0.051
$NaCl:Sr^{2+}$	0.008
$NaCl:Ba^{2+}$	-0.022
$KCl:Mg^{2+}$	0.184
$KCl:Ca^{2+}$	0.140
$KCl:Sr^{2+}$	0.107
$KCl:Ba^{2+}$	0.080
$KBr:Mg^{2+}$	0.169
$KBr:Ca^{2+}$	0.128
$KBr:Sr^{2+}$	0.097
$KBr:Ba^{2+}$	0.073

6. Conclusions

We have combined defect and lattice energy calculations in a study of the aggregation of impurity-vacancy dipoles and of Suzuki phase precipitation in alkali-halide crystals. The relative stabilities of nearest neighbour compared to next-nearest neighbour configurations of isolated dipoles is accentuated as the cluster size increases. Thus, aggregation is sensitive to the type of dipole which predominates. Nearest neighbour dipoles will complex to the {111} trimer while next-nearest neighbour species, which predominate in systems containing smaller dopant ions, will tend to form a Suzuki phase. Our conclusions corroborate the kinetic data of dipole decay and are also consistent with structural evidence for the presence of the Suzuki super-lattice in the systems studied.

Acknowledgements

We are grateful to Amdahl Ireland Ltd. and to the Computer Centre of University College Dublin for computational facilities.

References

1. Perlman, M.M. and Unger, S., J. Electrostatics $\underline{1}$, 231 (1975).

2. Hor, Ah Mee, Jacobs, P.W.M. and Moodie, K.S., Phys. Stat. Sol.(a), $\underline{38}$, 293 (1976).

3. Kirk, D.L. and Innes, R.M., J. Phys. C. $\underline{11}$, 1105 (1978).

4. Cook, J.S. and Dryden, J.S., Australian J. Phys. $\underline{13}$, 260 (1960).

5. Cook, J.S. and Dryden, J.S., Proc. Phys. Soc. $\underline{80}$, 479 (1962).

6. Chiba, Y., Ueki, K. and Sakamoto, M., J. Phys. Soc. Japan $\underline{18}$ 1092 (1963).

7. Dryden, J.S., J. Phys. Soc. Japan (Suppl. III), $\underline{18}$, 129 (1963).

8. Dryden, J.S., and Harvey, G.G., J. Phys. C. $\underline{2}$, 603 (1969).

9. Brown, N. and Jacobs, P.W.M., J. de Physique $\underline{34}$ C9-437 (1973).

10. Chapman, J.A. and Lilley, E., J. de Physique $\underline{37}$, C9-455 (1973).

11. Suzuki, K., J. Phys. Soc. Japan $\underline{16}$, 67 (1961).

12. Dryden, J.S., J. Phys. D $\underline{3}$, L51 (1970).

13. Dryden, J.S., Morimoto, S. and Cook, J.S., Phil. Mag. $\underline{12}$, 379 (1965).

14. Capelletti, R. and Benedetti, de E., Phys. Rev. $\underline{165}$, 981 (1968).

15. Crawford, Jnr. J.H., J. Phys. Chem. Solids $\underline{31}$, 399 (1970).

16. Kessler, A., J. de Physique $\underline{37}$, C7-291 (1976).

17. Dienes, G.J., Semiconductors and Insulators $\underline{4}$, 159 (1978).

18. Lilley, E., J. de Physique $\underline{41}$, C6-429 (1980).

19. Berg, G., Frohlich, F. and Siebenhuner, S., Kristall and Technik <u>10</u>, 1091 (1975).

20. Berg, G., Frohlich, F. and Siebenhuner, S., Phys. Stat. Sol.(a) 31, 385 (1975).

21. Berg, G., Frohlich, F. and Schneider, S., Phys. Stat. Sol.(a), <u>42</u>, 73 (1977).

22. Strutt, J.E. and Lilley, E., Phys. Stat. Sol.(a), <u>33</u>, 229 (1976).

23. Ramdas, S., and Rao, C.N.R., Crystal Lattice Defects <u>6</u>, 199 (1976).

24. Catlow, C.R.A., Diller, K.M. and Norgett, M.J., J. Phys. C. <u>10</u>, 1395 (1977).

25. Catlow, C.R.A., Corish, J., Quigley, J.M. and Jacobs, P.W.M., J. Phys. Chem. Solids <u>41</u>, 231 (1980).

26. Collins, W.C., and Crawford, Jnr. J.H., Solid State Commun. <u>9</u>, 853 (1971).

27. Ikeya, M. and Crawford, Jnr. J.H., Phys. Stat. Sol.(b), <u>58</u>, 643 (1973).

28. Boswarva, I.M., Phys. Stat. Sol.(a), <u>37</u>, 65 (1976).

29. Sors, A.I. and Lilley, E., Phys. Stat. Sol.(a), <u>32</u>, 533 (1975).

CHAPTER 18   COMPUTER SIMULATION OF FAST ION CONDUCTORS

by

M. Dixon* and M.J. Gillan
Theoretical Physics Division, A.E.R.E. Harwell, Oxon OX11 ORA

1. Introduction

Fast ion conductors are ionic crystals of a rather unusual kind, in which the ions of one type are in a disordered and highly mobile state[1,2]. The diffusion constant of the mobile ions in such materials is typically 10^{-5} cm^2 sec^{-1}, which is comparable with the diffusion constant of ions in molten salts. There has been vigorous activity in the field of fast ion conduction over the past ten years, both because of their technological usefulness and because of their intrinsic scientific interest[3,4]. From the theoretical point of view, these systems are very difficult to treat by simple analytical methods. Clearly, the usual harmonic theory of solids has little to contribute to an understanding of diffusion in a disordered state. The lattice statics methods[5] discussed in this volume by Catlow and Mackrodt are also of limited value, as they take no account of thermal motions of the lattice ions - clearly a significant omission for solids with high ionic mobilities. Almost the only theoretical technique which has the power to yield detailed results for particular fast ion conductors is that of molecular dynamics simulation[6]. In this method, one constructs, on the computer, a dynamical atomistic model, which is designed to reproduce, as nearly as possible, the real system of interest. The practicability of simulating fast ion conductors by the molecular dynamics method was first demonstrated by Rahman[7]. The purpose of the present chapter is to describe how molecular dynamics has been used to study one particular class of fast ion conductors, i.e, those having the fluorite structure.

Fluorite materials are well suited for simulation, firstly because they have a simple lattice structure, and secondly because the interactions between the ions are well understood. We shall try to show that molecular dynamics is capable of reproducing the fast ion behaviour and of giving valuable assistance in the interpretation of experimental results for these systems. In order to provide some context for the later discussion, we shall begin by summarising some important basic facts about fluorites. We shall then describe briefly how the molecular dynamics calculations are performed and how the required interionic potentials are obtained. Since the diffusion constant of the mobile ions is of central importance, illustrative results for this quantity will next be presented, and we go on to show how the simulations have shed light on the nature of the diffusion process and the spatial distribution of the disordered ions. We conclude by assessing the role of molecular dynamics simulation in the field of fast ion conduction - a topic which is also discussed in Chapter (5).

*Department of Theoretical Physics, University of Oxford.

2. Properties of fluorites

The fluorite crystal structure is shown in Figure 1. Many compounds of the type MX_2 crystallize in this structure[8], but for present purposes we shall be interested mainly in halides such as CaF_2, $SrCl_2$ and PbF_2; the fluorite structured oxides were, we note, discussed in Chapter ( ). The structure is best described in terms of the anions forming a simple cubic lattice and the cations occupying the centres of alternate cubes; the remaining cube centres are vacant. Most materials having this structure show a pronounced peak in the specific heat at a characteristic temperature T_c, which lies several hundred degrees below the melting point[9,10]; at the same temperature, the electrical conductivity rises continuously to the high values (~ 1 $ohm^{-1}cm^{-1}$) characteristic of molten salts[11]. Experimental results for $SrCl_2$, whose behaviour is typical, are given in Figure 2. The conductivity is due to the diffusion of anions[13], and it is the partial disordering of the anion sublattice which is responsible for the anomaly in the specific heat.

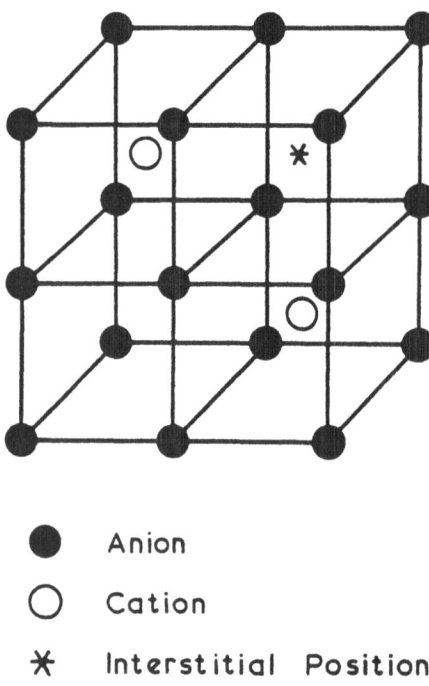

● Anion
○ Cation
✱ Interstitial Position

Figure 1   Fluorite lattice structure

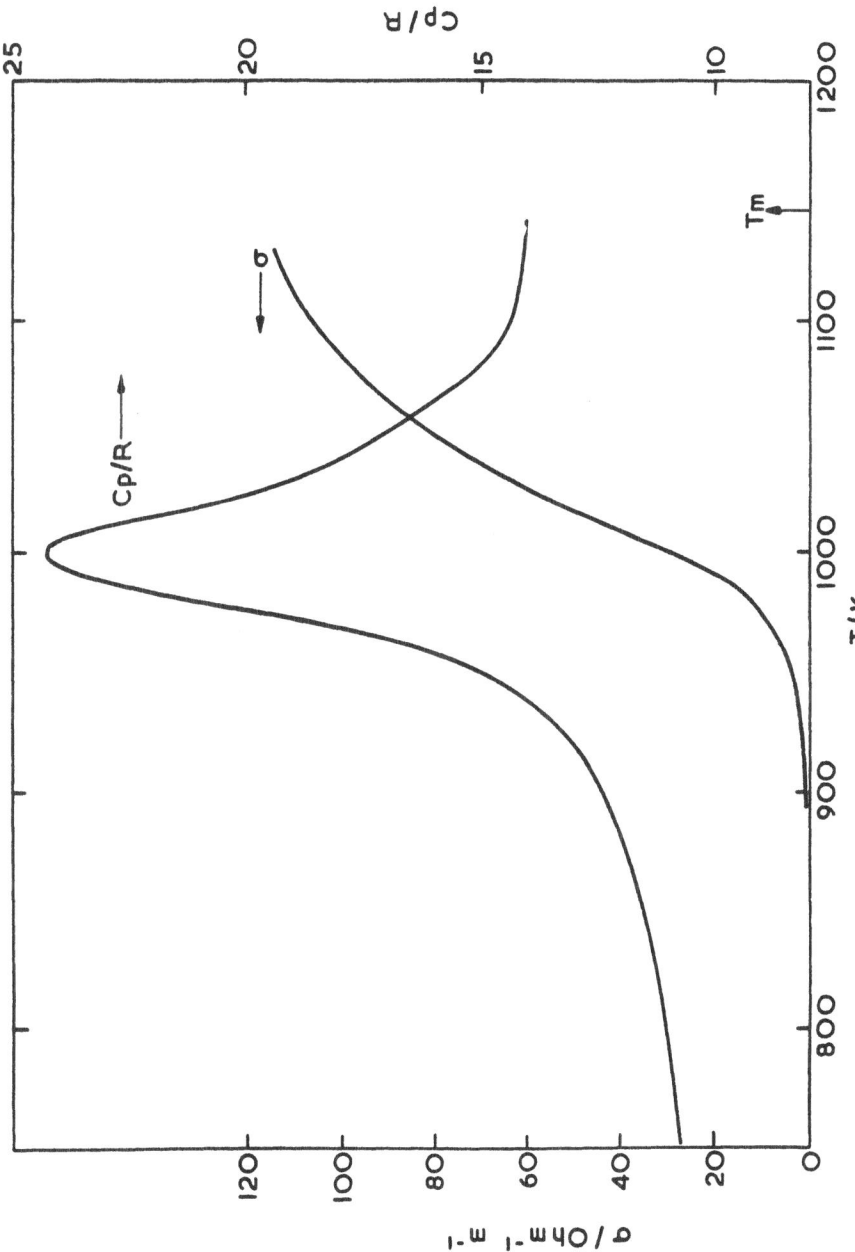

<u>Figure 2</u> The electrical conductivity and the molar specific heat C_p divided by the gas constant R for $SrCl_2$ as a function of temperature. The conductivity data are from Carr et al[12], and the specific heat data from Schroter and Nolting[10].

At temperatures well below T_c, the conductivity is considerably smaller than the values mentioned above, and the detailed mechanism is well understood[13]. Here, the conductivity is due to thermally generated anion Frenkel defects, i.e. to isolated anion vacancies and interstitials. It is well established that these interstitials reside at the cube centres not occupied by cations[14], as shown in Figure 1. As the temperature is raised towards T_c the concentration of defects approaches values of the order of one percent. One of the interesting questions about fast ion conduction in fluorites is whether the anion diffusion mechanism remains essentially the same as at low temperatures. This is clearly related to the nature of the spatial distribution of the anions, and in particular to the question whether, above T_c, one finds a substantial concentration of anions occupying the cube centre positions. As we shall see, the molecular dynamics calculations give some insight into this question, which has also been investigated with the aid of the static calculations[15,16].

3. Molecular dynamics

The aim of molecular dynamics simulation is to reproduce the detailed dynamical behaviour of a real system on an atomic scale. In the present short article, we can do no more than summarise the principles of the method. A full description of the modelling of ionic materials by molecular dynamics may be found in the review of Sangster and Dixon[6], and it is further discussed in Chapter (5).

The system of interest is represented by a collection of ions contained in a notional cube, which is periodically repeated in the three spatial directions, in order to eliminate the boundary effects. Initially, the ions are placed on their regular lattice positions and given random velocities. The system is then made to evolve in time, the motion of each ion being governed by Newton's equation of motion. The force on each ion is obtained by summing the forces exerted on it by all the ions in the system, these forces being derived from specified interionic potentials, about which more will be said shortly. In calculating the force on a given ion, the short range contributions need be summed only over nearby ions, but the Coulombic forces must be summed over all other ions and all their periodic images; this latter summation is performed by the Ewald technique[17]. The time evolution is produced by integration of the equation of motion using a numerical quadrature with a fixed value of the time step Δt.

In the first period of a simulation run, the positions and velocities of the ions become randomized, and the system attains a state of thermal equilibrium. During this period, kinetic energy is added or subtracted so as to bring the system to the required temperature. After this, the system is allowed to evolve undisturbed for a number of steps corresponding to a real time of a few tens of picoseconds. The first part of the run is discarded for the purpose of analysis. The second part, which is taken to represent the real system in thermal equilibrium, is analyzed to obtain the thermal average values of the quantities of interest.

In all our simulations on fluorites[18-21], we work with systems containing either 96 ions (= 32 cations + 64 anions) or 324 (= 108 + 216) ions. Although these are relatively small numbers, a comparison of the results obtained with the two sizes of system shows that even the smaller system is adequate for most purposes[21]. The step length Δt is typically 10^{-14} sec. so that the runs need to have a length of several thousand steps.

4. Interionic potentials

The materials in which we are interested are generally treated as fully ionic compounds. Accordingly, our calculations have been based on the Born-Mayer form of interionic potential, which is commonly used for such compounds[6,22], and which is discussed in detail in Chapter (10). The interaction energy of two ions of types α and β, is, we recall, given by

$$V_{\alpha\beta}(r) = z_\alpha z_\beta e^2/r + A_{\alpha\beta} \exp(-r/\rho_{\alpha\beta}) - C_{\alpha\beta}/r^6 \qquad (1)$$

where z_α is the charge on each ion of type α, in units of the electronic charge e.

The values of the parameters $A_{\alpha\beta}$, $\rho_{\alpha\beta}$ and $C_{\alpha\beta}$ of the short-range potentials have been determined using both the empirical and non-empirical techniques described in Chapter (10); details are presented in references (19-21). Molecular dynamics simulations almost always treat the ions as rigid; inclusion of polarisability using the shell model[6] described in Chapter (10) results in a great increase in computing time. In the case of CaF_2, where we have been able to compare the rigid-ion and shell-model simulations[21], we have found that the explicit inclusion of polarizability has very little effect on the quantities of interest here, so long as the rigid-ion potential is suitably constructed. In particular we ensure that the potential correctly reproduces the static dielectric constant which, as argued in Chapter (10) is essential for adequate representation of defect dependent properties.

5. Diffusion constant

Can simulation calculations of the kind we have described correctly reproduce the fast ion conduction effect in fluorite materials? The most straightforward way of answering this question is by examining the cation and anion self-diffusion constants, D_+ and D_-, as was done Rahman[7]. A simple way of determining a diffusion constant in a molecular dynamics simulation is to calculate the mean square distance travelled by a particle of type α in a time t[6]; we call this quantity $<r_\alpha(t)^2>$. Its asymptotic form for large $|t|$ is[6]

$$\langle r_\alpha(t)^2 \rangle \to B_\alpha + 6D_\alpha |t| \qquad (2)$$

where D_α is the self-diffusion coefficient.

In order to illustrate this, we show in Figure 3 our results for $\langle r_+(t)^2 \rangle$ $\langle r_-(t)^2 \rangle$ from a simulation[21] of CaF_2 at a temperature of 1585K, which is well above the transition temperature T_c. It is clear from this that the asymptotic form is rapidly attained (the fluctuations in the curves at longer times are the result of statistical noise). It is also clear that the anions are diffusing, while the cations are not, as required by experiment.

In Figure 4 we present results[21] for the anion diffusion constant as a function of temperature from a series of simulations on CaF_2 compared with values obtained, via the Nernst-Einstein relation[23], from the experimental electrical conductivity[11]. The comparison demonstrates that the molecular dynamics simulation does indeed reproduce fast ion conduction in this material and that the simulated diffusion constant is in at least qualitative agreement with experiment. Similar results have been obtained from simulations on $SrCl_2$[19] and PbF_2[24], though in the case of $SrCl_2$ the transition temperature comes out some 20% too high.

The specific heat peak mentioned in section 2 also has its counterpart in the simulations. What we actually calculate is the enthalpy as a function of temperature, which exhibits a diffuse step of about the correct magnitude at the temperature at which the diffusion constant rises to liquid-like values[20].

There can be little doubt, then, that the mechanism responsible for fast ion conduction in the real materials also operates in the simulated systems. It is therefore reasonable to use the simulation results to study these mechanisms.

6. Diffusion mechanism

We are interested in finding out how the mobile anions diffuse through the lattice. In order to provide some framework for the analysis, we have adopted the following procedure[18,19]. Around each regular anion site, we draw a sphere whose radius is arbitrarily taken to be one third of the anion-anion distance. We then construct, for a given simulation run, a catalogue of all events in which any anion leaves one of the spheres and passes to one of the others. In this way, the diffusive motion is broken down into a sequence of intersite hopping events. A different but related method has been employed by Jacucci and Rahman[25]. The analysis shows immediately that diffusion occurs almost entirely as a result of direct hops of anions between nearest-neighbour regular lattice positions[18,19]. The hopping event is rapid, and is achieved in a time of about 0.7 psec[19], i.e. about three or four vibrational periods. This is an order of magnitude smaller than the mean time for which an anion resides on a given lattice site[19].

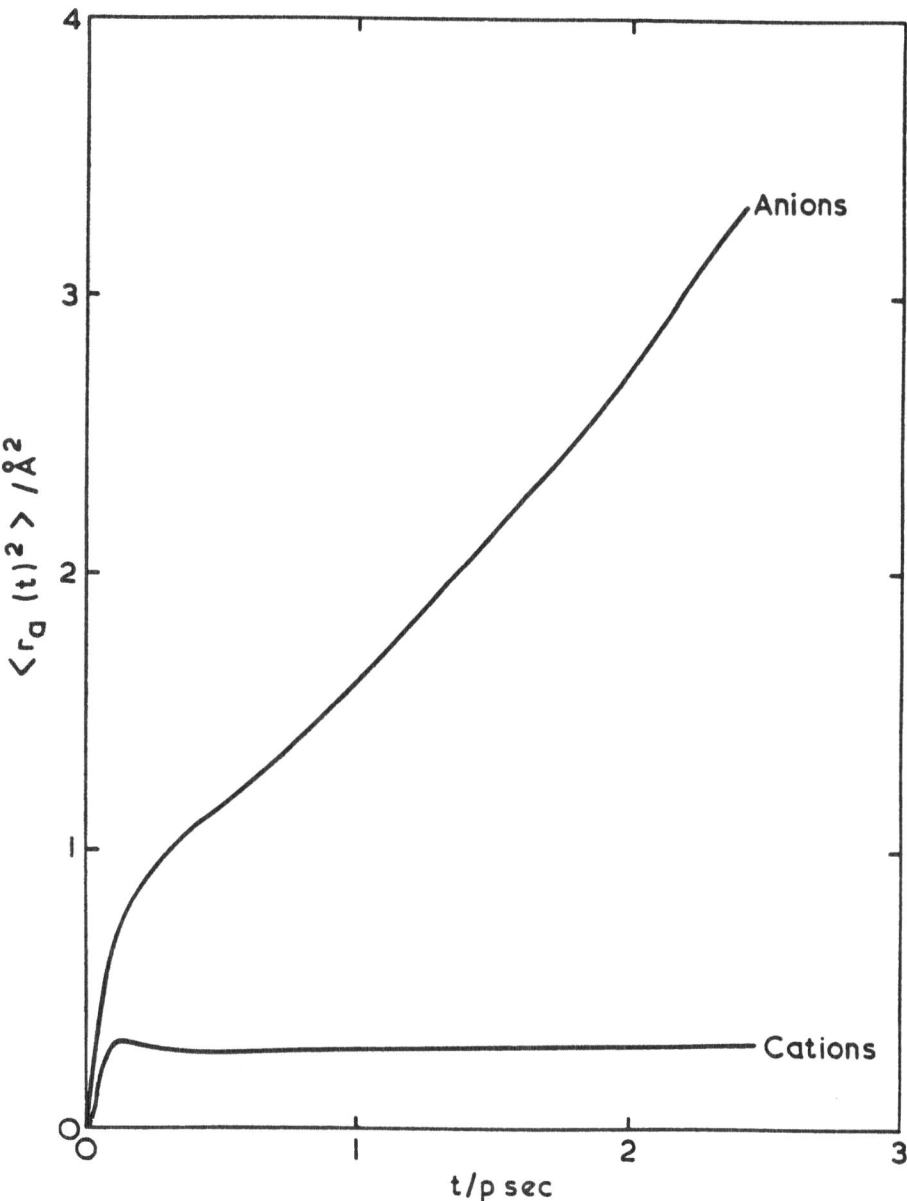

Figure 3   Cation and anion mean square displacements $\langle r_\alpha(t)^2 \rangle$ from molecular dynamics simulation of CaF_2 at T = 1585K using 324 ions.

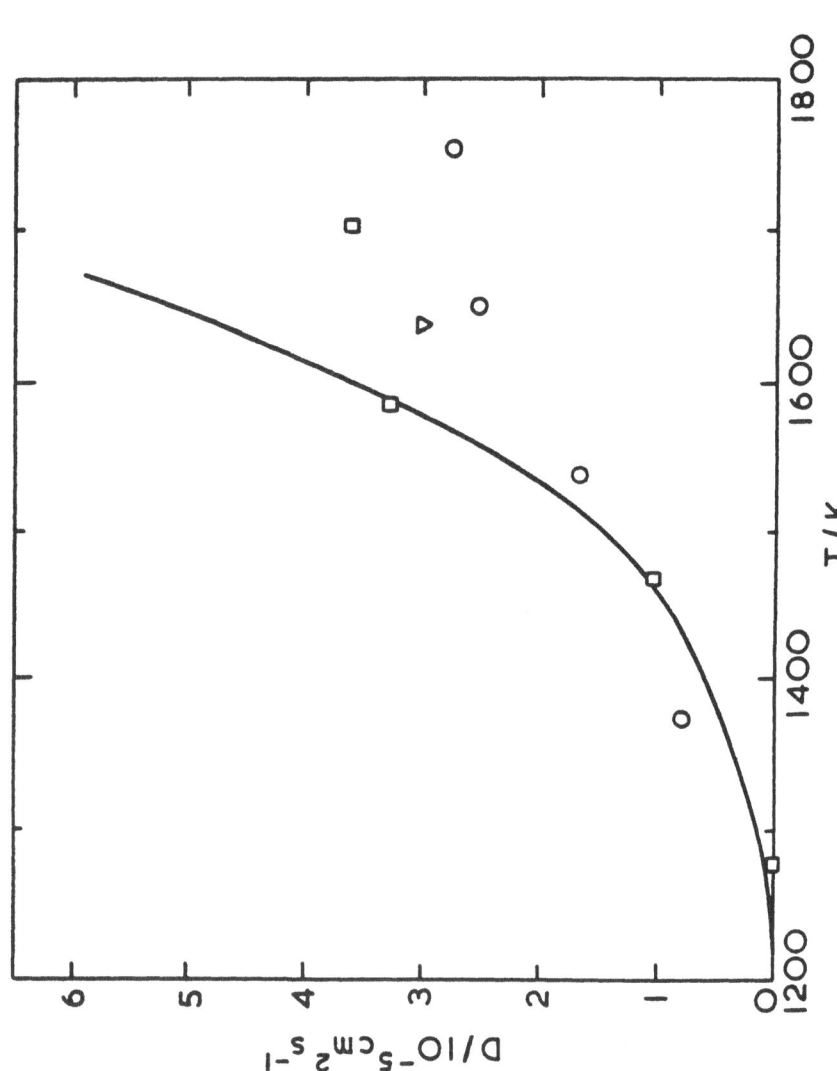

Figure 4 Simulation results for anion diffusion constant of CaF_2 (O and □ : 96- and 324- particle rigid-ion, ▽ : 96-particle shell-model) compared with values derived from conductivity measurements of Derrington et al[11].

The hopping analysis also enables us to determine whether any of the anions pass from their initial regular site positions to other positions of a different kind, where they remain for an appreciable length of time. The simulations show that this does not happen to any appreciable extent[19]. This indicates that the cube centre interstitial sites described in section 2 are not significantly occupied in the fast ion state. Thus, the picture which emerges from this analysis is that the anions spend most of their time vibrating about the regular lattice sites and only a small fraction (say 1/10) of their time passing from one site to another[19].

An important aspect of the analysis is that it yields results which are difficult to obtain by experiments on the real materials. However, one feature, namely the dominance of nearest-neighbour hops, does appear to be confirmed by preliminary results of incoherent neutron scattering measurements on $SrCl_2$[26].

7. Spatial distribution of ions

The picture outlined above is supported by an examination of the spatial distribution of the anions in the fast ion state. The quantity of interest here is the mean density $\rho_\alpha(\underline{r})$ of ions of type α, defined by

$$\rho_\alpha(\underline{r}) = A < \sum_{i=1}^{N_\alpha} \delta(\underline{r}-\underline{r}_{i\alpha})> \quad , \tag{3}$$

where $\underline{r}_{i\alpha}$ is the position of the ith ion of type α, the sum going over all such ions in the system; the angular brackets indicate the thermal equilibrium average and A is a normalization constant. The probability of finding an ion of type α at the position $\underline{r}$ in the unit cell is proportional to the quantity $\rho_\alpha(\underline{r})$.

In Figure 5 we show the mean ion densities for simulated CaF_2 in the fast ion state at a temperature T = 1651 K. As would be expected from the hopping analysis, the anion density is strongly concentrated about the regular anion lattice site, which lies in the centre of the plot. The large distortion of the distribution from spherical symmetry is mainly due to the strong anharmonicity of the anion vibrations, an effect which is present even below the transition temperature[27,28]. There is no evidence for an appreciable occupation by anions of the cube centre position, which is at the top corners of the plot. In fact, a more detailed examination suggests that the anion density has a local minimum at these positions.

In principle, it should be unnecessary to appeal to molecular dynamics results to find out about the spatial distribution of ions. Such information should be obtainable from the results of diffraction experiments, i.e. from the measurement of Bragg peak intensities. However, there has been a surprising degree of discord in the interpretation of diffraction experiments on fluorite materials. Some of

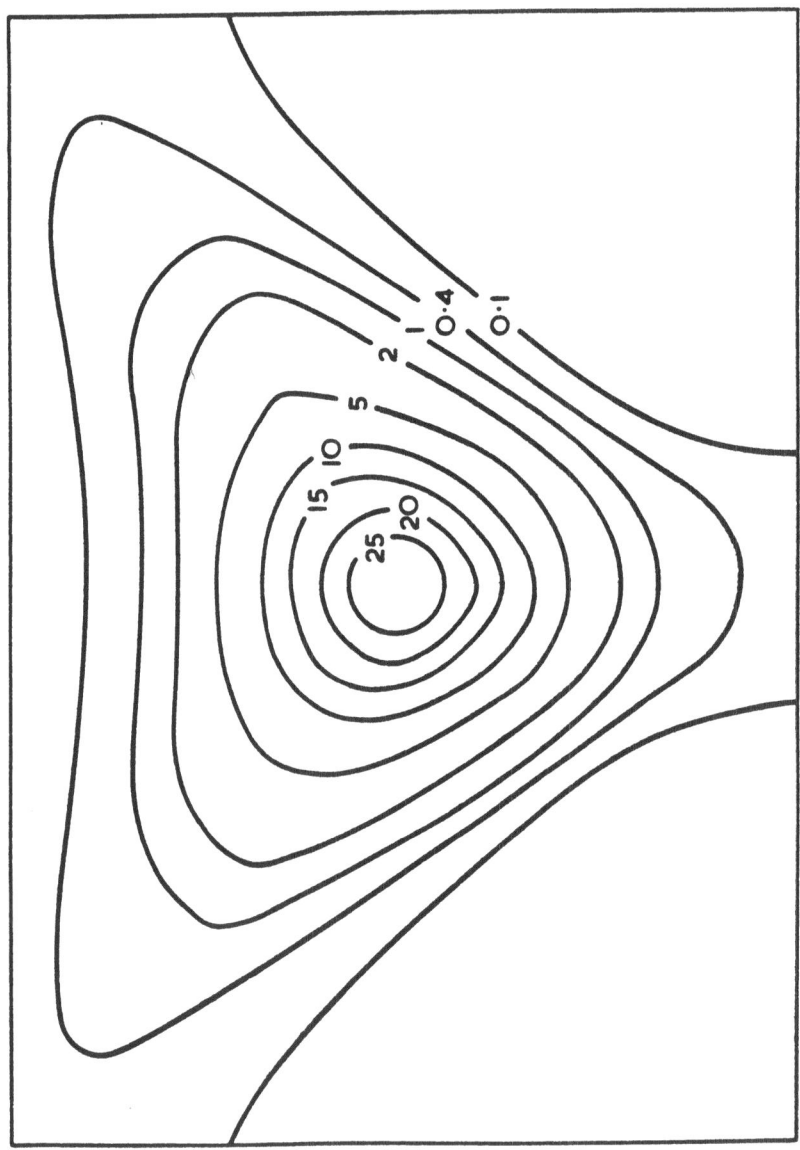

<u>Figure 5</u> Mean anion density on (110) plane from simulation of CaF$_2$ at T = 1651 K using 96 ions. Centre of plot is regular anion site, bottom corners are regular cation sites and top corners are cube-centre interstitial sites.

the first neutron diffraction measurements on BaF_2 and PbF_2 (Axe, Shapiro and Wakabayashi unpublished, cited by Shapiro[29]) appeared to give a large concentration of interstitials on the cube centre site. The more recent work has shown that this result was almost certainly erroneous. This later work is consistent with our finding that interstitials do not reside at the cube centre site in the fast ion state. Some of the diffraction results have been analyzed in terms of defect models[28,30,31]. Our recent simulation work on PbF_2[24] has suggested that this may be an unhelpful way of proceeding. Certainly our simulation results give no support to the existence of stable interstitial sites in the fast ion state in fluorites. It has been shown that there is no conflict between these results and the well-established stability of the cube-centre interstitial site at lower temperatures[32].

8. Conclusions

We wish to draw two main conclusions from the work we have described: first that molecular dynamics simulation based on simple, but realistic interionic potentials is capable of reproducing the main structural and dynamical features of superionic fluorites; second that the simulation results allow one to build up a detailed and coherent picture of the atomistic processes underlying the phenomenon. Our belief that the simulations are producing a faithful model of the real materials is based on the facts that the characteristic temperature dependence of the diffusion constant is correctly given and that we find an anomaly in the enthalpy content which also agrees qualitatively with experiment. These facts give one some confidence in the atomistic picture provided by simulation. We do not, of course, suggest that the molecular dynamics results should be accepted uncritically: it is particularly important to seek further ways of testing the results against experiment. However, the simulations do not at present appear to conflict with experiment in any important respect.

We have confined our discussion to the simulation study of fluorites since these are the systems on which we have worked ourselves. Similar dynamical studies on silver and copper ion conductors, such as AgI and CuI, have been reported by Rahman and coworkers[33,34]. In addition, molecular dynamics calculations have been carried out on a simple model of β-alumina[35]. Thus in conclusion, we believe that molecular dynamics simulation provides a practical and valuable aid in elucidating the behaviour of fast ion conductors.

References

1. Hayes, W., Contemp. Phys. 19, 469 (1978).

2. Salamon, M.B. (Ed.). Physics of Superionic Conductors. Springer, Berlin (1979).

3. Mahan, G.D. and Roth, W.L. (Eds.). Proc. Int. Conf. on Fast Ion Conductors. Schenectady. Plenum. New York (1976).

4. Vashishta, P., Mundy, J.N. and Shenoy, G.K. (Eds.). Fast Ion Transport in Solids. North-Holland. New York (1979).

5. Catlow, C.R.A. and Mackrodt, W.C. This volume (1981).

6. Sangster, M.J. and Dixon, M., Adv. Phys. 25, 247 (1976).

7. Rahman, A., J. Chem. Phys. 65, 4845 (1976).

8. Pauling, L. Nature of the Chemical Bond. 3rd edition. Cornell University Press (1960).

9. Dworkin, A.S. and Bredig, M.A. J. Phys. Chem. 72, 1277 (1968).

10. Schröter, W. and Nölting J. Proc. 3rd Europhys. Topical Conference on Lattice Defects in Ionic Crystals. J. de Physique (Paris), 41, Coll. C-6, 20 (1980).

11. Derrington, C.E., Lindner, A. and O'Keeffe, M., J. Solid St. Chem. 15, 171 (1975).

12. Carr, V.M., Chadwick, A.V. and Saghafian, R., J. Phys. C. 11, L637 (1978).

13. Lidiard, A.B., in Crystals with the Fluorite Structure. Ed. W. Hayes, pp.101-84. Clarendon, Oxford (1974).

14. Catlow, C.R.A. and Norgett, M.J., J. Phys. C. 6, 1325 (1973).

15. Catlow, C.R.A., Comins, J.D., Germano, F.A., Harley, R.T. and Hayes, W., J. Phys. C., 11, 3197 (1978).

16. Catlow, C.R.A. Comments Solid State Phys., 9, 157 (1980).

17. Woodcock, L.V. and Singer, K., Trans. Faraday Soc. 67, 12 (1971).

18. Dixon, M. and Gillan, M.J., J. Phys. C. 11, L165 (1978).

19. Gillan, M.J. and Dixon, M., J. Phys. C. 13, 1901 (1980).

20. Dixon, M. and Gillan, M.J., J. Phys. C. 13, 1919 (1980).

21. Dixon, M. and Gillan, M.J., Proc. 3rd Europhys. Topical Conf. on Lattice Defects in Ionic Crystals. J. de Physique (Paris), 41, Coll. C-6, 24 (1980).

22. Fumi, F.G. and Tosi, M.P., J. Phys. Chem. Solids, 25, 31 (1964).

23. Corish, J. and Jacobs, P.W.M. in Surface and Defect Properties of Solids, Vol. 2, Ch. 7, Chemical Society, London (1973).

24. Walker, A.B., Dixon, M. and Gillan, M.J., Solid State Ionics, 5, 601 (1981).

25. Jacucci, G. and Rahman, A., J. Chem. Phys. $\underline{69}$, 4117 (1978).

26. Hutchings, M.T. Private communication.

27. Dawson, B., Hurley, A.C. and Maslen, V.W., Proc. Roy. Soc. A$\underline{298}$, 289 (1967).

28. Dickens, M.H., Hayes, W., Hutchings, M.T. and Smith, C., J. Phys. C. $\underline{12}$, L97 (1979).

29. Shapiro, S.M., in Proc. Int. Conf. on Fast Ion Conductors. Schenectady. Ed. G.D. Mahan and W.L. Roth, p.261, Plenum, New York (1976).

30. Shapiro, S.M. and Reidinger, F. In Physics of Superionic Conductors. Ed. M.B. Salamon. Springer, Berlin (1979).

31. Dickens, M.H., Hayes, W., Hutchings, M.T. and Smith, C. Harwell Report MPD/NBS/153 (1980).

32. Gillan, M.J. and Richardson, D.D., J. Phys. C. $\underline{12}$, L61 (1979).

33. Vashishta, P. and Rahman, A., Phys. Rev. Lett. $\underline{40}$, 1337 (1978).

34. Vashishta, P. and Rahman, A., in Fast Ion Transport in Solids, p.527. North-Holland, New York (1979).

35. De Leeuw, S. and Perram, J., in Fast Ion Transport in Solids, p.345. North-Holland, New York (1979).

CHAPTER 19   COMPUTER SIMULATION OF IONIC CRYSTAL SURFACES

by

P.W. Tasker
Theoretical Physics Division, A.E.R.E. Harwell, Oxon OX11 ORA

1.   Introduction

The importance of crystal surfaces and interfaces in many poorly understood phenomena such as catalysis, corrosion and electrode processes has stimulated both experimental and theoretical studies. Computer simulation techniques analogous to those used in studying the bulk and defect properties of solids can be applied to the calculation of the properties of both perfect and defect surfaces. Much of the early work in this area was carried out by Benson and co-workers[1] who calculated the structure and thermodynamic properties of surfaces of crystals with the rock salt and fluorite structures. More recently, modern experimental techniques (e.g. LEED, ion scattering etc.) have renewed interest in the structure of "perfect" surfaces. Theoretical simulations can play a valuable role here in helping to interpret the experimental data. In general, the calculations can give information on the stable ionic configuration at the surface and calculate the basic thermodynamic properties, such as surface energy and stress, that are often difficult to determine reliably by experiment. Calculations on the perfect surface are necessary precursors to investigations into the defect structure including surface irregularities such as steps, point defect properties, adsorption and surface segregation of bulk impurities. In general terms, the techniques used in surface simulation are similar to these used in point defect calculations, discussed in Chapter (1). More details of the methods are given in the next section.

2.   Calculation

Since the work of Benson two theoretical developments have led to the increased reliability of surface calculations. First, interionic potentials have been derived for many ionic and semi-ionic materials. These potentials may have been derived ab initio[2] or empirically by fitting to bulk elastic and dielectric properties[3,4].   In both cases they give realistic estimates of the properties of the bulk and defect crystal and are suitable for the application to surface simulations. A compilation for many materials has recently been made[5]. Secondly, general methods for summing the long-range Coulomb interactions over two-dimensionally periodic structures have been derived that are both computationally efficient and completely general[6]. This means that codes can be developed that require only a crystal structure (i.e. lattice vectors and basis ions) and a specified surface in order to calculate the surface properties. There is no need to restrict calculations to simple crystal structures or low index faces.

These methods have been incorporated in the MIDAS (Minimisation for

Interfacial defects and Surfaces) code, which performs static lattice calculations
on surface and planar defects. The block structure for a surface calculation is
shown in Figure 1. The crystal is considered as a stack of planes parallel to the
surface and periodic in two dimensions. As is usual in lattice simulations (see
Chapter 1), the crystal is divided into two regions. Region I includes the defect,
in this case the free surface, and the ions are relaxed explicitly to equilibrium.
In region II more approximate methods are used, and in this case it is generally
sufficient for the lattice in this region to be rigid and unpolarisable. Most
calculations of this type[7-9] have employed the shell model treatment of polarisation,
discussed in Chapter (10). Dipole moments are created by the different
displacements of core and shell and the short-range interactions couple the intra-
atomic polarization with the inter-atomic forces. Computationally, the model
effectively doubles the number of ions in the calculations as cores and shells are
treated separately.

Static lattice calculations of this type can be related to the thermodynamic
properties and structure of the crystal at absolute zero temperature if we neglect
the zero point vibration. Calculations of the properties at higher temperature
strictly require inclusion of the dynamical effects[10-12].

3. Surface structure

Most low energy surfaces of ionic crystals have a structure close to a simple
termination of the bulk. Such systems are best discussed in terms of surface
relaxations; these may be distinguished from surface reconstructions where the
surface differs substantially from the bulk. Surface reconstruction is common in
semiconductors[13] where re-bonding is important at the surface; it is therefore
often associated with changes in the force law for the atomic interactions.

Most calculations on ionic crystals with both positive and negative ions in
the surface plane predict a rumpled structure. This is illustrated in Figure 2
for a rocksalt (100) surface. The cations relax inwards while the anions relax
outwards. Although any potential model that extends beyond first neighbours will
produce a difference in relaxation between anions and cations, the calculated
effect with outward relaxation of anions is mainly attributable to ionic polarization.
The anions and cations are polarized at the surface in opposite directions, but to
avoid increasing overlap with the ions on the plane below, the anions move outwards
while the cations are able to move inwards. Within the shell model, the shells do
not relax much due to the balance of the short-range forces while the cores relax
inwards or outwards to produce the appropriate dipole moment. Since the direction
of rumple is dependent on the sign of the shell charge, considerable care must be
taken with the choice of these parameters. It is usually assumed that potential
parameters derived for the bulk can be applied at the surface of ionic crystals. The
free ion polarizabilities are therefore unchanged while there is some difference in

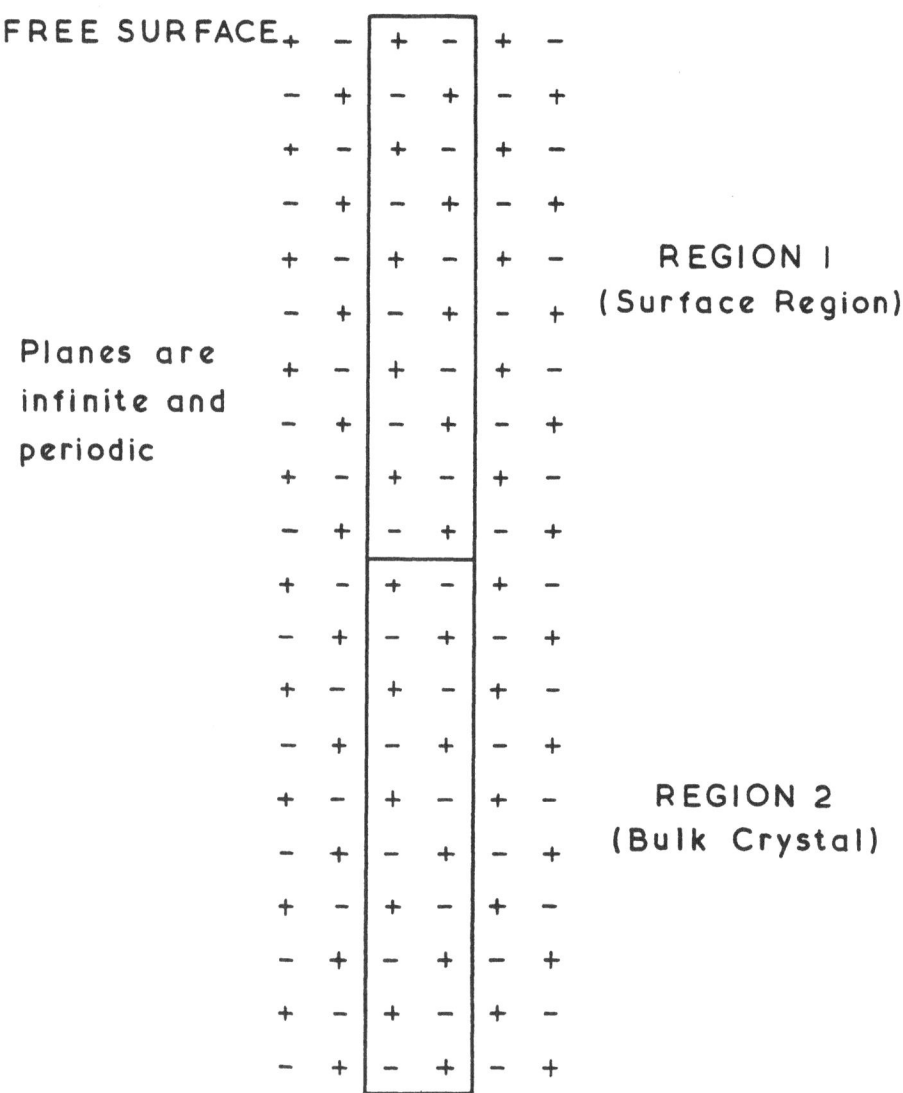

Figure 1   Schematic representation of the crystal regions for a surface calculation

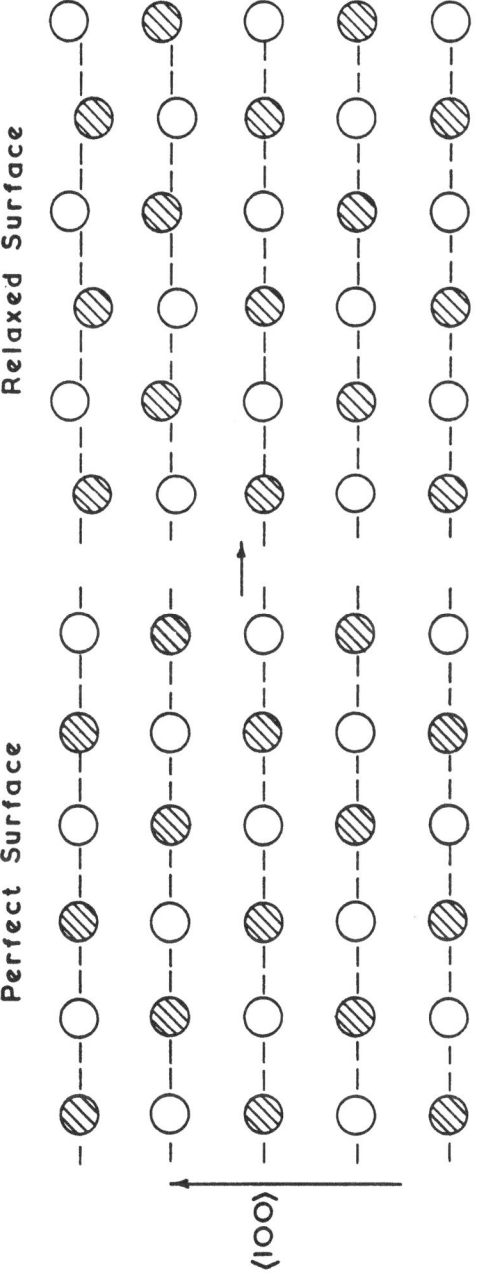

Figure 2 Rumpling at the (100) rocksalt surface

the effective polarizability at the surface due to the different environment. However, potentials derived from the bulk properties often have positive shell charges. This is usually ascribed to overlap effects[14] but is sometimes due to a non-unique fitting procedure. In either case, these potentials seem to provide a satisfactory description of bulk defect properties. However, at the surface, a positive shell charge will produce an opposite relaxation to a negative shell charge. An example of the problem is found with ZnO where a pronounced rumple is observed on the (10$\bar{1}$0) surface. Attempts to calculate this using available ionic potentials are doomed to failure since the positive shell charges are bound to produce the wrong direction of rumple.This emphasises the importance of using physically sensible potentials if useful information on surface structure is to be obtained. In some cases, suitably constrained fits to bulk properties may be adequate, whereas for other systems it may be necessary to permit different ionic parameters at the surface. Physically realistic changes in the potential, allowing for changes in polarizability, quadrupole contributions and overlap charge adjustments have been investigated by Martin and Bilz[9] for MgO.

Where the symmetry of the structure parallel to the surface is lowered, lateral distortions will be observed. These are still often small and need not alter the periodicity of the surface. An example of these relaxations is shown in Figure 3 calculated for the (110) surface of UO_2[15].

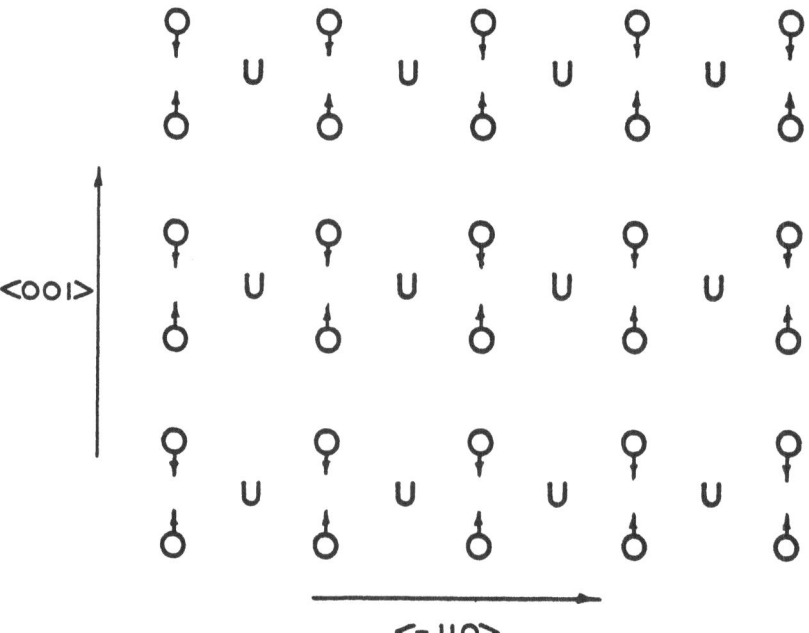

Figure 3 Lateral displacements of oxygen ions at a (110) uranium dioxide surface

293

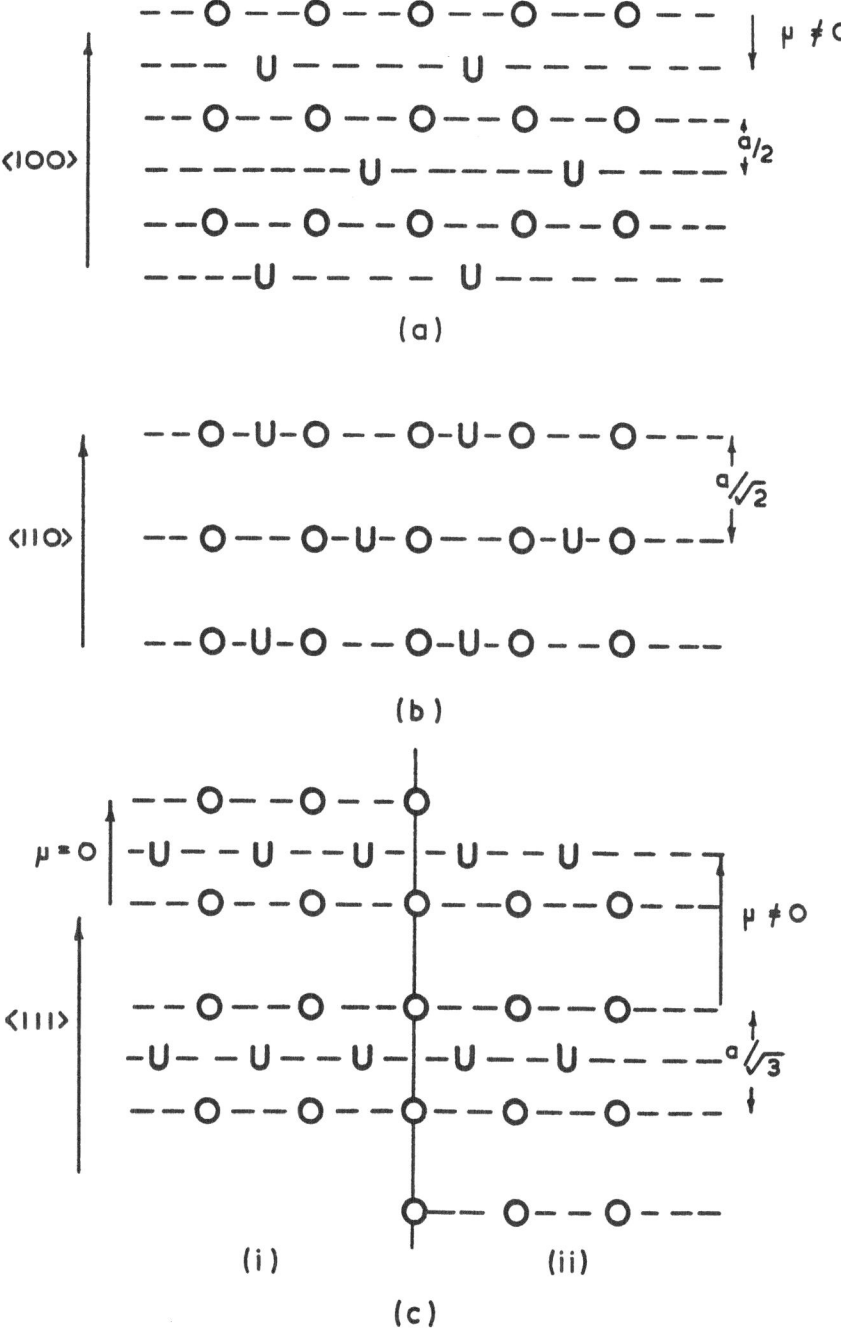

Figure 4   Schematic diagram of the stacking sequence in the low index directions in UO_2. (b) and (c) (i) have no dipole movement in the repeat unit; (a) and (c) (ii) have a dipole moment

Although small relaxations are usual for the low energy faces, there are some high energy surfaces which, if they exist, must show substantial reconstruction. It is readily shown that if there is a dipole moment in the repeat unit perpendicular to the surface, the lattice sums do not converge and the surface energy per unit area increases with increasing crystal size[16]. This is illustrated in Figure 4 for the fluorite structure adopted by UO_2. The (110) surface and the (111) anion surface do not have dipole moments in the repeat unit and are low energy surfaces. The (111) cation surface and the (100) face are dipolar and can only exist with restructuring to remove the internal electric field. The very high surface energy can be understood, as oppositely charged planes are being produced on generation of e.g. the (111) cation surface by cleavage; in contrast one is separating like charged planes to produce the lowest energy anion (111) surface. In general, therefore, (111) surfaces exist with anion termination and a (100) surface can only exist when stabilised by defects; the electrostatic instability is removed if the charge on the surface is halved, for example by 50% vacancies. This would be a natural process on cleavage since it requires less energy to split the plane leaving 50% of the the anions on opposite surfaces than to separately the adjacent oppositely charges faces. However, the surface energy of this face and analogous surfaces (e.g. (111) in the rocksalt structure) will be very high.

The (100) surface of UO_2 has been studied using a variety of surface scattering techniques[17]. The results indicate restructuring with a c(2x2) unit cell on the surface. Figure 5 shows the results of lattice simulation using an ionic-model potential. The shaded circles are the uranium ions on the plane below the surface, almost undistorted from their normal lattice sites. The open circles are oxygen ions (0^{2-}) in the surface plane. The structure has 50% vacancies and is substantially reconstructed with respect to the normal lattice sites, showing zig-zag chains of oxygen ions. The larger c(2x2) cell is marked on the diagram. This structure agrees very well with the experimental data although the ionic model predicts that a simpler, but also reconstructed (1x1) surface should have a lower energy. Nevertheless, these structures are not readily guessed from the 'perfect' configuration or predicted by billiard ball nodels; the calculations indicate how lattice simulations may be used to guide the interpretation of the experimental data.

4. Thermodynamic properties

The surface energy is generally calculated directly by subtracting from the energy of a crystal block terminated by a free surface, the energy of a block containing the same number of bulk ions. Experimentally, it is usually measured by contact angle or cleavage techniques. Both the latter methods are fraught with difficulties and there is often considerable variation in the experimental results.

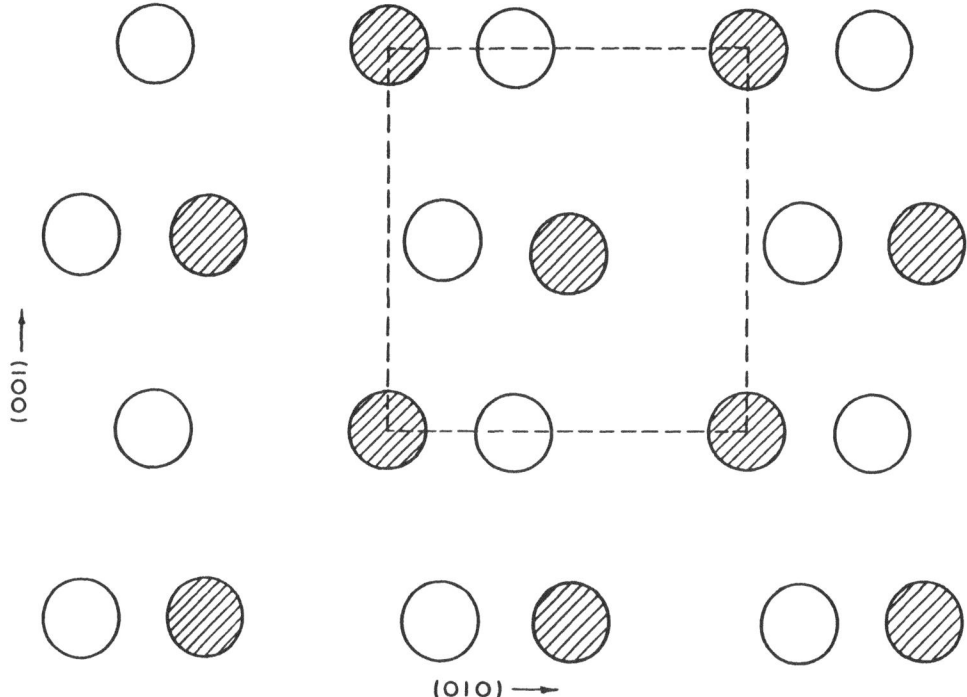

(001) →

(010) →

Figure 5   A calculated structure for the (100) UO_2 surface with c(2x2) unit cell.
Shaded circles are U^{4+} ions on plane 2, open circles are O^{2-} on surface
plane

The surface tension, γ, is related to the surface energy, E_s, by

$$\gamma = E_s + A \frac{dE_s}{dA}$$

and it can be evaluated directly by the appropriate lattice sums or differencing
the surface energy when the crystal is uniformly strained; it is a scalar
quantity that is related to the trace of the surface stress tensor.   In liquids
$dE_s/dA = 0$ and the surface tension and energy are equal and so the surface tension of
a liquid should be compared with the surface energy of the corresponding solid. The
experimental measurement of surface tension in a solid is even more difficult than
surface energy measurements.   The usual method is to deduce values from lattice
parameter changes in small crystallites,but the values obtained can differ not
only in magnitude but also in sign.

Surface energies of monovalent ionic crystals are typically ~ 0.2 J m^{-2} while
materials containing higher valence ions have proportionately higher energies. Indeed,
the correlation between surface energy and cohesive energy within a series of salts
is often very good, as is illustrated for the alkali halides in Figure 6[8].
Agreement between theoretical calculations and cleavage measurements is reasonable;

but a remarkable agreement is found between the calculated values and surface
tensions of the molten salts extrapolated to zero temperature. This agrees with
the observation of similarity between the short-range order and coordination number
of the molten and solid alkali halides[18]. Several cleavage measurements have
been made on the alkaline earth fluorides; which provide valuable tests of theory.
Table 1 shows the comparison: agreement is good. However the values quoted for
SrF_2 deduced from experiments employing both different crack velocities and different
theoretical approximations for the derivation of cleavage energies, indicate the
uncertainty of the measurements.

Table 1

Comparison of calculated surface energies with measured cleavage energies
for the alkaline earth fluorides (111) surface

Crystal	E_s Jm^{-2} calculated from lattice relaxation (19)	Experimental cleavage energies Jm^{-2} (20)	E_s Jm^{-2} static lattice unrelaxed (21)	
CaF_2	.476	.510, .450	.364	.543
SrF_2	.407	.420, .356, .260, .430, .465	.332,	.437
BaF_2	.349	.350, .280	.355,	.393

Surface tensions of ionic salts are usually calculated as positive, indicating
a compressive stress. This is borne out by most experiments on small
crystallites. The values are of similar order to the surface energies but may be
larger or smaller depending on dE_s/dA, whose value depends on the detailed structural
configuration of the surface and varies between the different faces of a particular
structure[8,19].

5.   The defective surface

So far we have been concerned with the properties of ideal crystal-vacuum
interfaces, but many important surface phenomena are dominated by the behaviour
of defects and impurities. Surface steps may be important in determining catalytic
activity and may be studied using the methods described here for high index
surfaces. Tsang and Falicov[22] have carried out calculations on the (1 0 11)

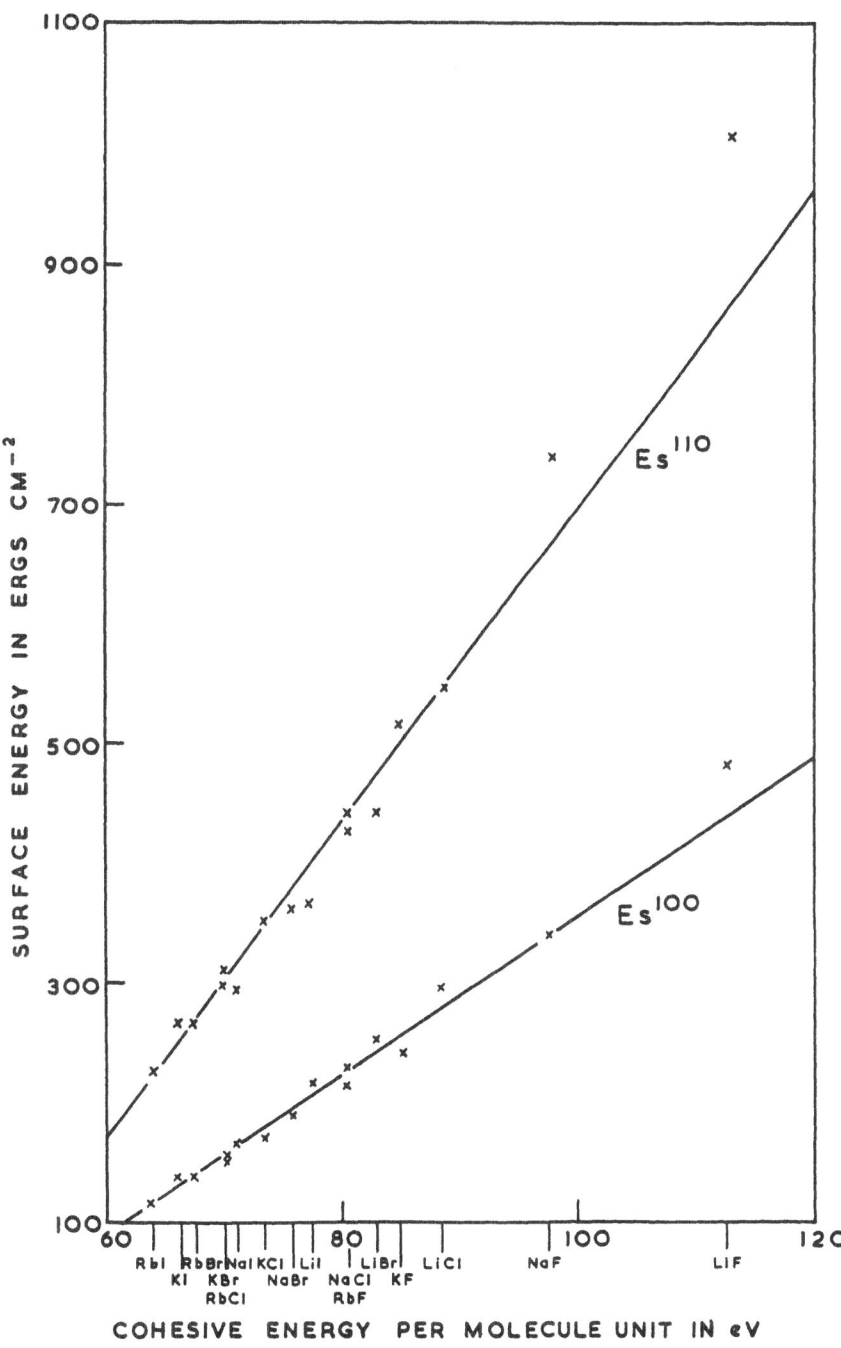

Figure 6   Calculated surface energy versus cohesive energy for the alkali halides

surface of NaCl which is equivalent to a regularly stepped (001) face. Point defect properties will vary at the surface and this has been investigated with lattice simulation techniques by Mackrodt and Stewart[23]. They calculate important variations in most defect energies suggesting, in particular, increased vacancy aggregation and reduced activation energies for diffusion at the surfaces of rock-salt structure crystals. Doping by foreign cations has also been studied showing significant differences in substitutional energies at the surface and bulk.

Gradients in the concentrations of impurities and defects at crystal surfaces have been observed; this has important consequences in catalyst materials[24] and nuclear fuels[25]. When surface concentrations are high, whole planes of defect or substituent ions may be present and the two dimensional periodicity may not be destroyed. In these circumstances, it is appropriate to use the methods described in section 2 with the crystal block containing the defects. As an example, consider calcium substitution in magnesium oxide. Table 2 summarises the results of the defect energy calculation including the isolated substitutional energy calculated using the HADES program. The substitutional energy per calcium for a (100) plane of defects in the bulk of MgO is ~ 0.2 eV higher than the isolated defect energy due to inter-calcium repulsions. This defect energy is constant at all sites except the surface layer itself where it is 0.9 eV lower.

Table 2

Energies for substituting (100) planes of Ca^{2+} ion in MgO as a function of distance from (100) surface. Plane 1 is the surface layer

Plane	Energy/Substitutional ion
1	5.74 eV
2	6.66 eV
3	6.64 eV
4	6.67 eV
5	6.65 eV

Energy of isolated Ca substitution = 6.45 eV

We would therefore expect that, at absolute zero, a doped crystal at equilibrium would have a saturated surface layer of Ca^{2+} ions with the rest of the dopant distributed in the bulk. To evaluate the surface segregation at higher temperatures we consider a simple thermodynamic model. Since the calculations indicate that only the top layer sites are different from the bulk we write the free energy, G, as

$$G = \sum_i n_i^B g_i^B + \sum_i n_i^S g_i^S - kT \ln \Omega$$

where g_i^B is the free energy of each of the n_i^B ions of type i in the bulk and g_i^S is the free energy of the n_i^S surface ions of type i. The third term represents the configurational entropy. Neglecting vibrational entropy contributions and assuming

that the number of surface sites is very much less than the number of bulk sites
we obtain for a two component system the simple Arrhenius form

$$\frac{n_1^S}{n_2^S} = f \exp(-H/kT)$$

where n_1^S/n_2^S is the ratio of two species in the surface plane, f is the dopant or
defect concentration n_1/n_2, and H is given by

$$H = \Delta h^B - \Delta h^S$$

where Δh^B is the bulk substitutional energy and Δh^S is the surface substitutional
energy. In order to compare different systems, it is convenient to define a
critical temperature, T_0, where there are equal concentrations of the two species

$$T_0 = \frac{-H}{k\ell nf} \quad .$$

Higher temperatures indicate stronger surface segregation. It is interesting to
note that the degree of segregation is not very sensitive to the dopant
concentration with only a logarithmic dependence in the critical temperature.
Figure 7 shows the results of this calculation for 1% Ca in MgO. The graph has
been extrapolated to high temperature above the melting point to indicate that the
surface fraction of Ca must reach the statistical limit of 1%. At lower
temperatures the ratio becomes infinite as the surface becomes saturated with the
substituent. In this case, the surface segregation persists to high temperatures
due to the large difference in defect energy.

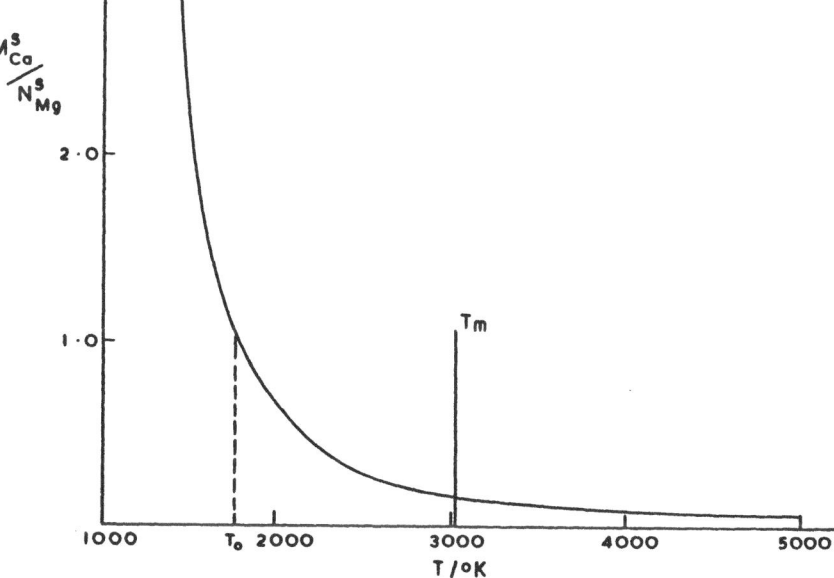

Figure 7 Ratio of Ca^{2+} to Mg^{2+} ions in the (100) surface plane of MgO doped with
1% Ca. T_m is the melting point

Recently, experimental measurements of the surface segregation in this system have been presented[26]. Measurements have been made on the ratio of calcium to magnesium on the (100) face of a magnesium oxide crystal as a function of temperature using Auger electron spectroscopy. The work finds an experimental heat of segregation of -0.8 ± 0.2 eV. This should be compared with the calculated difference between the surface substitutional energy and the isolated bulk substitutional energy of -0.71 eV (Table 2) and is in excellent agreement. Similarly good agreement is obtained between X-ray photoelectron spectroscopy measurements[27] on the NiO-MgO system and the earlier calculations of Mackrodt and Stewart[23] which predict a close to zero heat of segregation. A more detailed analysis of segregation in magnesium oxide has been given elsewhere[28].

6. Conclusions

This chapter has described the application of computer simulations to the study of crystal surfaces. The methods are general and can be applied to surfaces of any structure. Straightforward extensions of the techniques enable many two-dimensional defects and interfaces to be studied, including stacking faults[29], grain boundaries[30], shear planes and interfaces between different materials. Clearly, more work is required in applying these methods to the defective surface and solid-gas or solid-liquid interfaces. The problem of transferability of potentials to the surface, particularly in the less ionic materials needs more study. If suitable potentials can be derived for the more covalent materials, lattice simulation techniques could provide a valuable method for studying the more complex oxide and semiconductor surfaces.

Acknowledgement

I should like to thank Dr. A.B. Lidiard for valuable advice on the thermodynamics of surface segregation.

References

1. Benson, G.C. and Yun, K.S. in The Solid-Gas Interface. Ed. E.A. Flood (London: Arnold, 1967).

2. Mackrodt, W.C. and Stewart, R.F., J. Phys. C12, 431 (1979).

3. Catlow, C.R.A., Diller, K.M. and Norgett, M.J., J. Phys. C10, 1395 (1977).

4. Sangster, M.J.L. and Atwood, R.M., J. Phys. C11, 1541 (1978).

5. Stoneham, A.M., AERE Harwell Report R.9598 (1979).

6. Parry, D.E., Surf. Sci. 49, 433 (1975); ibid. 54, 195 (1976).

7. Welton-Cook, M.R. and Prutton, M., Surf. Sci. 64, 633 (1977); ibid. 74, 276 (1978).

8. Tasker, P.W., Phil. Mag. A39, 119 (1979).

9.   Martin, A.J. and Bilz, H., Phys. Rev. B19, 6593 (1979).

10.   Chen, I.S., Alldredge, G.P. and DeWette, F.W., Surf. Sci. 62, 675 (1977).

11.   Heyes, D.M., Barber, M., Clarke, J.H.R. and Faraday, J.C.S. II, 1469 (1979).

12.   Sawada, S. and Nakamura, K., J. Phys. C12, 1183 (1979).

13.   Duke, C.B., CRC Crit. Rev. Solid. St. Mater. Sci. 8, 69 (1978).

14.   Bilz, H., Buchanan, M., Fischer, K., Haberkorn, R. and Schröder, V., Solid Stat. Commun. 16, 1023 (1975).

15.   Tasker, P.W., Surf. Sci. 87, 315 (1979).

16.   Tasker, P.W., J. Phys. C12, 4977 (1979).

17.   Ellis, W.P. and Taylor, T.N., Surf. Sci. 91, 409 (1980).

18.   Edwards, F.G., Anderby, J.E., Howe, R.A. and Page, D.I., J. Phys. C8, 3483 (1975).

19.   Tasker, P.W., Proc. of 3rd Europhysics Conference on Lattice Defects in Ionic Crystals, J. Phys. (Paris) (1979), 41, C6-488 (1980).

20.   Becher, P.F. and Freeman, S.W., J. Appl. Phys. 49, 3779 (1978).
         Kratz, P. and Zoltai, T., J. Appl. Phys. 45, 474T (1974).
         Gilman, J.J., J. Appl. Phys. 31, 2208 (1960).

21.   Benson, G.C. and Dempsey, E., Proc. Roy. Soc. A266, 344 (1962).

22.   Tsang, T.W. and Falicov, L.M., Phys. Rev. B12, 3441 (1975).

23.   Mackrodt, W.C. and Stewart, R.F., J. Phys. C10, 1431 (1977).
         Stewart, R.F. and Mackrodt, W.C., J. Phys. (Paris), C7-247 (1977).

24.   Menon, P.G. and Prasada Rao, T.S.R., Catal. Rev. 20, 97 (1979).

25.   Matzke, Hj. and Lambert, R.A., J. Nuc. Mater. 64, 211 (1977).

26.   Wynblatt, P. and McCune, R.C., in Proc. of Berkeley Conf. on "Surface and Interfaces in Ceramic and Ceramic-Metal Systems" (Plenum Press, N.Y. 1981).

27.   Cimino, A., De Angelis, B.A., Minelli, G., Persini, T. and Scarpino, P., J. Solid St. Chem. 33, 403 (1980).

28.   Colbourn, E.A., Mackrodt, W.C. and Tasker, P.W. To be published.

29.   Tasker, P.W. and Bullough, T.J., Phil. Mag. A43, 313 (1980).

30.   Wolf, D., in Proc. of 3rd Europhysics Conf. on Lattice Defects in Ionic Crystals, J. Phys. (Paris), 41, C6-192 (1980).

CHAPTER 20   LONG RANGE ORDER IN NON-STOICHIOMETRIC OXIDES

by

A.N. Cormack
Department of Chemistry, University College London,
20 Gordon Street, London WC1H OAJ

1. Introduction

Strict translational periodicity in solids, necessarily, excludes the possibility
of non-stoichiometry. Nevertheless many binary and ternary inorganic compounds
exhibit bivariant thermodynamic behaviour whilst at the same time diffraction
techniques indicate them to be structurally monophasic. Historically, the concept
of non-stoichiometry was revived at the start of the 20th century, largely by
Kurnakov[1], and is now known to play an important role in many aspects of solid
state chemistry and physics.

The development of classical point defect theory has led to a thorough understanding of solids showing only small deviations from stoichiometry. Extensions of
point defect models to include the effects of clustering have led to the development
of successful treatments of more defective phases, notably $F_{1-x}O$ and UO_{2+x}[2],
although as noted in Chapter (15), the validity of the cluster approach becomes
questionable at very high deviations from stoichiometry. However, point defect
models fail entirely to account for the structural properties of certain non-
stoichiometric compounds, including the widely studied transition metal oxides
TiO_{2-x} and WO_{3-x}.

It is known[3,4] that in these latter materials extended rather than point
defects predominate, and that on reduction, oxygen loss leads to the formation of
crystallographic shear planes rather than anion vacancies or cation interstitials
(although, there is strong evidence that the point defect modes of reduction do
still operate in the very near-stoichiometric regions of the phase).   The formation
of shear planes alters the metal to oxygen ratio by changing the way in which the
MO_6 polyhedra are linked (e.g. corner is replaced by edge-sharing in the shear planes
in WO_{3-x}). A typical example is provided by the (102) shear plane which forms in
reduced WO_3, and which is discussed further in the next section.

Experimental observation, mainly by High Resolution Electron Microscopy, has
revealed two important properties of shear planes. First, with increasing reduction
their orientation changes; and second, with increasing concentration, the shear
planes order into regular or nearly regular arrays. These and other observations raise
several fundamental questions concerning the relationship between extended and
point defects and relative stabilities of the different shear plane orientations.
However, the aspect we shall consider here concerns the ordering of the extended
defects into arrays, which if entirely regular, gives rise to the well known Magneli
phases[5] in the homologous series of oxides Ti_nO_{2n-1} and W_nO_{3n-1}. The large

spacing between extended defects in those members of the series with large values of n, implies interactions between the shear planes over several hundred Angstroms; the explanation of the origin of these interactions clearly poses a major challenge to theoreticians.

Until recently, theoretical investigations of this problem had largely been confined to approaches based on continuum elasticity[6,7]. Recently, however, atomistic modelling techniques were successfully used to identify the factors that are important in stabilising shear-planes relative to point defects in TiO_{2-x}[8]. These methods have now been successfully extended to the study of CS plane formation in ReO_3 structured oxides - a group of compounds that includes the widely studied WO_{3-x} phase - and the present chapter concerns the application of the methods to a detailed study of the interaction between shear planes in ReO_3 structured oxides. We concentrate on the case of (102) shear planes, and show for this example how atomistic techniques can yield detailed information on the shear plane interaction energy and hence on the nature of extended defect ordering.

2. Crystallographic shear plane structures

In the aristotype ReO_3 structure which is symmetrically cubic, the cations are octrahedrally coordinated by the anions to give a 3-D periodic array of regular, corner-sharing octahedra, as illustrated in Figure 1. The loss of oxygen from the structure is accommodated by the formation of groups of edge-shared octahedra which are aligned into planes within the host crystal; the resulting extended defects are the crystallographic shear planes, which at relatively low deviation from stoichiometry are aligned along (102) as shown in Figure 2. In more grossly non-stoichiometric phases, shear plane orientations are found along (103) and (100). The present study, however, concentrates on the (102) planes which as noted in section 1, order to form well-defined super-cells.

The oxide ReO_3 itself does not form a non-stoichiometric phase of significant width. However, WO_3 which adopts a distorted ReO_3 structure does, as noted earlier, form shear planes on reduction. In addition MoO_3 on reduction forms a series of compounds essentially based on shear structures in an ReO_3 structured host, although the stoichiometric compound does not have the ReO_3 structure. However, the calculations discussed here were performed on ReO_3 itself; therefore, they should be taken as referring to the structure in general, rather than to specific materials.

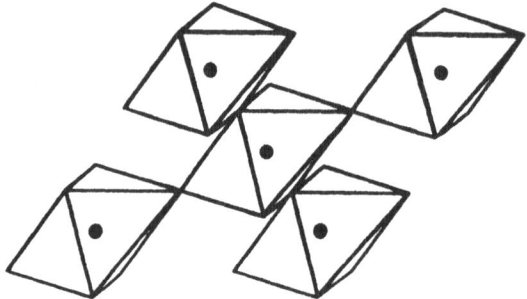

<u>Figure 1</u>   Perspective view of the ReO₃ structure. The cations lie at the
centre of the oxygen octahedra which are linked through their corners.

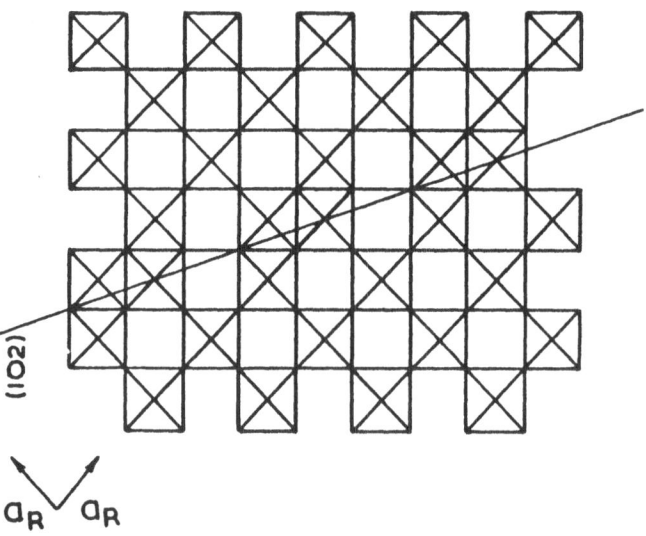

<u>Figure 2</u>   (102) CS plane in ideal ReO₃ structure projected on (010). a_R
represents the ReO₃ unit cell vectors (note the edge sharing
octahedra along the CS plane).

3. The energetics of crystallographic shear planes

The ordering of shear planes may be discussed quantitatively if we know the shear-plane interaction energy as a function of interplanar spacing; the energies must, of course, be expressed in terms suitable for comparison. Now, the composition of a crystal based on a regular array of (102) crystallographic shear planes is given by M_nO_{3n-1} where n denotes the number of MO_6 octahedra between the planar defects. The energy of the shear plane, E_{CSP}, in the array is thus given by

$$E_{CSP} = E_{ARRAY} - E_{IDEAL} \qquad (1)$$

where E_{ARRAY} is the lattice energy per formula unit of a crystal with composition MnO_{3n-1} and E_{IDEAL} is the corresponding lattice energy of n formula units of the ideal ReO_3 structure. Thus E_{CSP} is expressed in terms of the energy per eliminated oxygen atom; it is a function of n, i.e. a function of inter-planar spacing.

If the energy, E_{ISO} of an isolated shear plane is known, then the interaction energy may be written as

$$E_{INT}(n) = E_{CSP}(n) - E_{ISO} . \qquad (2)$$

Thus, by obtaining the lattice energy of the members of the homologous series M_nO_{3n-1}, the variation with n of $E_{INT}(n)$ may be derived. The methods used to calculate E_{ISO} are essentially the same as those used in the surface simulations discussed by Tasker in Chapter (19). The calculations on arrays of shear planes are performed using the PLUTO program discussed in Chapter (1) which, with the recent expansion of computer power, is capable of handling the resultant very large unit cells. The next section discusses the special problems posed by such calculations.

4. Techniques

Straightforward lattice energy calculations of shear-plane supercells, cannot in fact yield information on shear plane interactions; as the shear planar defects are neutral*, the interactions arise solely from weak quadrupolar terms (shown by Stoneham and Durham[6] to be negligible at observed interplanar spacings) provided the shear plane configuration does not vary with separation. Lattice relaxations are, however, known to play a vital role in stabilising shear planes in TiO_{2-x}[8]. They have an equally important, if more subtle influence on shear plane interactions, as it is the variation with interplanar spacing of the relaxation term which is responsible for the interaction. Thus atomistic calculations on shear plane interactions require equilibration of the whole super-cell; that is, it is necessary to adjust all unit-cell dimensions

---

*This is achieved by placing reduced cations in the sites neighbouring the shear-plane

and all atomic coordinates until the forces acting on the individual atoms in the unit cell and on the unit cell as a whole, are eliminated. We should note that calculations which do not adjust unit cell dimensions correspond to constant volume conditions, whereas the full equilibration procedure yields results appropriate to constant presssure conditions.

It is clear from the above discussion, that it is an essential pre-requisite of any reliable calculation on shear plane interactions that the host-lattice potential should be strain free (i.e. the potential should give zero forces acting on the host unit-cell and its components with the observed cell dimension). We therefore developed an ionic model potential for ReO_3 in which the parameter describing the short range potential were adjusted to achieve the necessary strain free condition; the parameters were also fitted to the elastic constants for ReO_3 using the type of procedure described in Chapter (10).

Once developed, the potentials may be implemented in the energy minimisation studies of the defect super-cells. The lattice equilibration procedure is iterative, and closely resembles that described in Chapter (15) for the study of pyroxenoid structures. Atomic coordinates are equilibrated first, followed by a relaxation of unit cell parameters, after which the process is repeated until the strains are zero.

Although in principle all the parameters could be varied at once, procedures are not available at present for such calculations and in practice it is convenient to use different minimisation methods for the coordinates and the unit cell parameters.

The coordinates are relaxed using a fast matrix technique based on a quasi-Newton method of optimisation, e.g. the Davidon-Fletcher-Powell method[10] discussed in Chapter (1) which has the following essential features

(i)   the new set of coordinates is obtained from the relationship:

$$\underline{x}_{k+1} = \underline{x}_k - \underline{\underline{W}}^{-1} \cdot \underline{g}_k \qquad (3)$$

where $\underline{\underline{W}}$ is the matrix of second derivative and $\underline{g}$ the gradient vector (of the lattice energy)

(ii)   the initial step is in the negative gradient direction, that is initially a steepest descent approach is adopted.

(iii)   In order to save computing time, lengthy line searches are avoided and a matrix update is employed. The update is essential since repeated calculation and inversion of the second derivative matrix would be extremely expensive in computer time, especially for large matrices. In this work, the largest matrix used had dimensions 500 by 500, a limit imposed by the available memory of the

CRAY-1s at the SERC Daresbury Laboratory where these calculations were performed. The particular update formula used is of the form

$$\underline{\underline{H}}_{k+1} = \underline{\underline{H}} + \frac{1}{\underline{y}_k^T \underline{s}_c} \left(\beta_k \underline{s}_k \cdot \underline{s}_k^T - \underline{s}_k \underline{y}^T \cdot \underline{\underline{H}}_k \underline{y}_k \cdot \underline{s}_T \right) \tag{4}$$

where

$$H_1 = W_1^{-1} \quad \text{and} \quad \beta = 1 + \frac{\underline{y}_k^T \cdot \underline{\underline{H}} \cdot \underline{y}_k}{\underline{y}_k^T \cdot \underline{s}_k}$$

$$s_k = \underline{x}_{k+1} - \underline{x}_k \quad \text{and} \quad \underline{y}_k = \underline{g}_{k+1} - \underline{g}_k .$$

This is known as the complementary DFP formula and it has been shown that its use avoids the necessity of line searches provided suitable step lengths are chosen, i.e. if one chooses

$$s_k = -\lambda_k \underline{1} + k \underline{g}_k \tag{5}$$

where λ_k is a suitable constant.

The lattice vectors are adjusted making use of elasticity theory. First we note that the calculations are done in a Cartesian coordinate system and so the lattice vectors (as well as the particle coordinates) are described in terms of three orthogonal basis vectors, which we write as $\underline{i}, \underline{j}$ and $\underline{k}$. Thus each lattice vector $\underline{\ell}^\alpha$ can be written as

$$\underline{\ell}^\alpha = c_i^\alpha \underline{i} + c_j^\alpha \underline{j} + c_k^\alpha \underline{k} . \tag{6}$$

Thus after the initial equilibration of the coordinates, with the original lattice vectors, the six bulk lattice strains $e_{ii}, e_{jj}, e_{kk}, e_{ik}, e_{ij}, e_{jk}$ (see Chapter (1)) are calculated. These indicate the amount of residual strain in the structure, and enable a new basis to be constructed as follows

$$i' = (1+e_{ii}) \underline{i} + e_{ij} + e_{ik} \underline{k}$$
$$j' = e_{ij} \underline{i} + (1+e_{jj}) \underline{j} + e_{jk} \underline{k} \tag{7}$$
$$k' = e_{ik} \underline{i} + e_{jk} \underline{j} + (1+e_{kk}) \underline{k}$$

Since the components of the lattice vector matrix in the new basis set $\underline{i}', \underline{j}', \underline{k}'$ are the same as those in the original basis set $\underline{i}, \underline{j}, k$ (i.e. $c_{i'}^\alpha = c_i^\alpha$ etc.) the new lattice vector matrix is obtained straightforwardly from equations (6) and (7). The iterative procedure is continued until the bulk lattice strains are negligible. It should also be noted that, as the frame of reference changes it is also necessary to alter the atomic coordinates after each adjustment of the lattice vectors, and indeed that these are generally re-equilibrated each time the lattice vectors are changed.

In practice, it is found that only three or four cycles of lattice vector adjustment are required, the bulk of the relaxation energy being obtained in the initial relaxation of the atomic coordinates.

5. Results

The interaction functions for (102) CS plans for both constant volume and constant pressure modes of equilibration are plotted in Figures 3(a,b). For both modes, E_{ISO}, was obtained, as discussed above, from calculations using the surface simulation techniques discussed in Chapter (19), which were developed by Tasker. In addition, it was possible for the case of the constant volume calculations (i.e. those without any lattice vector adjustment) to confirm that the shear-plane formation energies obtained from the super-cell calculations converged to the same values as those obtained from the direct calculation of E_{ISO}.[*]

The essential features of the shear plane interaction functions are apparent from Figure 3. For the constant volume calculations (Figure 3(a)) the interaction is everywhere repulsive with large magnitudes of the interaction energy. Such a function would imply that shear plane would separate with no tendency to form arrays - behaviour which obviously does not accord with experimental observation. However, for complete equilibration, the opposite behaviour is found. We see from Figure 3(b) that the interaction is attractive with a small but appreciable magnitude even at large separations, suggesting that shear planes will interact over large separations to form arrays.

The reason for this attractive term may be found in the changes in the lattice vectors on equilibration. The effect of this relaxation is essentially twofold. First, there is an expansion of the crystal so that the octahedra in the regions of crystal between the CS planes are restored to ideality, and secondly there is a lateral displacement of the two halves of the crystal on opposite sides of the CS plane, as shown in Figure 4. This displacement increases as the spacing between the CS planes decreases owing to the necessity of accommodating the greater strain fields around the CS planes at shorter separation. Thus, it is the strain fields that are mainly responsible for the interaction, which may therefore be said to be elastic in origin.

Although the results presented here are only for the (102) orientation, work in progress has shown that qualitatively similar results occur for the (103) planes.

---

[*] It is not possible to make a similar check for the constant pressure calculations as sufficiently large super-cells cannot be handled using available computer power.

(a)

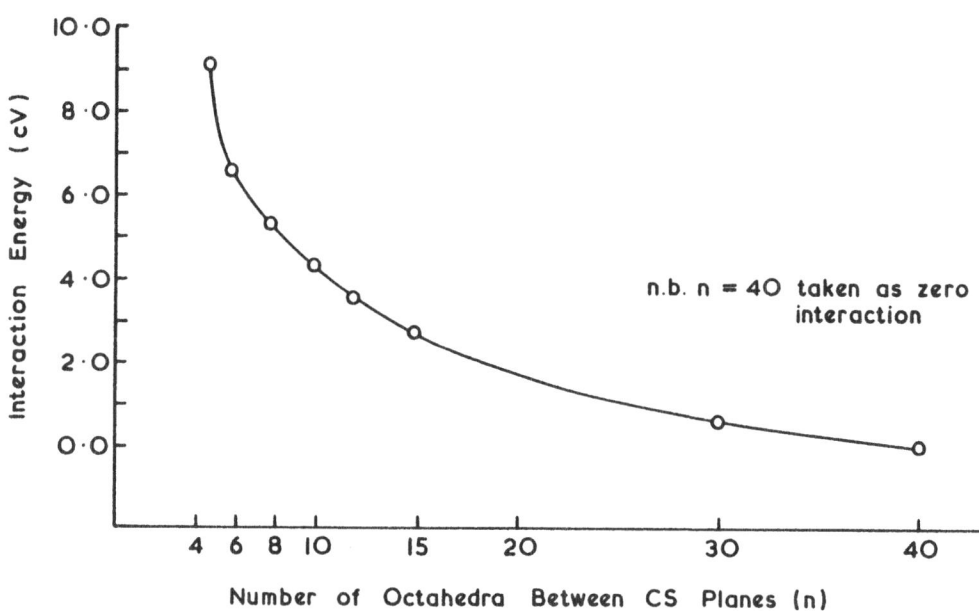

(b)

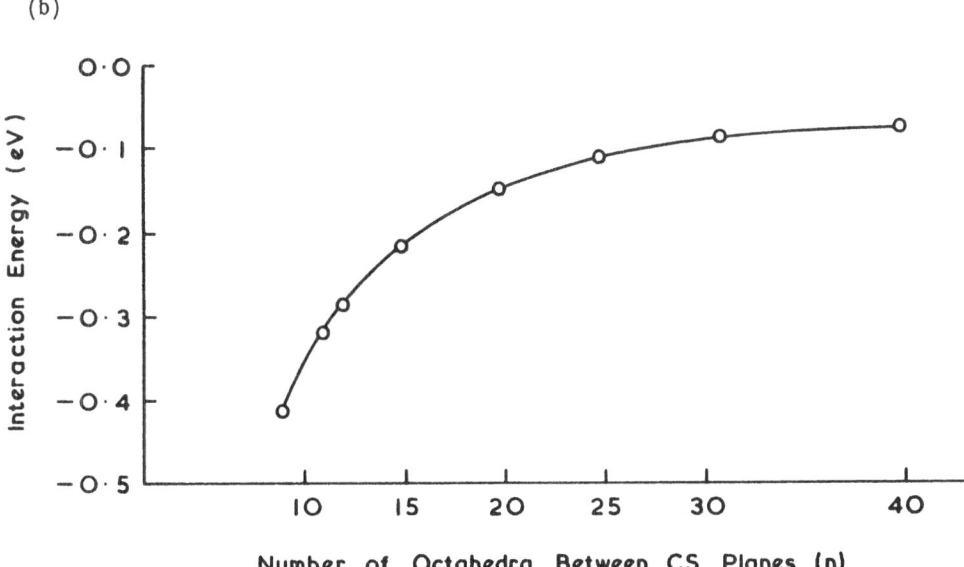

Plots of shear plane interaction energy vs. number of MO_6 octahedra between the planes. a) Function for contrast volume conditions, i.e. no adjustment of cell dimensions. b) Function for constant pressure calculation, i.e. cell dimensions equilibrated.

Figure 3(a,b)

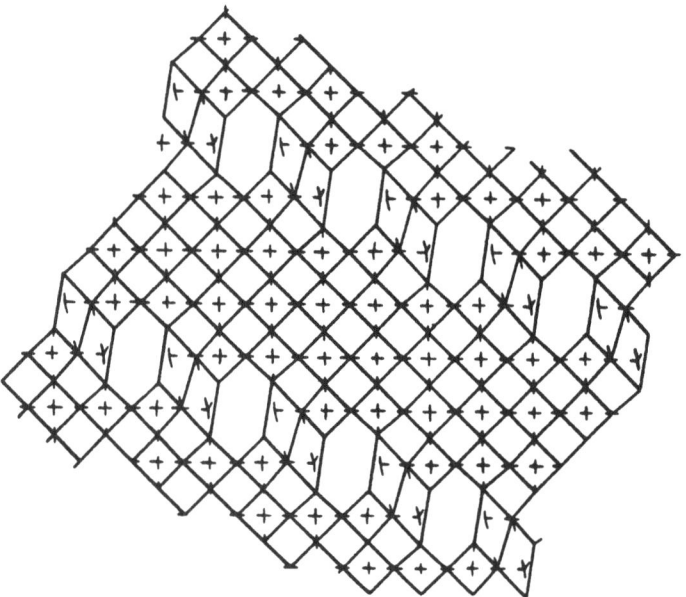

Figure 4 The completely equilibrated structure of the (102) CS plane, after
relaxation of both the atomic coordinates and lattice vectors. The
distortion of the structure around the CS plane is due to the dilation
and lateral displacements of the structure mentioned in the text.

6. Conclusions

We have shown that detailed atomistic computer modelling, which has enjoyed
considerable success in the field of point defect interactions as discussed in
Chapters (12) and (15), may profitably be applied to the considerably more complex
area of extended planar defect interactions, since computational resources are
now available to handle such large scale calculations.

Further, from these calculations, we have been able to identify lattice
relaxation as the dominant factor in determining the interaction between CS planes;
the relaxation of the unit cell parameters are found to be of particular
importance. The calculations we have described represent the first detailed
application of atomistic modelling techniques to the study of long range order.

7. Acknowledgements

Drs.C.R.A. Catlow and P.W. Tasker are thanked for their useful contributions.
Financial support from the SERC is acknowledged; and we are also grateful to the
SERC for the use of their CRAY-1s computer at Daresbury.

References

1.   Kurnakov, S.N., Z. Anorg. Allgem. Chem. 88, 109 (1914).

2.   Catlow, C.R.A. in 'Non-Stoichiometric Oxides' (ed. O.T. Sorensen), Academic Press (1981).

3.   Anderson, J.S. and Tilley, R.J.D. in Surf. Defects in Solids, Vol.3, p.1, (Spec. Periodic Report, RSC)(1976).

4.   Tilley, R.J.D. in Chemical Physics of Solids and their Surfaces, Vol.8, p.121 (Spec. Periodic Report, RSC)(1980).

5.   Magneli A. Acta Crystallogr. 6, 495 (1953).

6.   Stoneham, A.M. and Durham,P.J., J. Phys. Chem. Solids 34, 2127 (1973).

7.   Iguchi, I. and Tilley, R.J.D., Phil. Trans. Roy. Soc. A286, 55 (1977).

8.   Catlow, C.R.A. and James, R. Nature 272, 603 (1978).

9.   Catlow, C.R.A. and James, R. in Chemical Physics of Solids and their Surfaces, Vol. 8, p.108. Spec. Periodic Report of Royal Society of Chemistry, London 1980.

10. See for example Wolfe, H.A. Numerical Methods for Unconstrained Optimisation, Van Nostrand, Reinhold, New York (1978).

11. Cormack, A.N., Tasker, P.W., Jones, R.M. and Catlow, C.R.A., J. Solid State Chem. In press.

Materials Index

Subject Index

Lecture Notes in Physics

Selected Issues from
Lecture Notes in Mathematics